MATHEMATICAL IDEAS, MODELING AND APPLICATIONS

MATHEMATICAL IDEAS, MODELING AND APPLICATIONS

Volume II of Companion to Concrete Mathematics

Z. A. MELZAK

Department of Mathematics
University of British Columbia
Vancouver, Canada

A WILEY-INTERSCIENCE PUBLICATION

JOHN WILEY & SONS, New York · London · Sydney · Toronto

Library of Congress Cataloging in Publication Data (Revised)

Melzak, Z. A. 1926–
 Companion to concrete mathematics.

 (Pure and applied mathematics)
 "A Wiley-Interscience publication."
 Vol. 2 has title: Mathematical ideas, modeling and applications.
 Bibliography: v. 1, p. ; v. 2, p.
 1. Mathematics—1961– I. Title. II. Title: Mathematical ideas, modeling and applications.
QA37.2.M44 510 72-14171
ISBN 0-471-59338-9 (v. 1)
ISBN 0-471-59341-9 (v. 2)

Printed in the United States of America

10 9 8 7 6 5 4 3 2 1

ADAE PATRI FILIOQUE

Whatever is worth doing is worth doing badly

GILBERT KEITH CHESTERTON (**1874–1936**)

FOREWORD

Probably every mathematician is acquainted with the embarrassment caused by the simple question: What is mathematics? Representing the most abstract of sciences, we find ourselves unable to give an answer in general, abstract terms. We have to refer to a number of books that give samples of mathematical topics, as does the present book, but it answers a different, although equally puzzling and even more important question: What makes mathematics attractive? The answer has many components, and the book shows them beautifully in their natural interaction. I shall try to separate them by mentioning a few.

First of all, we are given a new awareness of the sources of mathematics. There is, of course, sight, that is, geometry, once synonymous with mathematics and still at the root of very sophisticated, very "modern" problems. But there are other sources: numerous basic tools like looms and hinges or basic crafts like casting, weaving, and measuring. It is both gratifying and surprising to see how many mathematical ideas are implicit in basic human activities. The surprises turn into near-miracles with the applications. To mention only one of many: Why should one expect the Moebius inversion to be connected with the counting of necklaces and therefore, with the counting of amino acids and comma-free codes?

Next, we see mathematics as a part of what Alfred North Whitehead called "adventures of ideas." At every turn of an argument, there is a new view and a new challenge. Starting with an elementary problem, we arrive at deep and general questions merely by looking at the underlying principle of the original problem. Many seemingly difficult problems can be solved by using ideas that are extremely simple once they are discovered. Above all, the ideas provide coherence.

Chapter 4 contains a highly original approach to computing and computability based, as the author puts it, on the method of computing by counting on one's fingers. It is a systematic chapter, and it illustrates another facet of the attractiveness of mathematics: the satisfaction of erecting a tall building on a narrow base.

This is an extraordinary book. It shows the infinite complexity of mathematics, and it enables us to see mathematics as a well-structured unit, as something like a living being in which every organ has a special function and yet serves the whole. The book provides an important service to all of those who want to pass on mathematics to the next generation, not because it is merely entertaining or intriguing, but because it is part of one of the most gratifying of human experiences in the search for insight and truth.

WILHELM MAGNUS

PREFACE

The general purpose of this, the second volume, remains the same as that of the first one. If anything, it is firmer now since, no matter how one interprets it, labor increases resolution. Neither volume is just a collection of problems or even of techniques, in spite of some reviews and other opinions. But then, perhaps the beads have hidden the string.

The concrete element is, of course, emphasized: linkages, casting of convex and other shapes, elliptic functions and the transmission of several telephone conversations on a single wire, an approach to computing and computability based on counting on one's fingers rather than with symbols, a most brutally involved functional equation which, however, arose from focusing the radiation from a point source to a plane beam, pursuit (as of rabbits by hounds) with an inverse which may apply to trendy salesmanship and government stability, wartime radar failure applied to looking at raindrops and giving rise to transport equations possibly generalizing those of the Lotka–Volterra theory of interspecies competition, some tentative extensions of topology arising out of a neurophysiological context, a set-theoretic approach to usury, and so on.

In addition, there is material on formal (not formalistic), manipulative and intuitive aspects of analysis, geometry, and combinatorics. Finally, there is an appendix treating, hopefully not too seriously, some of the "principles" used in volumes 1 and 2. If the subject matter is often elementary or looks antiquated, this is to compensate for the hypermodern mathematical training which builds the individual mathematical house solidly from the roof down, often stopping half-way. This might be due as much to the loss of pleasure in simple things as to the lack of a sense of proportion and reality. Against the second fault gods themselves contend unvictorious, but about the first one something might be done by men. Hence the Chesterton quotation opening this book.

In brief, what is presented here is a subjective distillation of topics found useful or amusing. As the title implies, this book is meant to accompany other texts, books, and instruction or self-instruction. However, several types of courses could conceivably be based on various parts of the combined two volumes:

A course for potential college teachers and instructors with an emphasis on perspective, background, and motivation, illustrating and interconnecting the material taught,

Another one on mathematical techniques and modeling for students of applied mathematics and related sciences or technological disciplines,

A third one, of rather variable nature, as an introduction for promising freshmen and sophomores, encouraging them to start formulating their own problems possibly early, and

A fourth one, for seniors or beginning graduate students, on the linking together of what they may have learned in different courses, including perhaps some supervised writing up of minor research projects on mathematical or historical topics.

One other potential use of the two volumes deserves perhaps a special mention. This concerns a type of retraining for those who have started their mathematical life on a steady pure-abstract diet. Until recently there was, perhaps, no need to change one's course, but we are witnessing a drying up of sources of support for mathematical work as well as an awakening of a sort of mathematical conscience. What should a hypothetical young mathematician do when he wishes to change toward something at least remotely relevant and beneficial in the social sense, and wants to avoid any traps of applied charlatanism as well as those of unbridled purism? The answer is simple and clear, though far from easy: he should develop sufficient breadth to be able to judge for himself. It is my sincere hope that my two books may perhaps help in this respect.

Friends, colleagues, students, well-wishers, correspondents, and others, too numerous to be named one by one, have offered help, suggestions, criticism, and advice. The University of British Columbia granted leave and the Killam Foundation awarded a Fellowship for the school year 1975–1976, during which a heavy part of the work was done. Grateful thanks are hereby tendered to all of them.

Z. A. Melzak

Vancouver, Canada
January 1976

CONTENTS

on the conservation of the number of intersections. Linear and nonlinear enumerative geometry.

CHAPTER 2 ANALYSIS   60

1 *Inequalities.* Young and Hölder inequalities. The arithmetic–geometric–harmonic inequality. A theory of the exponential function. Continuous and discrete means, applications. Iterated factorials. Sums of powers, related discrete and continuous inequalities. **2** *Differential equations and summation of series.* Summation of series and generating functions for certain orthogonal systems (Laguerre, Hermite, Bessel, Legendre). Truesdell unified treatment applied to Hermite polynomials. Abel's theorem and Polya's generalization, elliptic integrals. **3** *Discontinuous factors, random walks, and hypercubes.* Definitions and simple examples. Real zeros of random polynomials. Discontinuous factors and the geometry of n-cubes, areas and volumes. Connection with random walks in three dimensions, Polya's analysis. Normal law and sections of n-cubes. Order statistics and order factors, applications and illustrations. **4** *Calculus of variations.* Isoperimetric problems of the circle and the catenary. Shortest curves on surfaces, Bernoulli's theorem. Motion of light in medium with variable index of refraction, cycloid, Poincaré's model of hyperbolic geometry. Geodesics on cones, Clairaut's theorem. Maximization of convex hull volumes. A variational method for existence of periodic solutions. **5** *Vector, matrix, and other generalizations.* Gamma and Beta integrals of Euler with vector and matrix parameters. Volume of an n-sphere. Certain characteristic functions. The Ingham formula in statistics and the Siegel formula in analytic theory of quadratic forms. Mahler's theorem. Weierstrass and sub-Weierstrass programs in function theory. Binomial sequences of polynomials, vector sums of convex bodies, and mixed volumes. **6** *Continuity and smoothness.* C^n classification of functions. Baire scale. Some pathologies. Uses of continuity in certain proofs. Intermediate-value property and the n-dimensional regula falsi. An examination of $\exp(-1/x)$. Quasi-analytic functions. Synthesis of interpolatory and of singular C^∞ functions. Running averages and their iteration. Sequentially monotonic, or hyper-smooth, functions, Bernstein's theorem. Derivatives as limits and as linear operators. Derivatives from fast-convergent numerical procedures. **7** *Generalized functions.* Some background from physics and from electrical engineering. The Dirac δ-function.

Indirect and direct justifications, Laplace transforms. Examples of algebraic and other extension problems. Integration by parts and testing functions. Generalized functions as distributions, as equivalent sequences, and as fractional operators. Theorems of Lerch and Titchmarsh. Inverses of running averages.

CHAPTER 3 TOPICS IN COMBINATORICS,
NUMBER THEORY, AND ALGEBRA **151**

1 *Hilbert matrices and Cauchy determinants.* Ill-conditioned linear problems. Exact and asymptotic values for Hilbert matrices H_n. Round-off errors and decimal conversion. **2** *Multiplicative Diophantine equations.* The Bell–Ward method. Determination of all Pythagorean triangles. Other quadratic and higher problems. **3** *Gaussian binomial coefficients and other q-analogs.* Definitions and elementary examples. Analogy with permutations, k-dimensional subspaces of a vector space over $GF(q^n)$. Telescoping mirroring and its combinatorial uses. Identities for partitions. **4** *Some informal principles of combinatorics.* Statement of these principles. Elementary applications, simple combinatorial computations. Ordinary, exponential, Lambert, and Dirichlet generating functions, their relations. Borel transform, Voronoi regions. Plane packings of circles. Simple and multiple induction, generalizations. Ramsey's theorem. Computations associated with the combinatorics of branching. **5** *Möbius inversion.* Definitions and simple properties. The classical Möbius inversion. Analogies with binomial coefficients. Möbius function and the Riemann hypothesis. Counting of necklaces, amino acids, and comma-free codes. Combinatorial identities and unique representation, Galois fields $GF(q^n)$. Inclusion–exclusion principle. Summation and its inverting in partially ordered sets. Theorems of Dilworth and of Erdös–Szekeres. Generalized zeta and Möbius functions. Generalized Möbius inversion. **6** *The Lindelöf hypothesis.* Some number-theoretic background. Lattice points in convex and star-shaped regions. Equivalence of the Lindelöf hypothesis and a combinatorial counting problem.

CHAPTER 4 AN APPROACH TO COMPUTING AND
COMPUTABILITY **205**

1 *Descent of computers from the weaving machinery.* Control as part of manufacturing process. Jacquard loom and automatic control of computation. Some early history: Babbage, Menabrea, Augusta Ada Lovelace. Platonism and purity in mathematics. Echos in

CONTENTS

two and three dimensions, quadratic birth processes. Jacobian functions and filter design, pass-band and reject-band, Chebyshev approximation. **2** *Transport equations.* Radar meteorology and particle-size distributions. Random coalescence and breakdown. Laplace transforms and solution of simple scalar transport. Generalization of transport equations, approximation to nonlinear operators. Coagulation. Vector sums of sets, convolution and sets of positivity. Generalization to multiple transport, averaging transport. Connection with quadratic rate processes. Lotka–Volterra theory, the Volterra method of locus plotting. **3** *Branching.* Terminology, types of branching. Properties of branching structures, examples. Depth induction as method in branching. Taxonomy and glottochronology. Combinatorics of branching trees. Kinship distance and its generating function. Statistics of branching fibers, order-statistics regions. Hypothesis on the growth of dendrites. A nonlinear renewal equation. **4** *Pursuit and related topics.* Simple pursuit in the plane, linear and circular cases. Periodic orbits and limit cycles. Approximate solutions, error control. Capture, stability, spiralling, plane convex pursuit. Periodic and ergodic motions. Generalization of pursuit, tractrices, pseudo-sphere, hyperbolic geometry. Pursuit as a smoothing, antipursuit and applications to social sciences. **5** *Generalizations of Steiner's problem.* Effective and efficient solutions, embedded and abstract problems. Steiner's problem for set terminals. Spanning subtrees, Prim's algorithm, applications. A general plane graph minimization problem. Steiner's problem under the Minkowski metric, Minkowski circles, and Minkowski ellipses. Longest paths, self-intersections, layers.

CHAPTER 6 FUNCTIONAL EQUATIONS AND
            MATHEMATICAL MODELING**325**

1 *Certain types of functional equations.* Several meanings of the term. Reduction to differential equations. The Ramsey functional. Local and nonlocal equations, examples. Classical versus hereditary mechanics, Picard's and Volterra's work. **2** *Derivation of certain simple functional equations.* Differential equation for the nearest neighbor probability, some consequences, selective nerve-cell stains. Derivation of a coupled system of equations describing signals in cables. A coupled system of difference–differential equations for helical coil with leakage. Examples of linear difference–differential equations with constant coefficients. **3** *Geodesic focusing and the Mosevich functional equation.* Various methods of focusing. Variable index and the Luneburg lens, the Luneburg integral equation. Focusing and aberrations, scanning and wide-angle photography. The Schwarzschild system. Parallel-plate systems, an equivalent of a parabolic mirror. Geodesic focusing. The Mosevich equation, osculating tori, and numerical procedures. **4** *The Hartline–Ratliff equations and pattern perception.* Feedback and lateral inhibition. Schematic description of nervous system activity, interaction and noninteraction types. Investigations on the Limulus crab vision, the Hartline–Ratliff equations. Their continuous analog as a nonlocal example with a Urysohn operator. Uniqueness of solutions, translationally unique intersections. Possibilities for pattern discrimination. Curvature of the light–dark interface, Mach bands, disinhibition. Pattern perception and set-membership problem for disjoint unions. Problems of non-disjointness. Poincaré and Menger on nonrecognition of arbitrarily small differences. Statistical metric spaces of Menger. Set theory and some paradoxes, two roles of money and the resulting paradox.

MATHEMATICAL IDEAS, MODELING AND APPLICATIONS

1
FURTHER TOPICS IN GEOMETRY

1. ELEMENTARY GEOMETRY AND TRIGONOMETRY

(a) In this subsection we unite a certain amount of elementary geometrical material: It is shown that some distinguished points of triangles and certain distance-minimizing points can be handled together.

Let similar isosceles triangles with equal angles ω be built outward on the sides of a triangle ABC as shown in Fig. 1. It was mentioned in vol. 1, p. 140 that if $\omega = 60°$ then AA_1, BB_1, CC_1 all meet in P which is then called the Steiner point of the triangle. However, leaving the general ω, let us apply the law of sines to the triangles AKC, AKC_1, ACC_1, and so on; we get then

$$\frac{AK}{KB} = \frac{\sin(A+\omega)\sin B}{\sin(B+\omega)\sin A}$$

and analogous expressions for BL/LC and CM/MA. Multiplying out the three ratios we find that

$$AK \cdot BL \cdot CM = KB \cdot LC \cdot MA.$$

So, by Ceva's theorem (vol. 1, p. 8) the lines AA_1, BB_1, CC_1 are concurrent for *every* ω, not just for $\omega = 60°$. Let the common point of the three segments be $P(\omega)$. This is defined in the first place for $0 < \omega < \pi/2$, but by simple limiting procedures we get also $P(0)$ and $P(\pi/2)$. The points $P(\omega)$ include several important points of the triangle: $P(\pi/3)$ is the Steiner point, $P(\pi/2)$ is the orthocenter ($=$ intersection of the three heights), and $P(0)$ is the center of mass.

The appearance of $P(\pi/3)$ as the Steiner point suggests a connection with an extremum problem: Let $Q = Q(a)$ be that point which minimizes the expression

$$(|QA|^a + |QB|^a + |QC|^a)^{1/a}, \qquad 0 < a < \infty.$$

1

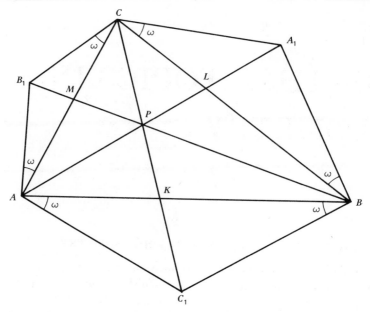

Fig. 1. Isosceles construction.

A limiting procedure shows that $Q(\infty)$ is the largest of the numbers $|QA|$, $|QB|$, $|QC|$. We find now that $Q(1) = P(\pi/3)$, $Q(2) = P(0)$, and $Q(\infty)$ is the circumcenter of the triangle ABC. Several problems suggest themselves; for example, the reader may wish to find the curves described by $P(\omega)$ and by $Q(a)$ as the parameters vary, and to find all their common points. Also, allowing for some modifications, we may extend $P(\omega)$ and $Q(a)$ to negative ω and a. Finally, a generalization to three dimensions might be attempted.

(b)   Since by elementary trigonometry

$$\sin(a+b)\sin(a-b) = \sin^2 a - \sin^2 b$$

we have

$$\sin 3x = 4 \sin x \sin(60° - x)\sin(60° + x), \tag{1}$$

and replacing x by $30° - x$

$$\cos 3x = 4 \cos x \cos(60° - x)\cos(60° + x). \tag{2}$$

Placing an equilateral triangle of side 1 with its three vertices on three parallel straight lines as in Fig. 2 we find by (1) and (2)

$$\sin 3x = 4ED \cdot CD \cdot CE, \qquad \cos 3x = 4AD \cdot AF \cdot DF.$$

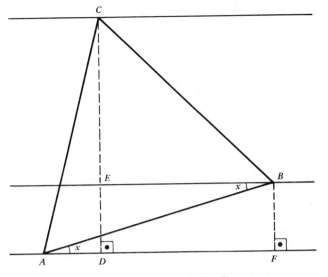

Fig. 2. An equilateral triangle.

Starting with (1) we can produce a simple proof of Morley's theorem: If the angles of a triangle are trisected and the trisectors adjacent to a side are allowed to intersect, then the three intersection points are the vertices of an equilateral triangle (Fig. 3). By taking the angles of the original triangle to be $3x$, $3y$, $3z$ and the circumcircle radius $\frac{1}{2}$, we get the sides to be $\sin 3x$, $\sin 3y$, $\sin 3z$ as shown in the figure. Also, $x + y + z = 60°$.

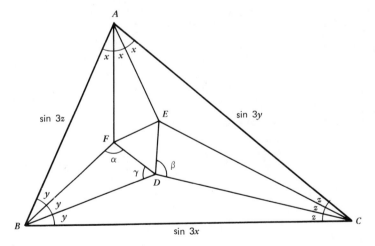

Fig. 3. Morley's theorem.

Applying the sine law to the triangle BDC we have therefore by (1)

$$BD = \frac{\sin 3x \sin z}{\sin (y + z)} = 4 \sin x \sin z \sin (60° + x)$$

and similarly

$$BF = 4 \sin x \sin z \sin (60 + z).$$

Hence

$$\frac{\sin \alpha}{\sin \gamma} = \frac{BD}{BF} = \frac{\sin (60° + x)}{\sin (60° + z)}.$$

But

$$\alpha + \gamma = 180° - y = 60° + x + 60° + z;$$

therefore

$$\alpha = 60° + x, \qquad \gamma = 60° + z,$$

and in the same way

$$\beta = 60° + y.$$

Now

$$\measuredangle FDE = 360° - \gamma - \beta - \measuredangle BDC$$

and

$$\measuredangle BDC = 180° - y - z$$

so that $\measuredangle FDE = 60°$. It follows that FDE is an equilateral triangle, since by symmetric argument all its angles are 60°. This method of proving Morley's theorem is essentially due to E. Borel [20].

Putting $x = 20°$ in (2) and (3) we get $\cos 20° \cos 40° \cos 80° = 1/8$, $\sin 20°$ $\sin 40° \sin 80° = 3^{1/2}/8$, and multiplying these we have

$$\prod_{k=1}^{8} \sin (10°k) = \frac{3}{256}.$$

This a special case $n = 18$ of

$$\prod_{k=1}^{n-1} \sin \frac{k\pi}{n} = 2^{1-n}n$$

which is itself a special case $\phi = 0$ of

$$\prod_{k=1}^{n-1} \sin \left(\phi + \frac{k\pi}{n} \right) = 2^{1-n} \frac{\sin n\phi}{\sin \phi}$$

given in vol. 1, p. 100.

(c)   The sums

$$C = \sum_{j=0}^{n-1} \cos (\alpha + j\beta), \qquad S = \sum_{j=0}^{n-1} \sin (\alpha + j\beta)$$

would be usually evaluated by using complex numbers, Euler's formula $e^{ix} = \cos x + i \sin x$, and the sum of a geometric series. But they can also be found by elementary geometry [76], exploiting the principle of evaluating one thing, here the length of a projection, in two different ways. Let a circle of diameter d pass through the origin O. Let OA_1, $A_1A_2, \ldots, A_{n-1}A_n$ be n chords of equal length $d \sin \beta/2$ where the angles α, β are as shown in Fig. 4. Projecting OA_1, $A_1A_2, \ldots,$ on the x-axis we find for the sum of the projections

$$Cd \sin \frac{\beta}{2}.$$

This must equal the projection of OA_n, which is $OA_n \cos (\measuredangle XOA_n)$. Since all angles in a circle on a chord, or on equal chords, are equal by a theorem of Thales, we have

$$\measuredangle A_1OA_2 = \cdots = \measuredangle A_{n-1}OA_n = \frac{\beta}{2}$$

and so

$$\measuredangle XOA_n = \alpha + (n-1)\frac{\beta}{2}.$$

Also, $OA_n = d \sin n\beta/2$ since OA_n subtends the angle $n\beta/2$ at the center

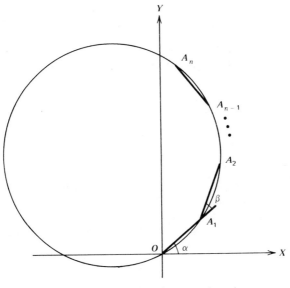

Fig. 4. Summation of a trigonometric series.

of the circle. Equating the two values for the quantity, we get

$$C = \frac{\sin n\beta/2}{\sin \beta/2} \cos\left(\alpha + \frac{n-1}{2}\beta\right).$$

Projecting similarly on the y-axis we find

$$S = \frac{\sin n\beta/2}{\sin \beta/2} \sin\left(\alpha + \frac{n-1}{2}\beta\right).$$

Thales of Miletus was an ancient Greek astronomer, geometer, philosopher, and speculator (640 B.C.–562 B.C.). Called one of the seven wise men of Greece, he is sometimes regarded as the father of European science and philosophy. With respect to the latter, he maintained that water was the principle of all things, being thereby the first recorded thinker who postulated a *natural* rather than a *supernatural* principle. As regards science, he was supposedly the first one to predict eclipses; by his geometrical work he laid the foundations of geometry and trigonometry. In particular, having "learned geometry from the Egyptians," [44], he did what they couldn't do: By his work on proportion and parallels he measured the height of a pyramid from the length of its shadow, and the height and the shadow length of a tree. He is reputed to have made a "killing" on the market by having bought out all available barrels before a good olive-crop. When asked, "what is difficult?" he said "to know oneself," "what is easy?"—"to give advice," "what is most pleasant?"— "success," "what is older, day or night?"—"night, by one day," "what is the strangest thing ever seen?"—"an aged tyrant," and so on.

(**d**)　If a (convex) quadrilateral Q has four fixed sides of lengths a, b, c, d we may regard it as a linkage of four bars joined by hinges, as in Fig. 5a. As the linkage shape changes it is not clear a priori when Q will be (a) inscribed into a circle, (b) circumscribed about a circle. It turns out that property (b) holds either always or never: The condition here depends on a, b, c, d alone. By reference to Figs. 5a and 5b the condition is

$$a + c = b + d (= u + v + w + z). \tag{3}$$

The condition for the existence of a circumcircle to Q is obtained from the previously mentioned Thales' theorem on the angles in a circle, based on the same chord: Moving D and B along the circumcircle to A, we find as in Fig. 5c, that the condition is $\alpha + \gamma = \pi$, or equivalently

$$\alpha + \gamma = \beta + \delta \tag{4}$$

providing a sort of dual to (3). A condition equivalent to (4) is Ptolemy's

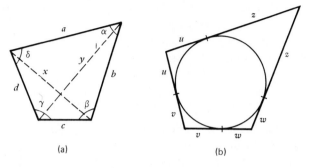

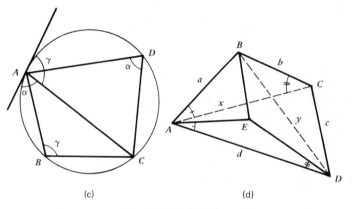

Fig. 5. Quadrilaterals.

theorem: The quadrilateral of Fig. 5a has a circumcircle if and only if $xy = ac + bd$.

We prove this by locating the point E of Fig. 1d so that

$$\sphericalangle DAE = \sphericalangle CAB, \qquad \sphericalangle BCA = \sphericalangle EDA.$$

Then, we conclude successively the similarity of triangles EDA and BAC, the equality $\sphericalangle DAC = \sphericalangle EAB$, and the similarity of BAE and DAC. By the cosine law

$$BD^2 = BE^2 + ED^2 - 2BE \cdot ED \cos(B + D) \tag{5}$$

since

$$\sphericalangle DEB = 2\pi - (\sphericalangle DEA + \sphericalangle BEA) = 2\pi - (B + D).$$

Multiplying (5) by AC^2 and using the previously established similarities of triangles, we get

$$x^2 y^2 = a^2 c^2 + b^2 d^2 - 2abcd \cos(B + D).$$

This is a generalization of Ptolemy's theorem for if there is a circumcircle then $B + D = A + C = 180°$ so that $xy = ac + bd$.

Ptolemy's theorem generalizes to Casey's theorem [88]: If t_{12} is the length of the common ipsilateral tangent to two circles C_1 and C_2, then four circles C_1, C_2, C_3, C_4 are tangent to a common circle (or line) if and only if

$$t_{12}t_{34} \pm t_{13}t_{24} \pm t_{14}t_{23} = 0.$$

We get Ptolemy's theorem when all four circles reduce to points.

Ptolemy's theorem is useful in a variety of situations and we give an application to elementary number theory: We shall show that the unit circle U can be arbitrarily well approximated by an inscribed polygon with all sides and all diagonals of rational length. The idea of the proof is to combine Ptolemy's theorem with the use of Pythagorean number triples $(m^2 - n^2, 2mn, m^2 + n^2)$:

$$(m^2 - n^2)^2 + (2mn)^2 = (m^2 + n^2)^2,$$

where m and n are integers. With reference to Fig. 6 we locate the points C and D corresponding in terms of the lengths AC, CB, AD, DB to, say, the Pythagorean triples 3, 4, 5 and 5, 12, 13, as shown. Now AB, AC, AD, BC, BD are rational, hence by Ptolemy's theorem CD is also rational. If a point E lying on U also corresponds to a Pythagorean triple, then we show in the same way that EC and ED, and hence all other necessary lengths, are again rational, and so on.

Leaving aside the trivial case of a straight line, we might conjecture that circles are the only curves arbitrarily well approximable by inscribed polygons with all sides and all diagonals of rational length. This is related

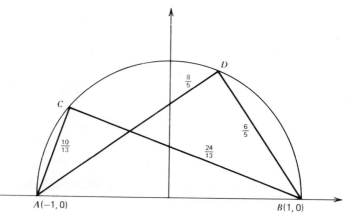

Fig. 6. Rational points.

to the conjecture that in the Euclidean plane there does not exist a dense set of points every two of whom are a rational distance apart. The two conjectures form the beginning of a whole hierarchy; we shall not go into this beyond stating just one more conjecture of that type. The idea here is that not only there cannot be a dense set with all distances rational, but that we depart, in a sense, further and further from rationality of distances the "denser" the set gets, and the departure is *extremely rapid.* To be specific, given a finite set X of points in the plane, let $R(X)$ be the field extension of rationals R obtained by adjoining all distances attained by pairs of points in X. Let Y be a compact plane set with nonempty interior and for small $\varepsilon > 0$ let $Y(\varepsilon)$ denote a finite set of points such that no point of Y is further than ε from a point of $Y(\varepsilon)$. Then our conjecture may be stated as follows:

$$\deg \frac{R(Y(\varepsilon))}{R} \geq 2^{c\varepsilon^{-2}},$$

where c is a positive constant depending on Y alone.

2. GENERATING SOLID SHAPES

In this section there will be considered several heavily idealized and simplified methods of generating solid shapes in, say, metal or plastic. These methods are selected to illustrate certain elements of the geometry of convexity, to provide a natural setting for a sample theorem of less elementary type, and to suggest a number of problems.

We consider first a version of the standard method of sand-casting. The solid S, of Fig. 1a, to be cast is first formed as a prototype in wood or plaster. It is then pressed partly into the sand in the lower box L as shown

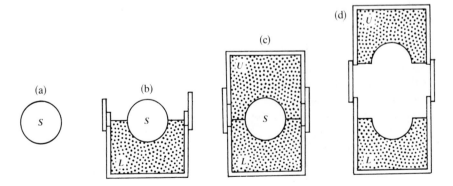

Fig. 1. Sand-casting.

in Fig. 1b. Next, an upper box U full of sand is placed tightly over S and fitting L, as shown in Fig. 1c. The sand used is of a special self-sticking variety so that the boxes can be taken apart and S removed, as in Fig. 1d, without the collapse of sand. The boxes are then put together again, a channel is poked into the cavity, and the molten metal or plastic is poured in. We suppose that in the process the boxes L and U can be moved only up or down, and that the interface between the two portions of sand, the so-called breaking surface, is in a horizontal plane. We suppose that when S is removed from the sand, this is done by simple lifting straight out. Thus, for instance, we cannot remove an S that is shaped like a corkscrew, even though a helical motion would lift it out of sand (of course, new problems arise when we admit such motions).

The natural question arises now of characterizing those shapes S that can be cast by this method. We suppose that S is connected (= of one piece, for the *present* purpose). Call a solid X a half-body if a part of its boundary ∂X lies in a plane P and the following is true: With a Cartesian coordinate system in which P is the *xy-plane*, X lies in the half-space $z \geq 0$, and for $a > b \geq 0$ the projection of the section P_a of P by the plane $z = a$, onto the plane $z = b$ lies in P_b. Now the reader may wish to show that S can be cast if and only if it is a union of two half-bodies whose base planes coincide and whose bases overlap partly. We observe an alternative description of a half-body X: A cavity in the shape of X can be drilled into a metal block with a flat face (assuming that the block is moveable only at right angles to the drill).

Next, we ask what solids S can be cast in arbitrary orientation with respect to the sandboxes. There is no difficulty in showing that a necessary condition is that S must be convex. To answer our question we need a theorem of Blaschke [16] which will require some machinery. Given a convex solid K and a direction u, the equator $E(K, u)$ of K along u is the locus of points in which lines parallel to u touch the surface ∂K of K. Briefly, $E(K, u)$ is the intersection of ∂K and the cylinder projecting K along u; we have clearly $E(K, u) = E(K, -u)$. If K is strictly convex, that is, when ∂K contains no straight segments, then every equator $E(K, u)$ is a simple closed space curve on ∂K which disconnects ∂K into two sets; by further analogy we can call these two sets the northern and the southern hemispheres of K along u. The reader may wish to prove conversely, that if every equator $E(K, u)$ is a simple closed curve on ∂K then K is strictly convex.

Now, the theorem we need is this: All equators $E(K, u)$ of K are plane curves if and only if K is an ellipsoid. An outline of the proof follows. Locate K in a Cartesian coordinate system. Let u_0 be a direction and suppose first that K has the weaker property $P(u_0)$ of having every equator $E(K, u)$ plane if $u \perp u_0$; choose u_0 along the z-axis. If the horizontal planes support K from above and below at p and q, say, then we may assume without loss of generality that p and q are on the z-axis, for, an affine transformation will achieve it, preserving $P(u_0)$. Suppose that planes $z = a$, $z = b$ cut ∂K in curves H_1 and H_2, let $u \perp u_0$, and consider the closed convex (plane) curve $E(K, u)$. This cuts each H_1 in two points and the tangents at the four points are parallel, being all in the direction u. Since u is arbitrary, subject only to $u \perp u_0$, it follows that the sections of ∂K by $z = a$, a arbitrary, are similar and similarly located curves, each with a center on the z-axis.

Hence $E(K, u)$ is not only crossing points p and q but, further, it is symmetric with respect to pq. Suppose now that $P(u_0)$ holds for every direction u_0 and consider a fixed equator $E = E(K, u)$. By the foregoing E is affine-symmetric with respect to every line $\perp u$; hence it must be an ellipse. Roughly, the last step is shown by the use of the conjugacy principle; it can be shown that there is an affinity A and a sequence of rotations R_n *converging to identity*, such that $A R_n A^{-1}$ are rotations; hence E is an affine image of a circle, that is, an ellipse, q.e.d.

Closely related theorems assert that for every direction u the midpoints of secants of K along u lie in a plane if and only if K is an ellipsoid, and that flat parallel sections of K are homothetic if and only if K is an ellipsoid.

Returning to our problem of castibility in arbitrary orientation, we show now that the shape S must be spherical. If the method of casting is changed so that the surface of breaking, while still plane, is not required to be horizontal, then every ellipsoid can be cast in arbitrary orientation. If nonplane breaking surfaces are allowed then any convex shape can be cast in arbitrary orientation.

The equators and half-bodies suggest some further questions in convexity. To begin with, it is easy to extend the definition of a half-body to n dimensions; by an affine half-body we shall understand simply the image under an affine transformation of a half-body. Given an n-dimensional convex body K we ask: What is the biggest volume of an affine half-body $H_u(K)$ which can be cut off K by a plane orthogonal to u? Here we may observe a restatement of our theorem on the ellipsoids: A convex body K is a union of two affine half-bodies with common base P, for arbitrary orientation of P, if and only if K is an ellipsoid. Thus, for an ellipsoid

$$\min_u \frac{\text{Vol } H_u(K)}{\text{Vol } K} = \frac{1}{2}$$

which is the largest value possible. Using the previously mentioned theorem of Blaschke on ellipsoids being the only convex bodies with plane equators, the reader may wish to prove the following characterization: Ellipsoids are the only convex bodies with the property that if any plane cuts them then of the two parts one is an affine half-body.

Given a convex body K we have defined the equator $E(K, u)$ in the direction u; $\partial K - E(K, u)$ is a union of two components $S(K, u)$ and $N(K, u)$ which may be called the southern and the northern (open) hemispheres; we have $N(K, u) = S(K, -u)$. We have now various problems referring to max's, min's, and minmax's of the thicknesses of parallel slabs which contain $E(K, u)$, $S(K, u)$, $N(K, u)$. We also note: The condition that for every convex body K there is a hemisphere $N(K, u_0)$ whose closure is in a union of some n open hemispheres $N(K, u_i)$, $i = 1, \ldots, n$, (n being the dimension of K), is equivalent to the still unproved Borsuk conjecture (see vol. 1, p. 157).

In another method of producing metal shapes S we take a solid metal block, put it on a drill press, and sink a thin drill to some prescribed depth; then we reorient the block and repeat. Here the question of which shapes S can be produced is simply answered: A necessary and sufficient condition is that the complement $E^3 - S$ of S with respect to its ambient space E^3 should be a union of straight rays. Equivalently, every point of the boundary ∂S must be freely visible in some direction from points arbitrarily far away from S.

Let us recall here that an equivalent definition of convexity is this: A solid K in E^3 is convex if and only if $E^3 - K$ is a union of half-spaces. Together with the preceding observation on producing shapes by drilling, this suggests the following generalization of convexity: Let F be a suitable family of sets in E^3 (or E^n); what is the family X of sets whose complement (in E^3 or E^n) is a union of members of F? For instance, if for F we take the set of all solid cones of semivertical angle α, then for $\alpha = \pi/2$ X is the family of convex bodies and for $\alpha = 0$ X is the family of sets generable by drilling. The reader may wish to consider the cases when F is the family of all spheres, or circular cylinders, or straight lines.

Finally, we consider producing a solid by laminating together shapes cut out of thin plane sheet. Here the problem is, roughly, to decide when two solids K and Q have the same plane sections $K \cap P$ and $Q \cap P$, for every plane P orthogonal to a given line L. This suggests the following. In the plane P we define a transformation T on plane figures, we take K and apply T to every section $K \cap P$ thus obtaining a new solid Q. Special cases include:

(a)　T is a rigid motion, Q is then composed of the same plane sections as K;

(b)　T is an area-preserving transformation, Q is then a solid of the same volume as K—that is the content of the Cavalieri principle (vol. 1, p. 150);

(c) T applied to a plane figure A produces a circle in the plane of A, of the same area as A, and centered on L; Q is then the symmetrization of K with respect to the line L (vol. 1, p. 47).

3. LINKAGES AND OTHER ANALOG MECHANISMS

Some examples of linkages have already been considered, for instance, those of Peaucellier and Hart, vol. 1, p. 13. These two devices perform the inversion of a plane locus and so, incidentally, by inverting a suitable circle they draw a straight line L. It must be emphasized that L is not drawn by reference to a physical law (plumb line, stretched elastic band, bent paper page) or to another straight line (ruler).

The simplest linkage is, of course, a rigid bar, one point of which remains fixed. In its various modifications this is known as compasses, lever, pry-bar, simple balance, oar, and so on. With reference to rulers and compasses, and the role of linkages as mechanical devices we may mention some work of the Italian mathematician L. Mascheroni (whose name is sometimes associated with that of Euler in the naming of the number

$$0.5772157\ldots = \lim_{n \to \infty} \left(\sum_{j=1}^{n} \frac{1}{j} - \log n \right)).$$

In his book, *La Geometria dell Compasso*, written in 1797 [108], Mascheroni set out to find which classical ruler-and-compasses constructions could be done by compasses alone. He found that *all* of them could be, provided that a straight line was considered as given once some two points of it were known. His aim was practical as well as theoretical—to try to help mechanics by replacing inaccurate workshop operations (i.e., those employing rulers and edges) by more accurate ones (i.e., those employing compasses and their modifications) [108]. We shall loosely define a linkage as a finite collection of rigid bars, connected by hinges and slides so as to have one degree of freedom. That is, once the linkage is set down in its working position with certain point(s) fixed, then a single parameter (length, angle, ratio, etc.) determines the position of every part. The importance of linkages derives from the fact that when some parts are moved in a certain way then other parts perform a type of motion that we wish. Briefly, linkages are geometrical transformers (much as certain devices with coils and electromagnets are electrical transformers). What such a geometrical transformer effects is an object well enough studied in mathematics: a geometrical transformation. Here we shall pay some attention to the transformer itself rather than the transformation. Five simple examples of linkages are given in Fig. 1.

In Fig. 1a there is the simplest of all linkages, a rigid bar with a point o fixed, the path of p is a circle about o. In Fig. 1b we have the inversion linkage of Peaucellier. By fixing o and moving p on a locus L the scriber s is made to trace out the locus which is the image of L under inversion in a circle about o. Fig. 1c shows the Roberval parallel weighing scales. The weighing itself could be performed without any multiple bars, but then the scales would not stay horizontal during their motion. Here we have five rigid bars hinged as shown, and the scales are always horizontal. In Fig. 1d we have the familiar fire tongs where the angular change in α results in the displacement d of the jaws. A somewhat similar arrangement in Fig. 1e produces the cable cutters in which a large angular motion α of the handles results in a small angular motion β of the blades, with the consequent mechanical advantage in shear.

Gilles Personne de Roberval (1602-1675) was a French mathematician, physicist, and inventor [7]. Along with B. Cavalieri he was the discoverer of an early form of integral calculus. In geometry he worked on the cycloid; this name was given to that very popular curve by Galileo through his student Torricelli; Descartes and Pascal used the name *roulette* while Roberval introduced the name *trochoid* (from the Greek: *kuklos*—circle, *trochos*—wheel). Roberval was professor of arithmetic,

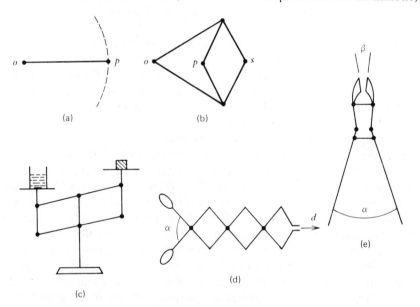

Fig. 1. Some linkages.

music, geometry, optics, mechanics, astrology, and geography at the Royal College of France.

We exhibit the use of a sleeve and slide arrangement on the example of a power linkage. As in Fig. 2a L, M, N are three rigid bars, L is fixed, and the sliding cross glides on L with its arms at right angles; as a result the movable bars M and N intersect at right angles on L. Recall that in a right-angled triangle the height belonging to the right angle is the geometric mean of the two segments into which it divides the hypothenuse. As in Fig. 2b we fix two bars X and Y at right angles and make the arrangement of sleeve–slides as shown. If the points A and B are at 1 and a units from the origin, we get a mechanical way of generating the first few powers 1, a, a^2, a^3, a^4, a^5, ... Alternatively, working in reverse, we can use the power linkage for an approximate *simultaneous* determination of the quantities $a^{1/5}$, $a^{2/5}$, $a^{3/5}$, $a^{4/5}$, given a (positive and fairly close to 1).

By using the power linkage together with the angle multisectors (to be described) the reader may wish to design a mechanical method for powers and roots of complex numbers.

An example of a sliding hinge is given in Fig. 3a. Here ABA_1 and ACA_2 are rigid rods, there are hinges at B and C, and $|AB| = |AC|$, $|BD| = |CD|$. The hinge D can slide freely along the rigid rod AD. The result is that we obtain a graphical device for angle bisection since clearly $\angle DAB = \angle DAC$. Putting together two such units in effect, as in Fig. 3b,

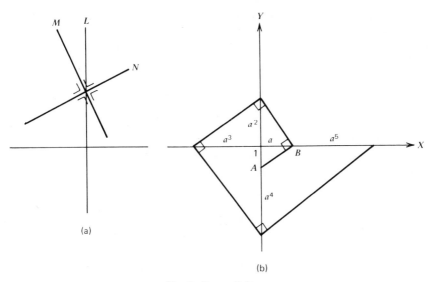

(a)

(b)

Fig. 2. Power linkage.

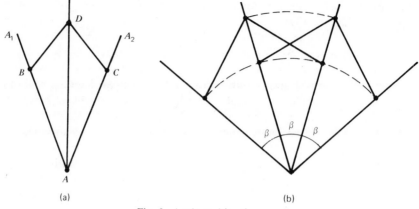

(a)   (b)

Fig. 3. Angle multisection.

will give us a mechanical angle trisector; it is clear that a general angle multisector (for any fixed multiple) can be similarly built up. The reader may wish to consider whether sliding hinges are necessary for trisection or whether a linkage with fixed hinges will do (see reference [77]); a matter which may, or may not, be connected is an attempt by Franz Reuleaux [142] to set up something like an algebraic theory of mechanisms, somewhat on the model of groups and semigroups presented in terms of generators and defining relations.

Combining together the sliding hinge arrangement of Fig. 3a and the sleeve and slide, we get a seven-bar linkage of Fig. 4 which can draw

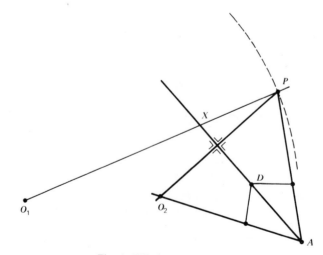

Fig. 4. Elliptic compasses.

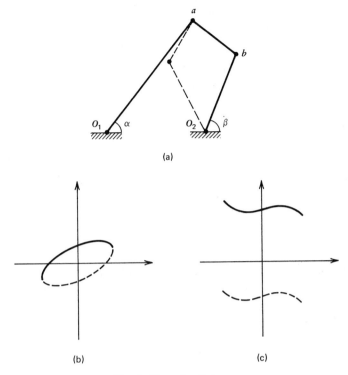

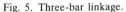

Fig. 5. Three-bar linkage.

ellipses and hyperbolas. O_1 and O_2 are fixed points, and the distance O_1P is fixed so that P describes a circle about O_1. It follows that X is equidistant from P and O_2; hence as P moves on the circle, X describes an ellipse. Similar arrangement but with the point O_2 outside the circle described by P, will lead to X tracing one branch of a hyperbola. The reader may wish to design a linkage to draw parabolas, and one to draw all three conics.

A special class is formed by the three-bar linkages; an example is shown in Fig. 5a. Here three rigid bars are hinged at a and b; O_1 and O_2 are fixed hinges. During the motion each point of O_1a and of O_2b describes a circle while points of ab describe rather complicated algebraic curves. One of the principal uses of three-bar linkages is to produce approximations to a required relation $\beta = f(\alpha)$ between the angles of Fig. 5a for various mechanical purposes [166]. Two of the many functions achievable in this way are shown (with the obvious provision for the ambiguity) in Fig. 5b and c. We observe that it is simple to produce mechanically the iterate $f(f(\alpha))$: Place two identical three-bar linkages in tandem as in Fig. 6b, starting with the linkage of Fig. 6a in which the

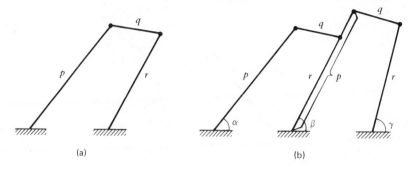

Fig. 6. Iterated linkages.

letters refer to the bar lengths. We find that $\beta = f(\alpha)$ and $\gamma = f(\beta)$ so that $\gamma = f(f(\alpha))$.

A slight modification of three-bar linkage which gives in effect the so-called antiparallelogram (Fig. 7a) is perhaps the simplest device to draw an ellipse. Assuming the rods to be hinged and with $|F_1F_4| = |F_2F_3|$, $|F_1F_2| = |F_3F_4|$, we fix F_1 and F_2; then the triangles F_1F_2X, F_4F_3X are congruent, hence during the motion X moves so that $F_1X + F_2X$ stays fixed, that is, on an ellipse.

Adding a rolling vertical rough-rim wheel to a bar of a linkage will change the motion so that the bar carrying the wheel stays normal to the curve C described by the point of contact of the wheel and the plane P on which it rolls (Fig. 7b). The wheel W is supposed to be free to turn on the bar but not to slide along it. Suppose now that the antiparallelogram linkage of Fig. 7a is provided with the two such wheels W_1 and W_2 shown, as viewed from above, in Fig. 7c. The points F_1 and F_2 are no longer fixed, the motion is now such that the point X is moving on a (sufficiently slowly curving) curve C. The reader may wish to show that the motion here is the same as though the ellipse E, imagined to be rigidly joined to the linkage, were rolling on C without slipping. Thus we have a simple mechanical arrangement for drawing compound roulettes of an ellipse (see vol. 1, p. 37) with respect to a plane curve C. When C is a straight line we get the simple roulette ($=$ elliptical equivalent of the cycloid); this was used by R. C. Yates to generate a minimal surface of revolution [186].

Similar device of a vertical wheel is also used in the Amsler planimeter, an analog linkage device for measuring plane areas, which we shall now describe.

Let a rigid straight bar p_1p_2 of length l move in the plane so that the end points are $p_1(x_1, y_1)$ and $p_2(x_2, y_2)$, and the bar makes angle θ with a

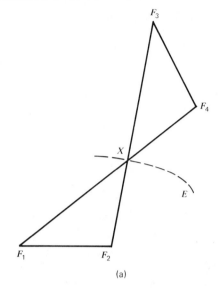

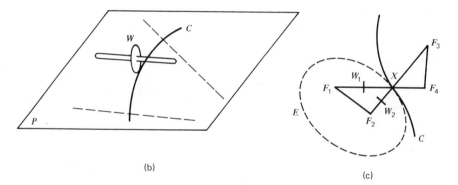

Fig. 7. Ellipses and their roulettes.

fixed direction in the plane. Then

$$x_2 = x_1 + l \cos \theta, \qquad y_2 = y_1 + l \sin \theta.$$

If p_1 describes a closed curve A_1 and p_2 a closed curve A_2, then by differentiating the preceding equations we find

$$x_2 \, dy_2 - y_2 \, dx_2 = x_1 \, dy_1 - y_1 \, dx_1 + l^2 \, d\theta$$
$$+ l(\cos \theta \, dy_1 - \sin \theta \, dx_1) + l(x_1 \cos \theta + y_1 \sin \theta) \, d\theta. \tag{1}$$

We also have the area formula

$$A_i = \frac{1}{2} \oint (x_i \, dy_i - y_i \, dx_i), \qquad i = 1, 2.$$

Hence, integrating (1), employing integration by parts, and observing the winding number

$$\oint d\theta = 2k\pi \tag{2}$$

we get

$$A_2 = A_1 + k\pi l^2 + l \oint (\cos\theta \, dy_1 - \sin\theta \, dx_1),$$

since after a complete revolution certain quantities will return to their original values.

Supposing now that the point p_3 of Fig. 8 describes a closed curve of area A_3, let s be the arc-length on the curve described by p_1 and α the angle which the tangent to the curve makes with $p_1 p_2$. Then we have the

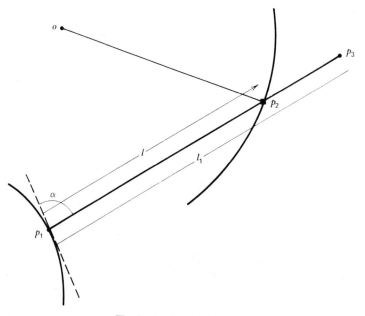

Fig. 8. Amsler planimeter.

equations

$$\cos \theta \, dy_1 - \sin \theta \, dx_1 = \sin \alpha \, ds$$

$$A_2 = A_1 + k\pi l^2 + l \int \sin \alpha \, ds \qquad (3)$$

$$A_3 = A_1 + k\pi l_1^2 + l_1 \int \sin \alpha \, ds.$$

Eliminating the integral we get

$$l_1 A_2 - l A_3 = (l_1 - l) A_1 + k\pi l l_1 (l - l_1),$$

which may also be written as

$$A_1 d_{23} + A_2 d_{31} + A_3 d_{12} + k\pi d_{12} d_{23} d_{31} = 0;$$

here d_{ij} is the signed distance $p_i p_j$. This generalizes the formula of vol. 1, p. 201. Eliminating A_1 from equations (3) we have

$$A_3 = A_2 + k\pi (l_1^2 - l^2) + (l_1 - l) \int \sin \alpha \, ds. \qquad (4)$$

The Amsler planimeter is an area-measuring device working on the basis of the above formula. It consists of a linkage shown in Fig. 8 in which o is a fixed point and op_2 a fixed distance. The point p_3 describes the closed curve whose area is to be measured. At p_1 we have a small circular rough disk D free to rotate on its shaft $p_1 p_2 p_3$. If θ is the rotation angle of D then

$$d\theta = c \sin \alpha \, ds.$$

During the complete motion p_2 describes a circle of fixed radius r and so we have

$$A_3 = \pi r^2 + k\pi (l_1^2 - l^2) + \frac{(l_1 - l)\theta}{c}$$

so that, with a suitable mechanization, we obtain the area A_3 in terms of the total angular turn θ of the dial on the disk D.

We return for the moment to the power linkage of Fig. 2. It is easy to design companion linkages which add and multiply; combining these with the power linkage, the reader may wish to prove a theorem credited in [5] to A. B. Kempe: A linkage can be given to draw any (real) algebraic curve. For its equation is of the form

$$\sum_{p,q} \sum a_{pq} x^p y^q = 0$$

so that, taking care of the constants a_{pq}, we can generate mechanically y from x by powers, multiplications, and additions.

A. B. Kempe was a nineteenth century English mathematician, known, i.a., as one of the first to attempt a proof of the four-color conjecture [93]; he also wrote a memoir on linkages, reprinted in [77].

4. POLES, POLARS, AND POLAR RECIPROCITY

Let a and b be two points on a line L; we take L for the x-axis, a for the origin, and $|ab|$ for the unit of length. If p is a point on L with the abscissa x then the ratio $x/(x-1)$ in which p divides ab is plotted in Fig. 1a and its absolute value $|x/(x-1)|$ in Fig. 1b.

From Fig. 1b it is clear, by considering the line $y = r$, that for any positive r other than 1, there are exactly two corresponding values of x. For $r = 1$ one value is $x = 1/2$, the other one is the "point at infinity." Once this is properly defined in projective terms, the quotation marks can be removed. Thus, there are generally two points c and d which divide ab in the given length-ratio r. The same thing can be shown geometrically if we recall that the locus of points with a given ratio of distances to a and b is the circle (of Apollonius) except when that ratio is 1. The points c and d are called harmonic conjugates of a and b.

The circle of Apollonius appears in the following problem: If two houses are polyhedral prisms standing on the horizontal plane P, with vertical walls and flat but not necessarily horizontal roofs, then the locus of points in P from which the houses appear equally tall is piecewise circular (or linear).

Let C be a fixed conic and p a point in its plane, and let a variable secant line L through p cut C in r and s. We consider the locus of the

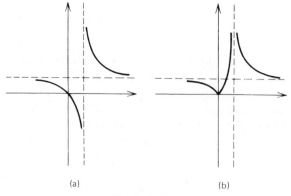

(a)   (b)

Fig. 1. Ratios.

point q, such that p and q are harmonic conjugates of r and s. It can be proved, by analytic geometry for instance, that this locus is a straight line P through the two points of tangency (real or imaginary ones) of C and the two tangents to C through p. P is called the polar of p, and p the pole of P. By means of this relation of polar reciprocity, to any figure F of lines and points there corresponds a dual polar figure F' of points and lines. The relation is involutory: $(F')' = F$. It is not hard to show that all poles of lines through a given point lie on the polar of that point. Conversely, all polars of points of a given line pass through the pole of that line.

If S is a plane curve we consider a variable tangent to S. Regarded as a moving polar, this induces a moveable pole that traces out the reciprocal polar curve of S.

Treated algebraically, the polars are quite simply obtained. Let $C = 0$ be the equation of the fixed conic, where C is some polynomial, of second degree in x and y. Let the point p have coordinates x_1, y_1; then (the equation of) the polar P is obtained by replacing in C

$$x^2 \text{ and } y^2 \text{ by } xx_1 \text{ and } yy_1,$$

$$xy \text{ by } \frac{xy_1 + x_1 y}{2},$$

$$x \text{ and } y \text{ by } \frac{x + x_1}{2} \text{ and } \frac{y + y_1}{2},$$

constant term by itself.

If p is on C so that $C(x_1, y_1) = 0$ we get an easy way of writing down the equations of tangents to conics.

With respect to the geometry of conics there are several uses of polar reciprocity which we shall illustrate on examples:

new theorems from old—given any general property of conics, transformation by reciprocal polars gives us a dual property;

hard theorems from easy ones—given a theorem to be proved for the general conic C, we arrange for a polar transformation under which C goes over into a circle, for which simple case the theorem may be much easier to prove;

new theorems—finally, polar reciprocity may be used to give elementary geometrical proofs and constructions for problems not easily tractable otherwise.

As our first example we consider Pascal's theorem: In a hexagon H inscribed into a conic C the three pairs of opposite sides intersect in three points which are collinear.

To prove this we let $C = 0$ be (the equation of) the conic and $L = 0$ (that of) a line cutting C. A conic through the intersection of C and L has the equation $C + k_1 LL_1 = 0$, where L_1 is a linear polynomial in x and y, and k_1 a constant. Let $C + k_2 LL_2 = 0$ be another such conic. Each two conics of the three intersect in four points; two of them are on L and two others determine a certain straight line. The three such lines are $L_1 = 0$, $L_2 = 0$, and $k_1 L_1 - k_2 L_2 = 0$; hence they have a common point.

We now let a, b, c, d, e, f be the vertices of H and we take the line ad as L. For the other two conics we take the degenerate cases of line-pairs: ab and cd, af and de. Applying the preceding we get Pascal's theorem.

Blaise Pascal (1623–1662) was a French mystic, philosopher, theologian, stylist, inventor (of calculating machines), natural philosopher, and mathematician. Son of a mathematician (Etienne), Pascal proved the above theorem at the age of sixteen, without much geometrical background except for some acquaintance with the previous work of G. Desargues. In spite of his short life Pascal had quit mathematics long before his death (of a huge brain tumor). Similar early transfer of interest from mathematics and natural philosophy to religion occurred with Sir Isaac Newton. Hermann Grassmann, one of the discoverers of multilinear algebra and tensor calculus, similarly relinquished mathematics, in this case for linguistic studies in Indoeuropean languages, and Rig-Veda in particular.

Returning to Pascal's theorem, we take reciprocal polars with respect to C and we find that the sides of H go over into the vertices of a hexagon circumscribed about c. We have, thus, a dualization of Pascal's theorem to Brianchon's theorem: If a hexagon circumscribes a conic then the three diagonals joining opposite vertices are concurrent.

For neither theorem is it required that the hexagon be simple, that is, not self-intersecting. We have, for instance, the configurations of Fig. 2a and 2b.

There are various special degenerate cases of the two theorems. For instance, the conic C itself may degenerate to a pair of (intersecting or parallel) straight lines. Pascal's theorem becomes then Pappus' theorem: If in a crossed hexagon H the vertices lie alternately on a pair of lines then the opposite sides of H intersect in three collinear points (see Fig. 3). This theorem is often taken as an axiom of projective geometry. Part of the reason is the following. If in a projective geometry one attempts to introduce coordinates by taking a line L and three points O, U, I on it, with O playing the role of the origin, U of the unit point, and I of the point at infinity, then we have the following projectively invariant construction for multiplication [78]. Let A be on L "a units away from O" and B similarly "b units away from O." Let P be a point off L, Q a point

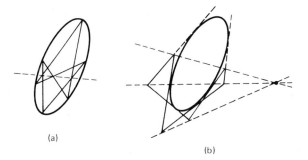

(a)

(b)

Fig. 2. Pascal's and Brianchon's theorems.

on *PI*, *R* the intersection of *UQ* and *AP*, *S* the intersection of *QB* and *OR*. Then *PS* cuts *L* in *T* which is "*ab* units away from *O*," (Fig. 4).

This construction is independent of the choice of *P* and *Q*. However, if with *P* and *Q* fixed the roles of *A* and *B* are interchanged then we get a new point T_1, which shows that the algebraic system with which we are coordinatizing our geometry is in general not commutative. But it can be shown that $T = T_1$ always if and only if Pappus' theorem holds in our geometry. Thus Pappus' theorem as a projective axiom is equivalent to commutativity.

Various special cases are obtained from Pascal's and Brianchon's theorems by merging two or more consecutive vertices *a*, *b* of the hexagon, to a single point *a*. As limit considerations suggest, the side *ab* of the hexagon goes over into the tangent *T* to the conic at *a* (it is interesting to speculate whether this played some role in the correspondence between Pascal and Fermat, which almost succeeded in establishing calculus years before Newton and Leibniz). For instance, the five-point version of Pascal's theorem is illustrated in Fig. 5. We note incidentally that this

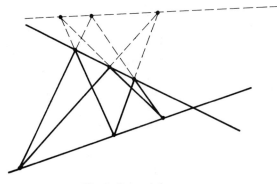

Fig. 3. Pappus' theorem.

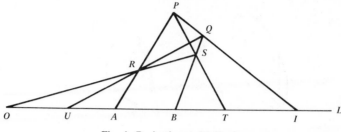

Fig. 4. Projective multiplication.

allows us to construct by ruler alone the tangent at a given point of a conic if four other points of the conic are known.

As a sample of degenerate Brianchon's theorem we consider an ellipse inscribed into a triangle *cea* as shown in Fig. 6. Regard now the triangle *cea* as the limiting case of the circumscribing hexagon *abcdef*. Applying Brianchon's theorem we find that the lines joining the vertices of the triangle to the opposite tangency points are concurrent.

Finally, using Pascal's and Brianchon's theorems, we find that a conic is uniquely determined by five points or, equivalently, by five tangents.

We give next an example of polar reciprocity applied toward deducing (relatively) hard propositions (about conics) out of easy ones (same ones but about circles). This applies in particular to focal properties [22]. Let *O* be a circle with the center *o*; from the definition of poles and polars it

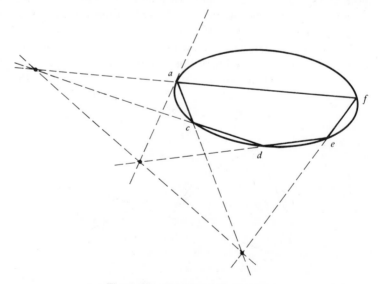

Fig. 5. Five-point Pascal's theorem.

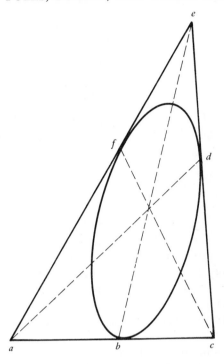

Fig. 6. Three-point Brianchon's theorem.

is not hard to show that the ratio of distances of two points from o is equal to the ratio of distances of each point from the polar of the other. Let C be a circle inside O centered at c, and let D be the polar of c with respect to O. To find the reciprocal polar curve C' of C with respect to O we let T be any tangent to C and t its pole. Applying the previous proposition on the ratio of distances, we find that the variable point t of C' has constant ratio of distances to D and to o. Hence C' is a conic with a focus o and the corresponding directrix D. Next, it follows from the definition of polarity that if x and y are the poles of X and Y then $\angle XY = \angle xoy$.

From the foregoing we deduce that if T_1 and T_2 are tangents to a conic C at points t_1 and t_2 on C, if o is a focus of C and p the point of intersection of T_1 and T_2, then op bisects $\angle t_1 o t_2$. This holds for C since it holds, trivially, for a circle (with center o); see Fig. 7.

In much the same way we start from the simple observation that the chords of a fixed length of a circle envelop a concentric circle. From this, by similar reasoning, we deduce the not so simple proposition that the family of all chords of a conic C which subtend a fixed angle at the focus o of C, envelops a conic with the same focus o and the same directrix.

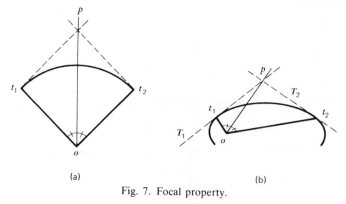

Fig. 7. Focal property.

This is illustrated in the Fig. 8 for the case when C is a parabola and the angle is 90°. The envelope E is an ellipse as shown in the figure.

Finally, as an example of polar reciprocity used in an essential way in a proof, we consider the solution due to Gergonne of the problem of Apollonius: to find the circle(s) tangent to three given circles C_1, C_2, C_3 in the plane. Allowing for the limiting cases when some of the C_i's become lines or points we have ten problems, some with several subcases. Here we shall consider only the case when C_i's are circles, external to one another, and of pairwise different radii.

Some easy solutions can be given here, which, however, involve the use of conics and so are not elementary in the traditional ruler-and-compasses sense. They are based on these observations: (a) the center of

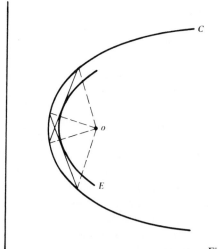

Fig. 8. Conics and envelopes.

a circle C tangent to all C_i's stays fixed when all C_i's have their radii changed by the same amount (plus or minus, depending on the type of tangency of C with C_i—external or internal); hence we can assume that one C_i reduces to a point; (b) the locus of points equidistant from a point and a circle is an ellipse or a hyperbola; (c) hence the center of C is obtained as the intersection of certain two conics. Two other ancient geometrical problems, the Delian problem of duplicating the cube, that is, solving (geometrically) the equation $x^3 = 2$, and the angle trisection, can be also solved by using conic sections. It may be noted that apparently Greeks started the theory of conic sections, not for any such majestic purposes as calculating planetary or cometary orbits, but just to solve the duplication of the cube (because it was thought that Apollo, appearing to a priest of his in a dream, required his cubical altar in Delos to be twice as big as it had been), the trisection of the angle, and the Apollonian problem.

Apollonius of Perga was a Greek geometer of the third century B.C., known best for his definitive work on conic sections. Among other things, he introduced the names ellipse, parabola, hyperbola (from Greek *elleipo, paraballo, hyperballo*—meaning roughly "I underthrow," "I throw as far as," "I overthrow"); these etymologies are with respect to the focus-directrix definitions of conics with the classification depending on which distance exceeds which. It was probably the preference of Greeks for spear as the weapon, and their fondness of javelin-cast and discus-throw as sports, that accounts for those single verbs to denote throwing not as far as, as far as, or further than; thus the names for the three conic sections in most of the modern languages may be ultimately derived from Greek weaponry and sports.

We observe first that an elementary solution of Apollonius' problem can be obtained by successive reductions. It was already shown how to reduce the three circles to two circles and a point. The next reduction is from two circles and a point to a circle and two points; this is based on Fig. 9a. If the circles are C_1 and C_2 and the point is A, let E be the (external) similitude center. If C is tangent to C_1 and C_2 at S and T, and passes through A, then it also passes through A_1 where $ET \cdot ES = EA \cdot EA_1$. Hence by a construction of similar triangles A_1 can be found. The final reduction is from a circle C_1 and two points A and A_1, to three points. With reference to Fig. 9b draw any circle through C_1, A, and A_1, cutting C_1 at, say, P and Q. If O is the intersection of AA_1 and PQ and the two tangents are drawn from O to C_1 touching it at T_1 and T_2 then the circle C through A, A_1 touching C_1 touches it at T_1 or at T_2.

Now we perform the three reductions in all possible ways eventually getting all the solutions to the Apollonian problem (8 at most) in an

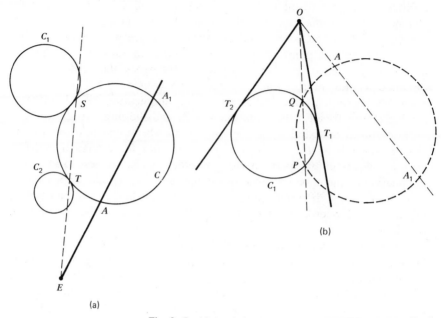

Fig. 9. Problem of Apollonius.

elementary though lengthy and inelegant fashion. To give Gergonne's solution of the problem it will be necessary to use some elementary notions of the radical axes, radical centers, and centers of similitude for circles. If p is a point outside the circle C and T is a tangent to C at x, passing through p, while S is a secant through p cutting C in y and z, then the triangles pxy and pzx are similar so that $py \cdot pz = (px)^2$. This quantity is called the power of p with respect to C. If $p = p(x_1, y_1)$ and $C(x, y) = 0$ is the equation of C then the power is $C(x_1, y_1)$. Given two such circles, the locus of points of equal powers with respect to the circles is then $C_1(x, y) = C_2(x, y)$. It is therefore a straight line—the so-called radical axis of the two circles. With three circles $C_i = 0$, $i = 1, 2, 3$, the three radical axes have equations $C_1 = C_2$, $C_1 = C_3$, $C_2 = C_3$; they are, therefore, concurrent and their common point is the radical center of the three circles C_i (here, as in what follows, we exclude certain special cases, parallel lines, etc.).

The radical center c is the center of the circle C which cuts each C_i orthogonally; the radius of C is the length of the tangent from c to any C_i. On inversion in C each C_i is mapped onto itself, the internally and externally tangent circles K_0 and K_1 of Fig. 10 get interchanged, and the lines joining the tangency points pass through c, as shown in the figure.

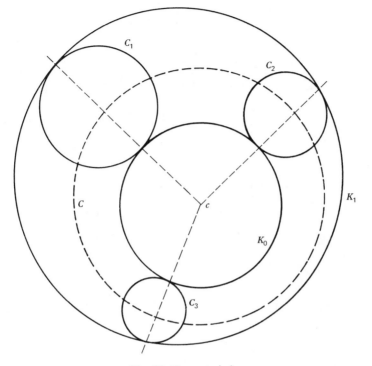

Fig. 10. Tangent circles.

We note that all constructions so far, or to follow, are of the classical ruler and compass type. For two external unequal circles the internal and external centers of similitude are shown in Fig. 11. Since the segment $c_1 c_2$ is divided by each similitude center in the ratio of the two radii, it follows that e, c_1, i, c_2 is a harmonic quadruple. From this it is not hard to show that for our three circles C_i the six similitude centers form the vertices of a complete quadrilateral and lie by threes on its sides (Fig. 12).

With reference to Fig. 10 let p be the point of intersection of the tangents to C_1 and K_0, and to C_1 and K_1. Since p is the radical center for K_0, C_1, K_1, it follows that the polar P and p with respect to C_1 passes through the tangency point of C_1 and K_1, and through the radical center c of the C_i's. Also, it can be shown that the radical axis of K_0 and K_1 passes through the three external ones of the six similitude centers of Fig. 12.

Now we have the Gergonne solution: (a) draw the line L through the three external similitude centers, (b) find its poles with respect to each C_i, (c) join these to the radical center c of the C_i's, (d) then these three lines cut C_i's in the points where the C_i's are touched by K_0 and K_1. Taking in

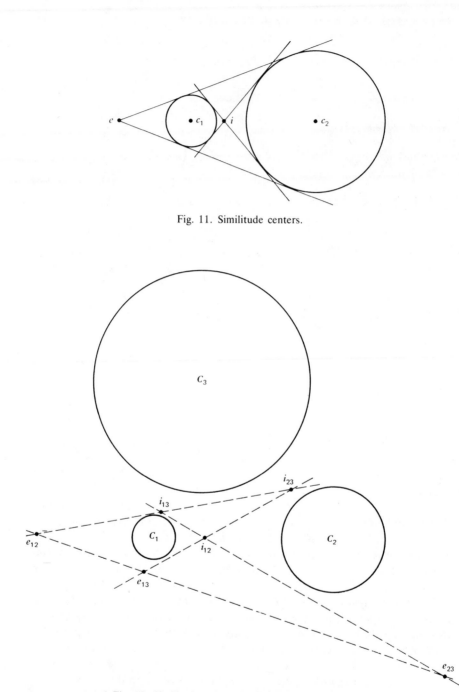

Fig. 11. Similitude centers.

Fig. 12. Similitude centers in Apollonian configuration.

succession for L each one of the four lines of Fig. 12 we can get as many as eight possible solutions to the problem.

5. NATURAL EQUATIONS AND INTRINSIC GEOMETRY

For a plane curve C let s be the arc length (measured from an arbitrary point), r the radius of curvature, and α the angle between the tangent and the x-axis. Then

$$r = \frac{ds}{d\alpha}, \qquad \frac{dx}{ds} = \cos \alpha, \qquad \frac{dy}{ds} = \sin \alpha. \qquad (1)$$

A natural equation of C is a relation between any two of the three quantities r, s, α; usually it is the relation $r = f(s)$. Its use to represent the curve dates to Euler, 1736 [54]. The relation displays C in a coordinate-independent, or intrinsic, fashion so that congruence or noncongruence of two curves follows from their natural equations at once. This is so since the matter depends on at most a shift in s; we may compare this with the relative difficulty of deciding on the congruence of, say, two ellipses by their Cartesian equations. Further, the natural equations are often simpler than their Cartesian counterparts. Relations (1) enable us to pass from the one to the other form:

$$\alpha = \int \frac{ds}{r}, \qquad x = \int \cos \alpha \, ds, \qquad y = \int \sin \alpha \, ds. \qquad (2)$$

The natural equation of a circle of radius a is $r = a$, for its evolute (= path of a point on a thread unwinding tautly off it) the equation is $r^2 = 2as$. The natural equation of a catenary is $r = a + s^2/a$, that of a "catenary of equal tension" is $r = a \cosh s/a$; the latter curve is mentioned in the section on the calculus of variations, in connection with the optimal cable-design. It will be noted that by an accident (?) the Cartesian equation of the one is the natural equation of the other. The equation

$$\frac{s^2}{a^2} + \frac{r^2}{b^2} = 1$$

represents an epicycloid for $a > b$, a cycloid if $a = b$, and a hypocycloid for $a < b$. Here, too, we note a sort of duality between the circle $(x^2 + y^2 = a^2)$ and the cycloid $(s^2 + r^2 = a^2)$. The equation $sr = a^2$ represents by (2) the curve given parametrically by

$$x = \int_0^s \cos \frac{u^2}{2a^2} \, du, \qquad y = \int_0^s \sin \frac{u^2}{2a^2} \, du;$$

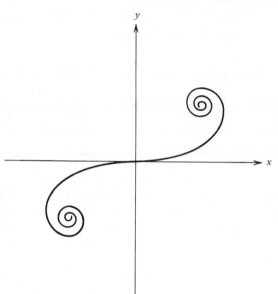

Fig. 1. Clothoid.

these are the Fresnel integrals occurring in physics (especially in connection with diffraction); the curve is called a clothoid or Cornu's spiral (Fig. 1).

After the problem of rectifying the ellipse it should not surprise us that an attempt to produce the natural equation of a conic leads to elliptic functions: We have

$$3s = \int (-1 + Ar^{2/3} + Br^{4/3})^{-1/2} \, dr$$

which is an elliptic integral, so that inverting for $r = f(s)$ we get an elliptic function. In this connection the reader may recall the previously made observation how the congruence of curves in their natural form reduces to a shift in arc length s—with this, the reader may attempt to obtain addition theorems for elliptic functions by using the congruence of ellipses. We get a parabola in the above equation if $B = 0$ and $A = 1$, and a rectangular hyperbola if $A = 0$ and $B = 1$. The relation of the constants A, B to the semiaxes a, b of the conic is

$$A = (ab)^{-4/3}(a^2 + b^2), \qquad B = (ab)^{-2/3}.$$

For details the reader is referred to [31]. In addition to representation of specific curves, it is important to note that natural equations allow us to pass fairly easily from a curve to some of its geometrical transforms: parallel curves, evolutes, pedals, roulettes and so on [31]; in this connection the reader may recall the contents of vol. 1, pp. 35–38. Further, natural equations might be considered in connection with pursuit. For instance, if C is an escape curve and C_1 the corresponding pursuit curve (see vol. 1, p. 241), we might be interested in the relation between the curvatures at the corresponding points of C and C_1, and for this purpose the natural equations are certainly better adapted than the Cartesian ones.

As an example using the natural equations we consider an extension of the prediction problem of vol. 1, p. 238. It was required there to find the exact region $R(t)$ which is the position at time t of a point moving with constant speed v, starting from the origin tangentially to the positive y-axis, and moving so that its radius of curvature is $\geq r_0$. We assume in the sequel that t is sufficiently small; as in Section 6, vol. 1, p. 236, it will be seen that if $R(t)$ can be found for small t, we can then "continue" to larger t. It turned out that the region $R(t)$ was bounded by two circle-evolute arcs in front, and by two cardioid arcs behind. Recalling that a cardioid is a special case of an epicycloid, with the rolling and stationary circles of equal radii, we see that the region $R(t)$ is given by a relatively simple system of inequalities in terms of the natural equations of its boundary.

Our new problem is to find the region $R(t, b)$ if the moving point is allowed a limited acceleration. That is, the moving point p starts at the origin, tangent to the positive y-axis at time $t = 0$ with speed v, there is a fixed tangential acceleration b, and the curvature limitation is now this: There is an upper bound c on the normal acceleration. If $r = f(s)$ is the natural equation of the path of p then our conditions give us

$$\left(\frac{ds}{dt}\right)^2 \leq cr, \qquad s = vt + \frac{bt^2}{2}.$$

Suppose that p moves with maximum allowable turning, then the inequality above changes to equality; if t is eliminated from the two equations we get

$$cr = v^2 + 2bs.$$

This is the natural equation of the "spiral of tightest turn"; it reduces to a circle if $b = 0$, is a tightening spiral if $b < 0$, and a straightening-out spiral if $b > 0$. The appearance of some sample curves $C(b)$ is given in Fig. 2.

We now tackle our problem of determining the region $R(t, b)$. The fixed value of b determines a unique right-bending spiral $C(b)$ and we consider an arc C of it, from the origin O to the point p corresponding to time t, assuming that t is small enough so that C does not turn too much.

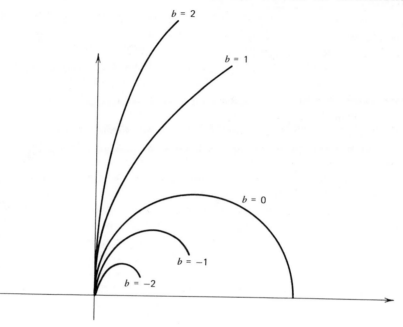

Fig. 2. Acceleration spirals.

We define a *simple rearrangement* of $C = \widehat{op}$ as follows: a point q is chosen on C and the subarc $\widehat{qp}$ of C is reflected in the tangent to C at q. A *rearrangement* of C is the result of a finite sequence of simple rearrangements. We are interested in the set of the end points of all rearrangements of C, $E(C)$. This set $E(C)$ can be defined for any sufficiently regular and slowly turning arc C. The reader may wish to prove that in the general case $E(C)$ is bounded, as before in vol. 1, p. 236, by arcs of evolutes and of roulettes of C (i.e., *by curves which lend themselves well to representation by natural equations*).

In our special case of the spiral $C = C(b)$, $E(C)$ is exactly the region $R(t, b)$ we want. The reader may wish to prove this; also, it may be asked what is the region $R(t, a)$ if we modify the conditions so that the acceleration is no longer fixed but allowed to vary between the limits $-a$ and a. In particular, the reader may wish to compare the sets $R(t, a)$ and

$$\bigcup_{-a \le b \le a} R(t, b).$$

Finally, the reader may wish to consider the problem of continuation of the region $R(t, b)$ to arbitrary values of t using, perhaps, as a model the considerations of vol. 1, p. 239.

So far we have been concerned with natural equations for plane curves. We consider next space curves C, limiting ourselves to three dimensions. As described in vol. 1, pp. 49–50, such curves have a curvature κ and a torsion τ; roughly, κ measures how fast C turns and τ measures how fast it twists away from planarity. Formally, if s is the arc length measured from some point on C, then κ is the rate of turning, with respect to s, of the unit tangent vector to C, and τ is the same with respect to the unit binormal vector (which is perpendicular to the osculating plane of C). A fairly precise analog of the natural equation $r = f(s)$ for plane curves is obtained now by taking κ and τ as given functions of s. The reciprocals $1/\kappa$ and $1/\tau$ are called the radius R of curvature, and the radius T of torsion.

Let us consider now the three-dimensional equivalent of equations (2), that is, the problem of determining the space curve C, given its curvature and torsion as functions of arc lengths s: $\kappa = \kappa(s)$, $\tau = \tau(s)$. We have the Frenet–Serret equations (vol. 1, p. 50):

$$t' = \kappa n, \qquad n' = \kappa t + \tau b, \qquad b' = -\tau n \qquad (3)$$

for the tangent, normal, and binormal unit vectors t, n, b; these play the role analogous to α in (2). To determine C it suffices to find the vector t. One method would be to eliminate n and b from (3) getting a third-order linear differential equation for t. However, the method of S. Lie and G. Darboux will reduce finding t to solving three first-order nonlinear differential equations of the Riccati type. To see this, let u, v, w be, say, the first components of t, n, b so that by (3)

$$u' = \kappa v, \qquad v' = -\kappa u + \tau w, \qquad w' = -\tau v, \qquad u^2 + v^2 + w^2 = 1. \quad (4)$$

We use the factorization of $u^2 + v^2 + w^2 - 1 = 0$ as

$$(u + iv)(u - iv) = (1 + w)(1 - w);$$

geometrically this corresponds to changing from the Cartesian equation of the unit sphere to its imaginary rectilinear generators. We then introduce the Darboux parameters σ, ω:

$$\frac{u + iv}{1 - w} = \frac{1 + w}{u - iv} = \sigma, \qquad \frac{u - iv}{1 - w} = \frac{1 + w}{u + iv} = \frac{-1}{\omega}$$

so that σ and $-1/\omega$ are conjugates. Then

$$u = \frac{1 - \sigma\omega}{\sigma - \omega}, \qquad v = i\frac{1 + \sigma\omega}{\sigma - \omega}, \qquad w = \frac{\sigma + \omega}{\sigma - \omega},$$

and from (4) it can now be verified that both σ and ω satisfy the same

Riccati equation

$$\frac{df}{ds} = \frac{i\tau}{2} f^2 - i\kappa f - \frac{i\tau}{2}.$$ (5)

More generally, we call a Riccati equation one of the form

$$f' = Af^2 + Bf + C,$$ (6)

where A, B, C are functions of the independent variable s. The same reduction to Ricatti equation is possible for a system somewhat more general than (4), namely for a third-order homogeneous system with a skew-symmetric matrix: $X' = MX$ where $M + M^T = 0$. For, writing it out by components, we have with

$$X = \begin{bmatrix} u \\ v \\ w \end{bmatrix}, \qquad M = \begin{bmatrix} 0 & r & q \\ -r & 0 & p \\ -q & -p & 0 \end{bmatrix}$$

the system analogous to Frenet–Serret's

$$u' = rv + qw$$
$$v' = -ru + pw$$
$$w' = -qu - pv;$$

introducing the Darboux parameters as before we find that these parameters satisfy the Riccati equation somewhat more general than (5):

$$f' = zf^2 + izf + \bar{z}$$

with $2z = -q + ip$.

Riccati's equation (6) is in general not solvable by elementary functions; it has also some claim to being the simplest nonlinear differential equation of some generality. There is here a certain slight analogy with the process of solving polynomial equations. We start with solving a first-degree equation $ax + b = 0$; that is, we learn to divide, as well as multiply, add, and subtract. Then we proceed to the quadratic equations, which are not solvable by the four arithmetic operations. Here we solve by completing the squares, and extracting square roots. Equivalently, in terms of the conjugacy principle, we use the arithmetic operations to make the quadratic polynomial $P(x) = ax^2 + bx + c$ conjugate under $L(x) = (2x - b)/2a$ to the "pure" quadratic:

$$L^{-1}PL = x^2 - A, \qquad A = \frac{b^2 - 4ac}{4} - \frac{b}{2}.$$

With differential equations we can solve the first-order linear ones, and the solution y is a linear function of the arbitrary constant:

$$y = Cy_1(x) + y_2(x). \tag{7}$$

On the other hand, we cannot in general solve the Riccati equation. However, just as with quadratic equations where knowing one root will enable us to find the other one by arithmetic operations, so here: If we know *one* solution of Riccati's equation then the complete solution can be effected by integration. This, and other properties, will be found to be closely related to the fact generalizing (7): the general solution is a bilinear function of the arbitrary constant:

$$y = \frac{Cy_1 + y_2}{Cy_3 + y_4}. \tag{8}$$

This is fairly easily shown since the substitutions $y \rightarrow y + h(x)$, $y \rightarrow h(x)y$, $y \rightarrow 1/y$ change Riccati equations into Riccati equations. The reader may wish to consider whether there exist first-order or other differential equations with other forms of dependence on the arbitrary constant than the linear form (7) and the bilinear form (8). It may be shown that the second-order linear equation

$$y'' + p(x)y' + q(x)y = 0 \tag{9}$$

is transformed by the substitution

$$y = e^{\int z \, dx}$$

into the Riccati form

$$z' + z^2 + p(x)z + q(x) = 0.$$

Conversely, any Riccati equation can be transformed into the linear second-order form (9).

Count Jacopo Riccati (1676–1754) of Venice was an Italian mathematician active in analysis. Aided by his sons Vicenzo (1707–1775) and Giordano (1709–1790), he conducted an extensive correspondence with many European scientists, and especially mathematicians (this was the equivalent of the publication of scientific journals of today).

For further details relevant to the determination of a space curve C from its natural equations $\kappa = \kappa(s)$, $\tau = \tau(s)$, see [51].

It is quite obvious that we can see at once from the natural equations of C whether C is a plane curve: A necessary and sufficient condition is $\tau = 0$. More generally, we can also tell whether C lies on a sphere. For, let C be represented in the Cartesian form by its radius vector as a function of s: $x = x(s)$. We produce a sphere of possibly high order of contact with C, its so-called osculating sphere. The equation of the sphere is written in vector form:

$$(u - c) \cdot (u - c) - r^2 = 0$$

so that c is its center and r its radius. We substitute x for u, and differentiate successively with respect to s; using the Frenet–Serret formulas (3) we get the equations

$$(x - c) \cdot (x - c) - r^2 = 0, \qquad (x - c) \cdot t = 0, \qquad \kappa(x - c) \cdot n + 1 = 0,$$

$$(x - c) \cdot (\kappa' n - \kappa^2 t + \kappa\tau b) = 0.$$

From these we determine c and r in terms of $R = 1/\kappa$ and $t = 1/\tau$:

$$c = x + Rn + TR'b, \qquad r^2 = R^2 + T^2(R')^2.$$

This determines completely the osculating sphere at a point of C. Now, a necessary and sufficient condition for C to lie on a sphere is that

$$r^2 = R^2 + T^2(R')^2 = \text{const.}$$

for we see then, by differentiating, that $c' = 0$ so that C has at all points the same radius and center of the osculating sphere, hence it lies on a sphere.

By using the Taylor series together with the Frenet–Serret formulas we can write down the local power-series for the Cartesian coordinate in the neighborhood of the point corresponding to $s = s_0$; this enables us to study the local shape of the curve:

$$x = (s - s_0) - \frac{\kappa^2}{6}(s - s_0)^3 - \frac{\kappa\kappa'}{8}(s - s_0)^4 + \cdots$$

$$y = \frac{\kappa^2}{2}(s - s_0)^2 + \frac{\kappa'}{6}(s - s_0)^3 + \frac{\kappa'' - \kappa^3 - \kappa\tau^2}{24}(s - s_0)^4 + \cdots$$

$$z = -\frac{\kappa\tau}{6}(s - s_0)^3 - \frac{2\kappa'\tau + \kappa\tau'}{24}(s - s_0)^4 + \cdots$$

(here x is used as a coordinate whereas it was used before as the radius vector).

Of the many further properties which can be read from the natural equations of a space curve C we mention that of being a curve of constant slope (or a generalized helix). Here C is called of constant slope if its tangent vector t makes a constant angle with a fixed vector a in space (if we regard the latter as being vertically up we have the fixed slope). This generalizes the ordinary circular helix for which both κ and τ are constant. It turns out that a necessary and sufficient condition for C to be of constant slope is somewhat similar: The ratio κ/τ is constant. For if

$t \cdot a = \text{const.}$ then by the Frenet–Serret formulas $a \cdot n = 0$ so that $a = t \cos \alpha + b \sin \alpha$, and differentiating this, we get $\kappa \cos \alpha - \tau \sin \alpha = 0$; converse is proved similarly.

As a simple application we consider the following problem arising in practice in connection with staircases on large spherical tanks (known in the trade as Horton spheres). Let the equation of a sphere S be $x^2 + y^2 + z^2 = a^2$; find the shortest curve on S joining a point on the equator, say $P(a, 0, 0)$, to the north pole $N(0, 0, a)$ and having the slope $\leq b$ at all points. It is not hard to show that this shortest curve C has initially the slope exactly b and continues so till it gets sufficiently close to the pole; then it runs to the pole "straight up," that is, along a meridian arc. Therefore the shape of C is as follows: It consists of two distinct but smoothly joined arcs PL and LN where L is a point on the circle of latitude at which the tangent plane has the slope b, LN is an arc of the meridian from L to the pole N, and PL is an arc of the spherical helix, that is, a spherical curve of constant slope b. Using the natural equation (in the plane) of the epicycloid, the condition for a curve to lie on a sphere, and the constancy of κ/τ for curves of constant slope, the reader may wish to prove that the projection of PL onto the xy-plane is half an arc of an epicycloid.

6. DEFORMATIONS AND RIGIDITY

In this section we consider deformations of various geometrical structures; our considerations link together such diverse things as mechanical rigidity of frameworks, bending invariants of classical differential geometry, and the elements of embedding abstract geometrical structures into ordinary spaces.

(**a**) In connection with the linkages of Section 3 let us consider a plane $n - \text{gon } P_n$ as a framework of rigid sides smoothly joined at the vertices by hinges. It is clear that a triangle P_3 is rigid: it can only be moved as a whole, by rigid motions and reflections (we always stay in the plane). For $n \geq 4 P_n$ is not rigid: Parts of it can be moved relative to other parts with the preservation of all side lengths. This preservation of length is characteristic of the subject; it is called isometry (from the Greek: *isos*—same, *metron*—measure). Somewhat more formally, we could say that there is a polygon Q_n, not homothetic to P_n, and a continuous family $I(t)$ of isometries, with $I(0)P_n = P_n$ and $I(1)P_n = Q_n$. That is, $I(t)$ acts on P_n, transforming it into another n-gon; the action is continuous, and the distances are preserved: $|xy|$ is the distance from $I(t)x$ to $I(t)y$ for $x, y \in P_n$. Here it must be emphasized that we measure the distances, say from x to y, not via the ambient space (here the Euclidean plane) but intrinsically: via P_n (or its image). We speak here of an infinitesimal (past and (possibly) future word) or a continuous isometric deformation.

The reader may wish to show that P_n becomes rigid as soon as we add to it any $n-3$ diagonals as "braces." This, though simple, is not entirely trivial to show. The trouble is that for $n \leq 5$ we always get a rigid subtriangle to begin with, whereas for $n \geq 6$ we must use a slightly more sophisticated argument (based on decreases and increases of lengths of various members of the framework) since there may be no obvious rigid substructure. We define now a self-strained configuration as follows: The polygon P_n is a closed perfectly elastic rubber band, there are $n-3$ straight rigid rods (the braces) and the arrangement of $n-3$ rods kept together by the rubber band is in equilibrium under no other forces. Since a rod must pass through each vertex we have $n-3 \geq n/2$ so that $n \geq 6$—there are no self-strained configurations of less than 6 vertices. Using static equilibrium conditions and Pascal's theorem of Section 4, the reader may wish to show, following M. W. Crofton [39], that a hexagonal frame is self-strained if and only if the six vertices lie on a conic. The reader may also wish to investigate three-dimensional analogs of self-strained frames, as well as possible extensions of the above result of Crofton.

A type of nonrigidity other than that of the polygon P_n for $n \geq 4$ is provided for two-dimensional structures by a convex polyhedron P consisting of two tetrahedra based from opposite sides on the same triangle T. Under suitable restrictions one of the tetrahedra may be reflected in the plane of T giving us a concave polyhedron P_1, isometric in every reasonable sense to P. But there is here no continuous family of isometric deformations "joining" P and P_1. It is reasonable to inquire whether a similar pair P and P_1 of isometric *convex* polyhedra might exist. The negative answer is provided by the theorem of Cauchy on the rigidity (in the large, rather than infinitesimally) of convex polyhedra, which we shall take up later.

(**b**)   To observe a simple situation with both one- and two-dimensional parts, and to indicate how the lack of rigidity can be sometimes exploited, we consider the Banach table [9]. First, it is noted that any rigid table with three legs is *statically determinate*: Given a heavy object on the table, as in Fig. 1a, we can calculate the reactions R_1, R_2, R_3 of the legs (for instance, by setting the moments about the three edges to 0).

On the other hand, the same is not true for the rigid rectangular table with four legs, shown in Fig. 1b. Here, if the structure is rigid, the four reactions R_i cannot be found by the (three) conditions of static equilibrium. However, the static indeterminacy disappears if we suppose the tabletop to be rigid, while the legs are compressible and the distortion is

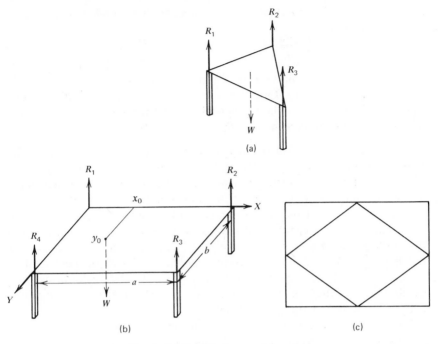

Fig. 1. Tables with three and four legs.

proportional by Hooke's law to the acting force. We must also assume "infinitesimally small" deformations to assure that the legs do not displace from the vertical, since the top is rigid. Now we have the three equilibrium equations

$$R_1 + R_2 + R_3 + R_4 = W \qquad \text{(net force 0)}$$

$$b(R_3 + R_4) = y_0 W, \qquad a(R_2 + R_3) = x_0 W \qquad \text{(net torque 0)},$$

(1)

and additionally the consequence of tabletop rigidity:

$$R_1 + R_3 = R_2 + R_4.$$

(2)

This condition is easily obtained by calculating the infinitesimal displacement of the center of the tabletop in two different ways (from the distortions of diagonally opposite legs) and equating. Solving equations (1) and (2) we have

$$R_1 = \frac{W}{4}\left[3 - 2\left(\frac{x_0}{a} + \frac{y_0}{b}\right)\right], \qquad R_2 = \frac{W}{4}\left[1 + 2\left(\frac{x_0}{a} - \frac{y_0}{b}\right)\right],$$

$$R_3 = \frac{W}{4}\left[-1 + 2\left(\frac{x_0}{a} + \frac{y_0}{b}\right)\right], \qquad R_4 = \frac{W}{4}\left[1 - 2\left(\frac{x_0}{a} - \frac{y_0}{b}\right)\right].$$

Since these must be by their nature nonnegative we find that

$$-\frac{1}{2} \le \frac{x_0}{a} - \frac{y_0}{b} \le \frac{1}{2}, \qquad -\frac{1}{2} \le \frac{x_0}{a} + \frac{y_0}{b} \le \frac{1}{2}.$$

This means that the point (x_0, y_0) of application of W must lie on or within the rhombus of Fig. 1c. Otherwise, if (x_0, y_0) is inside one of the four triangles, the reaction of the leg opposite to the triangle is 0 and we have in effect a statically determinate three-legged table then.

(c) As shown in Section (a), a polyhedron may fail to be rigid in the large. This may be rephrased: If all the faces of a polyhedron are given as well as their interconnection scheme (i.e., the prescription which edges and which vertices coincide) then the polyhedron itself is not necessarily uniquely given yet. This, in turn, suggests the problem, important for the abstractionists, of concrete realizations of an abstract geometrical structure that we shall illustrate on a simple example of an abstract convex polyhedron. Such an abstract convex polyhedron will be described in terms of its net, the idea here being that we can cut out the faces from cardboard and then glue them together at certain edges, to obtain a cardboard model of the polyhedron.

Specifically, taking (convex) polygons for granted, we define a net to be a finite set of k polygons $P_1, \ldots, P_k$ such that

(a) each side is common to exactly two P_i's (whose coincident edges are of equal length) and each vertex is common to at least three P_i's;
(b) the numbers v and e of vertices and edges (*after* the identifications of (a)) satisfy the Euler relation $v + k = 2 + e$;
(c) any two P_i's of the net are connected by a chain, starting with the one and ending with the other, in which successors share edges;
(d) the sum of all face angles at a vertex is $< 2\pi$.

Under these conditions we have the realizability theorem of A. D. Alexandrov [3]: The abstract convex polyhedron (or net) can be realized as an actual convex polyhedron in the Euclidean space E^3. Alexandrov's theorem constitutes an existence proof: showing that an entity of a certain type or description actually exists. It is natural to ask also whether it is unique; in our case, whether the actual convex polyhedron realizing the abstract one, is unique. *But this is precisely asking whether convex polyhedra are rigid in the large.* The affirmative answer—Cauchy's theorem—can be expressed in several equivalent ways:

(A) the convex realization of an abstract net satisfying conditions (a)–(d) is unique;
(B) if the faces of a convex polyhedron and their arrangement together

are both given then the convex polyhedron itself is given, up to a rigid motion or a reflection;

(C) if two convex polyhedra have faces that are congruent and similarly located, then the corresponding dihedral angles are equal.

Before outlining a proof of Cauchy's theorem we note a possible difficulty of such a proof: necessity of connecting together local with global properties. A very much simpler example of similar difficulty is the rigidity, previously mentioned in Section (a), of an n-gon with $n-3$ diagonal braces.

The proof of Cauchy's theorem depends on a sequence of lemmas, some of which are not quite as simple as they seem to be.

LEMMA 1.   Let P_n and Q_n be two convex n-gons with vertices $v_1, \ldots v_n$ and $w_1, \ldots, w_n$. Suppose that $|v_i v_{i+i}| = |w_i w_{i+1}|$ for $i = 1, 2, \ldots, n-1$. If $\angle v_i \le \angle w_i$ for $i = 2, \ldots, n-1$, with at least one strict inequality, then $|v_1 v_n| < |w_1 w_n|$.

This says that if a length of wire is bent into the shape of a convex polygon and a side ab of it removed, then on decreasing some of the angles the distance from a to b is diminished. Hence, this lemma generalizes the simple geometrical fact that if a triangle has two sides fixed in length and subtending an angle α, then the length of the third side increases or decreases together with α. Thus, this lemma acts as a generalized, though only qualitative, law of cosines for polygons.

LEMMA 2.   Let P_n and Q_n be as above. Suppose now that all sides are equal: $|v_i v_{i+i}| = |w_i w_{i+i}|$ for $i = i, \ldots, n$, with v_{n+1} and w_{n+1} taken to be v_1 and w_1. If P_n and Q_n are not congruent then there are some four values i_1, i_2, i_3, i_4 such that $1 \le i_1 < i_2 < i_3 < i_4 \le n$ and either

$$\angle v_i < \angle w_i \quad i = i_1, i_3; \qquad \angle w_i < \angle v_i \quad i = i_2, i_4$$

or

$$\angle w_i < \angle v_i \quad i = i_1, i_3; \qquad \angle v_i < \angle w_i \quad i = i_2, i_4.$$

This can be expressed simply in terms of the n-gonal framework of rigid sides hinged at the vertices, mentioned at the beginning of this section: If the shape of the framework changes then (at least) four angles must be changed, so that as we go round the vertices in cyclic order, the changes will alternate in sign: increase of angle, decrease, increase, decrease. This lemma is proved without too much difficulty, by using the preceding lemma.

LEMMA 3.   If a convex solid angle is deformed into another convex solid angle so that all vertex angles are preserved, then at least four dihedral angles change and the changes, in cyclic order, alternate in sign.

This lemma is proved by observing that the first two lemmas hold for spherical polygons as well as for plane ones. Here it is assumed that each spherical polygon lies in some open hemisphere; it is called convex, or spherically convex, if every two points can be connected within by an (smaller) arc of the great circle. Then Lemma 3 becomes, essentially, the spherical version of Lemma 2.

Now we can prove Cauchy's theorem, in the form (C). Suppose that it is false so that there are two isometric convex polyhedra P and Q, which are not congruent. We assign signature $+1$, 0, or -1 to each edge of P according as to whether its dihedral angle exceeds, is equal to, or is less than the dihedral angle of the corresponding edge of Q. Removing from P all edges marked 0 we obtain a net N, possibly disconnected, and we get a contradiction by counting in two different ways the number T of sign changes.

Suppose that the net has n_k k-gons, $k = 3, 4, \ldots$ Then, counting the sign changes, along consecutive edges of a constituent polygonal region of N, say a k-gon, we note that there must be an even number of such changes. Hence the number of changes is $\leq k$ if k even, $\leq k - 1$ if k odd. Therefore

$$T \leq 2n_3 + 4(n_4 + n_5) + 6(n_6 + n_7) + \cdots. \tag{3}$$

Let the net N have r regions, e edges, and v vertices. We count T for the second time by going, in cyclic order, over the edges with a common vertex, for each vertex. By Lemma 3 there is a contribution ≥ 4 at each vertex, hence

$$T \geq 4v. \tag{4}$$

The generalized Euler equation gives us

$$v + r - e \geq 2, \tag{5}$$

and we have

$$r = n_3 + n_4 + n_5 + \cdots, \qquad 2e = 3n_3 + 4n_4 + 5n_5 + \cdots.$$

Hence

$$4e - 4r = 2n_3 + 4n_4 + 6n_5 + \cdots$$

so that by (5)

$$4v - 8 \geq 2n_3 + 4n_4 + 6n_6 + \cdots. \tag{6}$$

Now, by (3), (4), and (6)

$$4v \leq 4v - 8,$$

which is a contradiction.

From Cauchy's theorem it follows that a convex polyhedron does not admit a continuous isometric deformation (images being at all times likewise convex polyhedra) since it does not even admit a single such deformation. However, the reader may wish to construct an independent proof. The reader may also wish to consider polyhedra from the point of view of edge rigidity: Here we consider only the edges, treating them as rigid rods smoothly hinged at the vertices. (Similarly, a higher-dimensional polytope might be considered with respect to the rigidity of its two-dimensional skeleton).

(**d**) On the basis of Cauchy's theorem one may try for an extension to a general closed convex surface which is not necessarily polyhedral. It is quite natural to conjecture that a complete closed convex surface S is uniquely given by its *distance-structure*. Somewhat more precisely, we may conjecture that S is isometrically unique (up to rigid motions and reflections); alternatively, we might say that S is both infinitesimally and finitely rigid. An idea of a proof would be to approximate to S by a sequence of suitable polyhedra. This was the method used by S. Cohn-Vossen [36], who was the first to prove the conjecture, for analytic S. A different method and much weaker assumptions were used by G. Herglotz in another proof [72], 1943. In still greater generality the proof was given by A. Pogorelov in 1949 [134]. A generalization of Cauchy's theorem to certain nonconvex polyhedra, as well as other extensions of Cauchy's result, are given in [164].

Closed convex surfaces being rigid, that is, indeformable by isometric mappings, we may ask whether *parts* of convex surfaces are deformable. The answer is not quite simple as may be seen from the following examples. Let T be the ordinary torus and C its convex part: The set of points at which the plane tangent to T has only one point in common with T. Then C is rigid. The same is true of other incomplete surfaces with several plane closed boundary curves, provided that at all points of each boundary curve there is one and the same tangent plane to the surface. We note that in all such examples, the spherical image of the surface (=set of end points of all unit outward-drawn normal vectors, when translated so as to start at the origin) is a sphere with merely a finite number of points removed.

To gain a better understanding of the rigidity of these, let us consider some opposite examples, of *deformable* incomplete surfaces. We shall consider certain subsets of the unit sphere S given by the equation $x^2 + y^2 + z^2 = 1$. Complete S is of course indeformable but if we draw on it a simple closed curve C and remove from S either of the two regions bounded by S, no matter how small it may be, then the remainder of S is deformable. The same is true even if we remove from S a simple arc.

The reader may wish to find out whether such a slit sphere can be turned inside out by a continuous family of isometries.

We shall now consider a deformation of a part of S to a surface of revolution, the details of which can be easily followed. From S we remove a lune $L(\alpha)$ of angle α consisting of the part of S lying between two half-planes, meeting on the z-axis, and making the angle α, $0 < \alpha < 2\pi$. The remainder R of S can be continuously deformed into a football-shaped surface of revolution. Moreover, at all intermediate times the image of R is part of a surface of revolution (we conjecture that this can happen only if the part removed from S includes a pair of antipodal points). To show this, we let n be a large positive integer, put $\varepsilon = (2\pi - \alpha)/n$, and slice up R into n congruent lunes of angle ε. Let L_1 be one of them and let P_1 and P_2 be the half-planes containing it between them. We "open out" the region between P_1 and P_2 so that the angle ε increases to $2\pi/n$. We suppose that during this opening out the lune L_1 remains at all times in contact with P_1 and P_2; its end points move away from each other on the z-axis, and it becomes flatter. Repeating the same procedure symmetrically on each one of the n lunes and letting $n \to \infty$ we deform R into a football-shaped complete surface of revolution that has, however, two singular points. The reader may wish to verify that the new surface is obtained by rotating the curve $y = f(x)$ where

$$y'^2 = (1 + y^2)\left[\left(1 - \frac{\alpha}{2\pi}\right)^2 + y^2\right]$$

so that y can be expressed in terms of the Jacobian elliptic functions [48].

The above describes an isometric deformation of the part R of the sphere S into a more tightly curved surface. Suppose now that we wished to study a deformation into a less tightly curved surface. For this purpose we remove from S all points (x, y, z) such that $|z| > b$, for some b, $0 < b < 1$. This amounts to removing two symmetric polar caps from S; we slit the remainder R_1 along a half-plane through the z-axis. As before, we cut up R_1 into lunes, although we have really only parts of lunes now. We consider the region between half-planes corresponding to P_1 and P_2 but instead of opening the wedge $P_1 P_2$ out, as before, we squeeze it, decreasing its angle. Under the same conditions as before, the part of R_1 in the contracting wedge will curl up, all of its points receding from the z-axis. This goes on until the end point tangent planes are at right angles to the z-axis; beyond that point no isometric deformation can be applied.

The interested reader, if possessed of some experimental bent, may actually perform the above two deformations of the sphere by operating (delicately!) on a ping-pong ball with a sharp knife or a fine fret-saw, and then applying force to the remainder. In the second case the remainder of the ping-pong ball opens out to what looks rather like a part of the torus, in fact, a part of the convex "outside" of the torus. However, this is only apparent. By the *theorema egregium* of Gauss, the Gaussian curvature K is a bending invariant (since it depends only on the coefficients E, F, G of the first fundamental form

$$dx \cdot dx = E(du)^2 + 2F\,du\,dv + G(dv)^2$$

and on their derivatives; however, $dx \cdot dx$ is the arc-length element squared, $(ds)^2$, hence the invariance since the isometric deformations by definition preserve arc length). On the other hand, it is easy to see that while the Gaussian curvature of a sphere is constant, that of a torus is not. Hence our deformation being an isometry of a part of a sphere, could not result in a part of a torus.

7. SPHERICAL TRIGONOMETRY

The usefulness of this small but nowadays nearly forgotten and untaught subject is undoubted: We live on the surface of a sphere, and the celestial objects we see in the sky are projected upon the imaginary celestial sphere. In this section we give a very brief introduction to spherical trigonometry.

(**a**) The subject matter of spherical trigonometry are spherical triangles (and spherical polygons) and quantities associated with them. A spherical triangle ABC consists of three arcs of great circles of a fixed sphere S, as shown in Fig. 1. Its six *elements* are the three vertex angles, or briefly angles, A, B, and C, and the three side angles, or briefly sides, a, b, c. The angles are formed by the tangents to the circular arcs, and the sides are subtended at the center of the sphere, as shown. Alternatively, in terms of solid geometry, we have here a trihedral angle $OABC$ with face angles a, b, c and dihedral angles A, B, C. We suppose that the spherical triangle

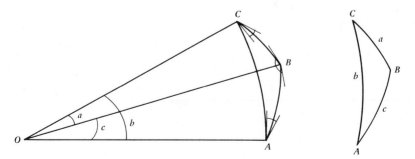

Fig. 1. Spherical triangle.

ABC is (spherically) convex. This means that the cone which it subtends at *O* is a convex cone; as a consequence each angle of the triangle is $\leq \pi$ and the whole triangle $T = ABC$ always lies in some closed hemisphere *H* of *S*. Excluding some degenerate cases, we suppose that each angle is $< \pi$ and *T* lies in an open hemisphere *H* of *S*.

If the three great circles forming the sides of *T* are regarded as equators then each one has just one pole in the hemisphere *H*. If A_1, B_1, C_1 are the poles corresponding to the arcs *BC*, *AC*, *AB* in *T*, then $A_1 B_1 C_1$ is another spherical triangle, also in *H*, which we call the polar triangle of *T*. The reader may wish to show that the relation is involutory: The triangle polar to $A_1 B_1 C_1$ is the original *T*. If the six elements of *T* are *a*, *b*, *c*, *A*, *B*, *C* then the corresponding elements of T_1 are

$$a_1 = \pi - A, \quad b_1 = \pi - B, \quad c_1 = \pi - C,$$
$$A_1 = \pi - a, \quad B_1 = \pi - b, \quad C_1 = \pi - c.$$

This polarity is a spherical case of just one of the dualities of convexity. In spherical trigonometry it enables us to dualize any relation between the elements of a spherical triangle to a new relation.

No such duality obtains in the plane. Further, *any* three of the six elements determine *T*. This, of course, comes from the fact that the notion of similarity, of plane geometry, does not apply to spherical geometry.

(**b**)   With reference to Fig. 1 let *u*, *v*, *w* be the vectors represented by *OA*, *OB*, *OC*, normalized to length 1. Starting with a standard vector identity

$$(m \times n) \cdot (p \times q) = (m \cdot p)(n \cdot q) - (m \cdot q)(n \cdot p)$$

we take $m = p = u$, $n = v$, $q = w$, getting

$$(u \times v) \cdot (u \times w) = v \cdot w - (u \cdot w)(u \cdot v). \tag{1}$$

Recall that *u*, *v*, *w* are unit vectors so that the scalar product of any two is the cosine of the angle between them, and the magnitude of their cross product is the sine of that angle. Hence $u \times v$ is a vector of magnitude sin *c*, at right angles to the plane *OAB*; similarly, $u \times w$ has the magnitude sin *b* and is at right angles to *OAC*. Therefore the angle between $u \times v$ and $u \times w$ is the angle between the normals to *OAB* and to *OAC*, that is the angle *A*. Hence the L.H.S. of (1) is just sin *b* sin *c* cos *A*. The R.H.S. is cos *a* − cos *b* cos *c* and so we get the spherical law of cosines

$$\cos a = \cos b \cos c + \sin b \sin c \cos A, \tag{2}$$

with analogous expressions for cos *b* and cos *c*. Dualizing by means of polar triangles we get

$$\cos A = -\cos B \cos C + \sin B \sin C \cos a \qquad (3)$$

and like expressions for $\cos B$ and $\cos C$. Formulas (2) and (3) enable us to solve the spherical triangle (that is, find all its six elements) when either three sides, or three angles, or two sides with the included angle, or a side and two adjoining angles, are known. To solve the remaining cases we derive first the spherical law of sines. From (2) we have

$$\sin^2 A = 1 - \cos^2 A = \frac{\sin^2 b \sin^2 c - (\cos a - \cos b \cos c)^2}{\sin^2 b \sin^2 c},$$

and replacing sines by cosines

$$\sin^2 A = \frac{1 - \cos^2 a - \cos^2 b - \cos^2 c + 2 \cos a \cos b \cos c}{\sin^2 b \sin^2 c}. \qquad (4)$$

Hence also

$$\frac{\sin^2 A}{\sin^2 a} = \frac{\sin^2 B}{\sin^2 b} = \frac{\sin^2 C}{\sin^2 c} \qquad (5)$$

since the numerator in (4) is symmetric in a, b, c. We check that $\sin A$ and $\sin a$ have the same sign and so we get from (4) and (5) the spherical law of sines:

$$\frac{\sin A}{\sin a} = \frac{\sin B}{\sin b} = \frac{\sin C}{\sin c}$$
$$= \frac{(1 - \cos^2 a - \cos^2 b - \cos^2 c + 2 \cos a \cos b \cos c)^{1/2}}{\sin a \sin b \sin c}. \qquad (6)$$

This, together with (2) and (3), enables us to solve any spherical triangle.

We got the cosine law (2) by applying a simple vector identity to the unit vectors u, v, w; we now reverse the procedure and obtain a vector inequality from the spherical trigonometric formula (4). Namely, we observe that $\sin^2 A \geq 0$ and so the numerator of (4) is ≥ 0. Expressing the cosines back by scalar products gives us

$$(u \cdot v)^2 + (u \cdot w)^2 + (v \cdot w)^2 \leq 1 + 2(u \cdot v)(u \cdot w)(v \cdot w),$$

and replacing the unit vectors u, v, w by $U/|U|$, $V/|V|$, $W/|W|$ where U, V, W are arbitrary vectors, we get

$$|W|^2 (U \cdot V)^2 + |V|^2 (U \cdot W)^2 + |U|^2 (V \cdot W)^2 \leq |U|^2 |V|^2 |W|^2$$
$$+ 2(U \cdot V)(U \cdot W)(V \cdot W)$$

with the equality if U, V, W are linearly dependent, and only then. Alternatively, we can proceed thus: express the numerator in (4) as a determinant, getting

$$
\begin{vmatrix}
1 & \cos a & \cos b \\
\cos a & 1 & \cos c \\
\cos b & \cos c & 1
\end{vmatrix} \geq 0
$$

and replace the cosines by scalar products getting eventually

$$
\begin{vmatrix}
U \cdot U & U \cdot V & U \cdot W \\
V \cdot U & V \cdot V & V \cdot W \\
W \cdot U & W \cdot V & W \cdot W
\end{vmatrix} \geq 0.
$$

This is merely a special case of the positivity of the Gram determinant: For any n vectors $v_1, \ldots, v_n$ in E^n the $n \times n$ determinant $G_n = |v_i \cdot v_j|$ is ≥ 0 with equality for n linearly dependent vectors, and only then. This, of course, is geometrically obvious once we recall that G_n is the square of the volume of the parallelopiped formed by the n vectors.

(c) Let us consider now the heights, the circum-radius, and the in-radius of a spherical triangle. These have obvious applications to simple problems of navigation on earth, assuming that the ships or planes in question move along the great-circle routes. For instance, given three cities A, B, C with given latitudes and longitudes, we may want to know the closest approach to A by a plane flying from B to C. This calls for calculating the height h_a in a spherical triangle with given sides a, b, c. We first use the cosine law (2) and calculate the angle C, then we apply the sine law to the right-angled spherical triangle with one side b, the angle C, and right angle opposite b, to get $\sin h_a = \sin b \sin C$.

Again, under the same circumstances we may suppose that planes fly between any two of the three cities A, B, C and we ask: What is the point X which minimizes the maximum of the three distances from X to the three lines of flight? The point X could be, for instance, the site of an omnidirectional beacon. Simple considerations show that X is the center of the circle inscribed into the spherical triangle ABC, and that just as in the plane case, the (great-circle) bisectors of the angles intersect at X. Hence we find X by starting with the known sides a, b, c, calculating A and B from the cosine law (2), and then solving the triangle whose side is c and two adjoining angles are $A/2$ and $B/2$.

Taking r to be the (angular) radius of the circle inscribed into ABC, the reader may wish to verify that

$$\tan r = \frac{\sqrt{1 - \cos^2 a - \cos^2 b - \cos^2 c + 2\cos a \cos b \cos c}}{2\sin (a + b + c)/2}$$

$$= \frac{\sqrt{-\cos S \cos (S - A) \cos (S - B) \cos (S - C)}}{2\cos \dfrac{A}{2} \cos \dfrac{B}{2} \cos \dfrac{C}{2}}, \quad 2S = A + B + C.$$

If, instead of minimizing the maximum distance from X to the three lines of flight, we wish to minimize the largest of the three distances XA, XB, XC, then X becomes the center of the circle that circumscribes the spherical triangle ABC. We could find X directly as the intersection of perpendicular bisectors of sides. But we can here exploit the duality by polar triangles, showing that the pole of the inscribed circle is also the pole of the circle circumscribing the triangle which is polar to the given one. This reduces our problem to the previous one. If R is the (angular) radius of the circumcircle of the triangle ABC then the reader may wish to verify that

$$\tan R = \frac{\sin (s - a) + \sin (s - b) + \sin (s - c) - \sin s}{\sqrt{1 - \cos^2 a - \cos^2 b - \cos^2 c + 2\cos a \cos b \cos c}}, \quad 2s = a + b + c,$$

$$= -\frac{\cos(A + B + C)/2}{\sqrt{-\cos S \cos (S - A) \cos (S - B) \cos (S - C)}}.$$

The reader may also wish to consider the spherical variant of the simplest case of Steiner's problem: to find X so as to minimize the sum $XA + XB + XC$.

Given a spherical triangle ABC, if the arcs b and c enclosing the angle A are produced beyond a, then their continuations intersect again at the point A' antipodal to A. The triangle $A'BC$ is called the triangle colunar with ABC with respect to the side a; its six elements are simply determined from those of ABC. The reader may wish to recall the spherical excess formula for the area of a spherical triangle (vol. 1, p. 5) and to prove the theorem of Lexell: Let a spherical triangle T have a fixed base and a fixed area, then the vertex opposite the base lies on a small circle whose pole is the circumcenter of the triangle colunar with T with respect to the base. In the limiting case of a *plane* triangle (which we get by letting the radius of S tend to infinity) the small circle becomes merely a straight line parallel to the base.

Given a spherical triangle ABC, let a great circle arc through A intersect the opposite side a and divide it into two (angular) segments a_1 and a_2. Similarly, let two other great circle arcs through B and C divide the opposite sides into b_1, b_2 and c_1, c_2. We suppose that the order of segments round T is a_1, a_2, b_1, b_2, c_1, c_2. Then the three great circle arcs are concurrent if and only if

$$\sin a_1 \sin b_1 \sin c_1 = \sin a_2 \sin b_2 \sin c_2,$$

in complete analogy with the corresponding plane case of Ceva's theorem (vol. 1, p. 7). Again, the plane case may be obtained as the limit $R \to \infty$ of the spherical case, R being the radius of the sphere.

As the foregoing examples suggest, many planar theorems and problems have spherical counterparts, and conversely. As an exercise in formulation (and perhaps also partial solution) the reader may wish to recall the plane problem of the traveling salesman (vol. 1, p.138) and to convert it into the spherical "problem of the stationary astronomer" (with the earth either fixed or rotating).

For further material on spherical trigonometry and related matters, the reader is referred to [111].

8. CONDITIONS AND DEGREES OF FREEDOM

In geometry, as in other branches of mathematics, we meet sometimes the following pattern of reasoning. A certain entity E is to be produced, built up of points, lines, and so on, and meeting certain conditions. We count the number of degrees of freedom describing the selection of the constituent points, lines, and so on, and we subtract the number of conditions. If the difference d is positive we conclude that infinitely many E's exist; more precisely, the E's form a d-parameter family. If $d < 0$ then in general there is no such E. If $d = 0$ then a finite number of E's exist. In the present section we consider several situations of the above type.

(**a**)   In vol. 1, beginning on p. 139, we have solved Steiner's problem: how to connect n given points in the plane by the shortest possible network. The solution given there depended on the following: If $n = 3$ and the triangle defined by the three points, A_1, A_2, A_3, has no angle $\geq 120°$ then the minimum net is obtained by joining the unique point X inside $A_1 A_2 A_3$ to the vertices so that every two segments form at X the $120°$ angle. In other words, for $n = 3$ Steiner's problem reduces to minimizing the sum of distances

$$S = \sum_{i=1}^{3} |A_i X|.$$

Consider now the following analog: Given four noncoplanar points A_1, A_2, A_3, A_4 in the Euclidean three-dimensional space, to find the point X

which minimizes the sum of distances

$$S = \sum_{i=1}^{4} |A_i X|. \tag{1}$$

Introduce the four unit vectors a_i given by

$$a_i = \frac{XA_i}{|XA_i|} \tag{2}$$

starting at X and pointing toward the A's. On the basis of a superficial analogy with the plane problem for three points it might be thought that S is minimized when the (a_1, a_2, a_3, a_4) configuration is *regular*, that is, when every two vectors a_i, a_j make the same angle. However, it is simple to show that with the general tetrahedron $A_1 A_2 A_3 A_4$ no point X gives rise to the regular configuration. We might reason as follows. Take four straight rays emanating from a point and forming the regular configuration. Each ray contains a vertex of a tetrahedron, hence we get the four-parameter family of tetrahedra admitting the regular configuration. However, the family of all tetrahedra has six parameters, not four. One way to show it is to observe that three coordinates per vertex give us $3 \cdot 4 = 12$ parameters; now we remove six parameters by factoring out the rigid motions. This can be done either in terms of the six parameters describing the matrices of the orthogonal group or, simply, by counting off three parameters for translations, two parameters to fix a straight line in the tetrahedron, and one parameter for rotation about that line. Thus, not all tetrahedra admit the regular configuration (a_1, a_2, a_3, a_4).

If such a configuration is to exist, then we might expect that two $(= 6 - 4)$ conditions must be satisfied by the tetrahedron. This could also be seen as follows. Let $\alpha = 109° 28' 15'' \cdots$ be the angle made by any two vectors a_i, a_j (so that $\cos \alpha = -1/3$ and for the n-dimensional case $\cos \alpha = -1/n$). Then the locus of points at which a segment subtends the angle α is a sphere. We get thus six such spheres, one for each edge of the tetrahedron. What we want is the existence of the regular configuration, which means that the six spheres should have a common point. In general, any three edges of the tetrahedron forming a triangle have spheres with a common point G (provided that no angles are too big). If *two* other edges also have spheres passing through G then the sphere of the sixth edge also passes through G.

We return now to the minimization of S in equation (1). Applying any one out of a variety of possible perturbations we show that minimization of S leads to the condition

$$a_1 + a_2 + a_3 + a_4 = 0. \tag{3}$$

This could be shown, for instance, by the principle of the minimum perturbation. We keep $|A_1X| + |A_2X|$ fixed so that X lies in the surface of the ellipsoid E_{12} of revolution. This ellipsoid is obtained by rotating about the line A_1A_2 the ellipse through X whose foci are A_1 and A_2. We define in similar way the ellipsoid E_{34}. Then the two ellipsoids are externally tangent at X since otherwise X could be moved into the interior of their intersection and so S would be decreased. Now (3) follows by some simple geometry: The common normal at X to the two ellipsoids must bisect both $\measuredangle A_1XA_2$ and $\measuredangle A_3XA_4$. It might be suspected that there exist perhaps some further minimizing conditions for S besides equation (3), which we could obtain by some cleverer perturbations; the reader may wish to show that this is not the case.

Thus we have exhibited here on a very simple example an instance of what could be called partial generalizability. What is meant is this: If the *plane* case of Steiner's problem with three points were considered in the same way then we could easily prove the same relation as (3):

$$a_1 + a_2 + a_3 = 0. \tag{4}$$

Now, the "partialness" is this: in the plane relation (4) for the three unit vectors implies that they form the regular configuration so that every two make the angle $120°$. Similar regularity does not follow from (3) for three-dimensional case.

A problem entirely different from minimizing S in (1) could be raised: Given the tetrahedron with vertices A_1, A_2, A_3, A_4, find the shortest net joining these four vertices. Here it is not difficult to show that depending on the angles of the tetrahedron we get one or another of the three possible solutions. The type of the solution goes together with the number of Steiner points ($=$ vertices of the net other than the four original A_i's) which is $\leq 4 - 2 = 2$. The three cases are shown in Fig. 1a–c. For the case of Fig. 1a to occur we must have $\alpha \geq 120°$ and $\beta \geq 120°$, the number of Steiner points is 0. Similarly, in the case of Fig. 1b there is one Steiner point, S_{12}, and we must have $\gamma \geq 120°$. It will be observed that the minimal net is carried by the edges of the tetrahedron in Fig. 1a and by the faces in Fig. 1b. For the case of Fig. 1c we have two Steiner points, S_{12} and S_{34}, and the minimal net is in the interior of the tetrahedron (except for the points A_i).

(b)   The natural question arises: How do we determine the minimal net of Fig. 1c; in particular, how do we find the positions of the Steiner points S_{12} and S_{34}? As in the plane case of solving Steiner's problem, we show that there are points A_{12} and A_{34}, with the following properties: $A_1A_2A_{12}$ and $A_3A_4A_{34}$ are equilateral triangles, and the four points A_{12},

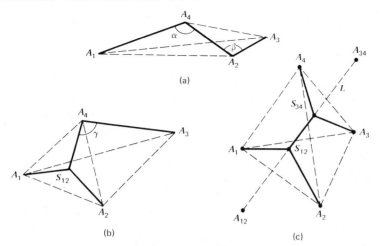

Fig. 1. Minimal tetrahedral net.

S_{12}, S_{34}, A_{34} are on a line L. This line, which we call the axis, intersects the edges A_1A_2 and A_3A_4. In the plane case of Steiner's problem with four points we would be finished now (as is shown in Fig. 3 in vol. 1, p. 141) since the points A_{12} and A_{34} are then constructible from the initial data.

Here we have no unique A_{12} or A_{34}; we only know their loci which are circles C_{12} and C_{34} lying in the planes bisecting perpendicularly the edges A_{12} and A_{34}, from which these edges subtend 60°. Actually, we do not need all of C_{12} (or of C_{34}) but only the arc of it which is antipodal to the arc lying inside the tetrahedron. Hence we get the following condition to determine L with: L is a straight line intersecting the circles C_{12} and C_{34}, and the straight edges A_1A_2 and A_3A_4.

Are those conditions sufficient to determine L? To determine it uniquely? We are faced here with an argument on the number of degrees of freedom and of the conditions, again. In three dimensions a straight line is determined by four parameters $(3+3-1-1$: three degrees of freedom to fix a point, three to fix another, subtract 1 for the distance between the points and 1 for sliding the line along itself). Since there are also four conditions to be met—the line we want is to intersect four one-dimensional loci—we could expect a finite number of solutions.

We have thus formulated a simple problem in what might be called nonlinear enumerative geometry. The general problem of the (linear) enumerative geometry is as follows: To determine the number N of Euclidean flats (and to determine the flats) that have dimensionally prescribed incidence with a finite number of given Euclidean flats $F_1, \ldots$,

F_s, provided that the numbers of conditions and of degrees of freedom are equal. Let us take a simple nontrivial example in three-dimensional space: The four flats F_1, F_2, F_3, F_4 are straight lines L_1, L_2, L_3, L_4 and the line L we want is to intersect them (clearly $L \cap L_i$ is a point, hence there is no need to prescribe the dimensionality of $L \cap L_i$). The determination of the number N of such lines L is greatly facilitated by the principal theorem of enumerative geometry, due to H. Schubert [153, 154]: The above number $N = N(F_1, \ldots, F_s)$ remains the same when the flats F_i are specialized or restricted in various ways, provided only that the specialization does not make N infinite.

This is referred to as the principle of special position, or the principle of the conservation of the number; since Schubert's proof was not general enough, Hilbert has placed the problem of producing a fully general proof as the fifteenth on his list of 23 problems [75]. We exploit the special position principle as follows. Of the four lines L_i let L_1 and L_2 intersect, similarly let L_3 and L_4 intersect. Now L must either contain the point $p_{12} = L_1 \cap L_2$ or else lie in the plane P_{12} of L_1 and L_2. By the symmetric argument L either contains P_{34} or lies in P_{34}. Neither $p_{12} \in P_{34}$ nor $p_{34} \in P_{12}$ is allowed, for then the number of lines L becomes infinite. Hence there are just *two* lines L which meet every L_i: the line $p_{12} p_{34}$ and the line $P_{12} \cap P_{34}$. Therefore there are just two lines intersecting four lines L_1, L_2, L_3, L_4 in the general position.

This is a special case of the following result of Schubert [163]: The number $N(n, k)$ of k-flats in E^n that intersect $(k+1)(n-k)$ given $(n-k-1)$–flats in general position, is

$$N(n, k) = [(k+1)(n-k)]! \prod_{j=0}^{k} \frac{j!}{(n-j)!}.$$

The two lines cutting four given ones on E^3 are obtained when $n = 3$, $k = 1$: $N(3, 1) = 2$. We could obtain this result, that there are two such lines, without employing Schubert's specialization principle, by an alternative method. This has the advantage of giving us the two lines as well as their mere number.

The idea is as follows: To produce the line(s) L cutting L_1, L_2, L_3, L_4 find first the locus of all lines cutting L_1, L_2, L_3 and intersect it with similar locus of all lines cutting L_2, L_3, L_4. First, we choose the line (rather, any one of the lines) making equal angles with L_1, L_2, and L_3. Then we translate it suitably, into the position U, which has the following property: When L_1 is rotated about U the locus S contains L_2 and L_3. S is of course a one-sheeted hyperboloid of revolution on which L_1, L_2, and L_3 belong to one system of generators. Hence, the lines intersecting L_1, L_2, and L_3 form the other system of generators of S. Now the two lines L

cutting all four lines L_1, L_2, L_3, L_4 are obtained by intersecting in pairs certain hyperboloids of revolution.

(c) We return finally to the problem of determining the minimal net corresponding to the case of Fig. 1c. This hinges on finding the axis L. In the first place, the question arises which of the three opposite pairs of the edges of the tetrahedron are to be paired together as A_1A_2 and A_3A_4 in our Fig. 1c. Once the axis L is found, we can show that $|A_{12} A_{34}|$ is the length of the whole net; this is proved exactly as in the plane case of Steiner's problem with four points. We therefore compare the distances $|A_{12} A_{34}|$, $|A_{13} A_{24}|$, and $|A_{14} A_{23}|$; the shortest of these tells us which edges of the tetrahedron to pair up. The reader may wish to develop a criterion for correct pairing, which avoids the need to determine the points A_{12}, A_{34}, and so on, in advance.

Supposing that the pairing is as in Fig. 1c, how do we find the minimal net, that is, the axis L? Let us recall the condition that L must satisfy: it intersects A_1A_2, A_3A_4, the circular arc C_{12}, and the circular arc C_{34} (these have been defined above). Let Q be a point of C_{12} and let $T(Q)$ be the tangent to C_{12} at Q. Similarly, we let R be a point of C_{34} and $T(R)$ the corresponding tangent. The four straight lines A_1A_2, A_3A_4, $T(Q)$ and $T(R)$ possess common straight secants. Now we set up the argument by the intermediate-value-property that, as Q and R traverse their circular arcs, there will occur the position when a common secant line of the four lines cuts $T(Q)$ at Q and $T(R)$ at R. This common secant is then the axis L that we wished to find. By a similar employment of tangent structures the reader may wish to set up and solve other problems of nonlinear enumerative geometry.

2
ANALYSIS

1. INEQUALITIES

(a) Since $e^t > 0$ for all real t, the curve $y = e^t$ lies above every one of its tangents, except for the tangency point itself. In particular, we have

$$e^t \geq 1 + t \tag{1}$$

by taking the tangent at $(0, 1)$. From (1) it follows, without any further use of calculus, that the function $y = x^{1/x}$ attains its maximum at $x = e$: We apply (1) with $t = (x - e)/e$; after some cancellation and simplification we get $e^{x/e} \geq x$ or $e^{1/e} \geq x^{1/x}$, with the equality at $x = e$ only. We recall that $\lim\limits_{x \to \infty} x^{1/x} = 1$; actually more can be proved without difficulty. Let $x > 1$ so that $x^{1/x} = 1 + a$ with $a > 0$; hence $x = (1 + a)^x$ so that, using the binomial theorem, $x > [x(x-1)/2]a^2$. Therefore, for $x > 1$ we get a bound on a which leads to the bracketing

$$1 < x^{1/x} < 1 + \left(\frac{2}{x-1}\right)^{1/2}.$$

Similarly, taking $x > k$ and using further terms in the binomial expansion, we get

$$1 < x^{1/x} < 1 + \left[\frac{k!}{(x-1)(x-2)\cdots(x-k+1)}\right]^{1/k}.$$

Although better bounds are simply obtained in this case by using logarithms (we write $x^{1/x} = \exp(x^{-1}\log x)$, etc.) we observe here a method of estimating which may be useful sometimes: With infinitely many estimates as in the last inequality ($k = 2, 3, \ldots$) we get the "enveloping" estimate

$$1 < x^{1/x} < 1 + m(x)$$

where

$$m(x) = \min_{2<k<x} \left\{ \left[\frac{k!}{(x-1)(x-2)\cdots(x-k+1)} \right]^{1/k} \right\}.$$

This could also be adapted to the case of continuous parameter k.

(b) Let $f(x)$ be a continuous increasing function with $f(0) = 0$. and let a, b be positive. With the areas as in Fig. 1 we find that

$$A_1 = \int_0^a f(x)\, dx, \qquad A_2 = \int_0^b f^{-1}(x)\, dx, \qquad A_1 + A_2 \ge ab,$$

and the inequality is strict except when $b = f(a)$. Hence Young's inequality

$$ab \le \int_0^a f(x)\, dx + \int_0^b f^{-1}(x)\, dx. \tag{2}$$

There is a slight generalization: If $f(x)$ and $g(x)$ both satisfy the conditions of Young's inequality then

$$ab \le \int_0^a \max\left[f(x), g(x)\right] dx + \int_0^b \min\left[f^{-1}(x), g^{-1}(x)\right] dx.$$

The reader may wish to convert (2) into a two-sided inequality by suitably

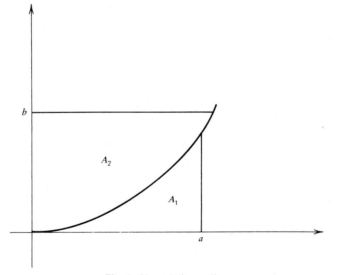

Fig. 1. Young's inequality.

bounding the difference between $A_1 + A_2$ and ab, for instance

$$ab \leq \int_0^a f(x)\,dx + \int_0^b f^{-1}(x)\,dx \leq ab + [f(a) - b][a - f^{-1}(b)]; \qquad (3)$$

better bounds might be attempted, based perhaps on the Euler–McLaurin formula (vol. 1, p. 125). As an application of (2) we take $f(x) = x^{p-1}$ with $p > 1$ and $p^{-1} + q^{-1} = 1$. Then (2) yields at once

$$ab \leq \frac{a^p}{p} + \frac{b^q}{q};$$

with $a_1, \ldots, a_n; b_1, \ldots, b_n$ nonnegative we have, therefore,

$$\sum_{i=1}^n a_i b_i \leq \frac{1}{p} \sum_{i=1}^n a_i^p + \frac{1}{q} \sum_{i=1}^n b_i^q. \qquad (4)$$

Suppose that not all a's or b's are 0 and put

$$A_i = a_i \left(\sum_{i=1}^n a_i^p \right)^{-1/p}, \qquad B_i = b_i \left(\sum_{i=1}^n b_i^q \right)^{-1/q}, \qquad i = 1, \ldots, n;$$

using (4) with A_i and B_i in place of a_i and b_i we obtain the Hölder inequality

$$\sum_{i=1}^n A_i B_i \leq \left(\sum_{i=1}^n A_i^p \right)^{1/p} \left(\sum_{i=1}^n B_i^q \right)^{1/q}.$$

A useful extension of (2) is obtained by generalizing the argument from 2 to n dimensions. Instead a continuous increasing function $y = f(x)$ with $f(0) = 0$ we take in the Euclidean n-dimensional space a curve given parametrically by

$$x_i = f_i(t), \qquad i = 1, \ldots, n, \qquad f_i(0) = 0$$

where each $f_i(t)$ is a continuous increasing function of t. Let the sets $A_i(t)$ be given by

$$A_i(t) = \{(x_1, \ldots, x_n) : x_i = f_1(t), 0 \leq x_j \leq f_j(t) \text{ for } j \neq i\}$$

and let $X_i(T)$ be the sets swept out by $A_i(t)$ as t varies from 0 to T:

$$X_i(T) = \bigcup_{0 \leq t \leq T} A_i(t) \qquad i = 1, \ldots, n.$$

Let $t_1, \ldots, t_n$ be n positive numbers; then the sets $X_1(t_1), \ldots, X_n(t_n)$ have disjoint interiors, and their volumes correspond precisely to the areas A_1 and A_2 of Fig. 1. Repeating now the same reasoning as before (essentially, the whole is no less than the sum of *disjoint* parts), this time on the volumes of the sets $X_i(t_i)$, we get

$$\prod_{i=1}^{n} f_i(t_i) \le \sum_{i=1}^{n} \int_0^{t_i} f_i'(t) \prod_{\substack{j=1 \\ j \ne i}}^{n} f_j(t) \, dt. \tag{5}$$

The reader may wish to obtain the two-sided equivalent of the type (3) here. If the passage from 2 to n dimensions is too sudden the case $n = 3$ may be considered first.

As an application let us take in (5)

$$f_i(t) = A_i t^{a_i}, \qquad A_i \text{ and } a_i > 0, \qquad i = 1, \ldots, n.$$

We have then

$$\prod_{i=1}^{n} A_i t_i^{a_i} \le \frac{\displaystyle\sum_{i=1}^{n} A_i a_i t_i^{\sum_{i=1}^{n} a_i}}{\displaystyle\sum_{i=1}^{n} a_i} \tag{5a}$$

which is a generalized arithmetic mean–geometric mean inequality. The generalization consists of different weights; by specializing to equal weights A_i and equal exponents a_i we have

$$a_1 = \cdots = a_n = \frac{1}{n}, \qquad A_1 = \cdots = A_n = 1$$

which yields the ordinary arithmetic–geometric inequality

$$(t_1 \cdots t_n)^{1/n} \le \frac{1}{n} \sum_{i=1}^{n} t_i. \tag{6}$$

Given n positive numbers $x_1, \ldots, x_n$ let us write $A(x) = A(x_1, \ldots, x_n)$ and $G(x) = G(x_1, \ldots, x_n)$ for their arithmetic and geometric means, and $H(x_1, \ldots, x_n) = H(x)$ for their harmonic mean

$$H(x) = n \left(\sum_{i=1}^{n} x_i^{-1} \right)^{-1}.$$

Employing simple self-explanatory terminology, we find that

$$H(x) = \frac{1}{A(1/x)}, \qquad G(x) = \frac{1}{G(1/x)}. \tag{7}$$

Hence we see that $G(x) \le A(x)$ implies $H(x) \le G(x)$; moreover, the inequality or equality occurs in both together. We get thus the full arithmetic–geometric–harmonic inequality

$$H(x_1, \ldots, x_n) \le G(x_1, \ldots, x_n) \le A(x_1, \ldots, x_n) \tag{8}$$

and the equality holds (on both sides together) if and only if $x_1 = x_2 = \cdots = x_n$. The reader may wish to complete the generalized arithmetic–geometric inequality (5a) to a full generalized arithmetic–geometric–harmonic case as in equation (8). A further generalization of (8) comes by introducing the function

$$F(t) = \left(\frac{1}{n} \sum_{i=1}^{n} x_i^t\right)^{1/t} \tag{8a}$$

and showing, by calculus for instance, that $F(t)$ is strictly increasing except when all x_i's are equal. Using limits to define $F(-\infty)$, $F(0)$, and $F(\infty)$ we find that with x_i's not all equal

$$F(-\infty) < F(-1) < F(0) < F(1) < F(\infty); \tag{8b}$$

this is a generalization of (8), for it can be written as

$$\min_i x_i < H(x) < G(x) < A(x) < \max_i x_i. \tag{8c}$$

Of the many uses of (8) we outline one: development of a small theory of the exponential function, based on nothing beyond the elements of limits and continuity. Let $0 \le x \le 1$ and apply the G–A part of (8) to the $n+1$ numbers: 1, $1 + x/n$, $1 + x/n$, ..., $1 + x/n$; we get then

$$\left(1 + \frac{x}{n}\right)^n \le \left(1 + \frac{x}{n+1}\right)^{n+1}. \tag{9}$$

Next, we apply the H–G part of (8) to the $n+2$ numbers: 1, $1 + x/n$, $1 + x/n$, ..., $1 + x/n$, getting

$$\left[(n+2)\bigg/\left(1 + \frac{n(n+1)}{n+x}\right)\right]^{n+2} \le \left(1 + \frac{x}{n}\right)^{n+1}. \tag{10}$$

Since $0 \le x \le 1$ it is easy to verify that the expression in the square brackets is no less than $1 + x/(n+1)$, hence (10) gives us

$$\left(1 + \frac{x}{n+1}\right)^{n+2} \le \left(1 + \frac{x}{n}\right)^{n+1}. \tag{11}$$

Consider now the two sequences

$$\left\{\left(1 + \frac{x}{n}\right)^n\right\}, \quad \left\{\left(1 + \frac{x}{n}\right)^{n+1}\right\}, \quad n = 1, 2, \ldots;$$

by (9) and (11) the first one is nondecreasing and the second one is nonincreasing, the first sequence is dominated term-by-term by the second one, and the ratio of the two nth terms tends to 1 as n increases.

Hence there exists the limit

$$\phi(x) = \lim_{n \to \infty} \left(1 + \frac{x}{n}\right)^n, \qquad 0 \le x \le 1,$$

and we get the crucial two-sided bracketing

$$\left(1 + \frac{x}{n}\right)^n \le \phi(x) \le \left(1 + \frac{x}{n}\right)^{n+1}, \qquad 0 \le x \le 1, \qquad n = 1, 2, \dots. \quad (12)$$

From this it follows that

$$\phi(x) \text{ is increasing} \qquad (13)$$

since it is bracketed between two increasing functions whose difference

$$\left(1 + \frac{x}{n}\right)^{n+1} - \left(1 + \frac{x}{n}\right)^n \le \frac{\phi(1)}{n} \qquad (14)$$

can be made arbitrarily small. Next,

$$\phi(x) \text{ is continuous in } x. \qquad (15)$$

This is proved by smuggling in, so to say, the idea of uniform convergence in the form which is perhaps easier and more natural than in some other contexts, and which lends itself well for introductory purposes. We have

$$\phi(x+h) - \phi(x) = \phi(x+h) - \left(1 + \frac{x+h}{n}\right)^n$$

$$+ \left(1 + \frac{x+h}{n}\right)^n - \left(1 + \frac{x}{h}\right)^n$$

$$+ \left(1 + \frac{x}{h}\right)^n - \phi(x)$$

so that by (12) and (14)

$$|\phi(x+h) - \phi(x)| \le \frac{2\phi(1)}{n} + \left|\left(1 + \frac{x+h}{n}\right)^n - \left(1 + \frac{x}{n}\right)^n\right|.$$

The right-hand side can be made arbitrarily small: We fix $\varepsilon > 0$ and choose first n so large that $2\phi(1)/n < \varepsilon/2$. Then, keeping this large n fixed, we use the continuity of polynomials to make the second term on the right-hand side also $< \varepsilon/2$ by taking h sufficiently small. Next, we prove the functional equation

$$\phi(x+y) = \phi(x)\,\phi(y), \qquad 0 \le x, y, \qquad x+y \le 1. \qquad (16)$$

In fact, by the definition of ϕ

$$\phi(x)\,\phi(y)=\lim_{n\to\infty}\left(1+\frac{x+y+xy/n}{n}\right)^n$$

and since xy/n can be made arbitrarily small, (16) follows from (15). To handle the derivative we have by (16)

$$\frac{\phi(x+h)-\phi(x)}{h}=\phi(x)\,\frac{\phi(h)-1}{h}. \tag{17}$$

Using (12) with $x=h$ and with $n=1$ on the left and $n=k$ on the right, we get

$$1+h\le\phi(h)\le\left(1+\frac{h}{k}\right)^{k+1}$$

and after subtracting 1 and dividing by h

$$1\le\frac{\phi(h)-1}{h}\le\frac{(1+h/k)^{k+1}-1}{h}.$$

We use the squeeze principle on the above, letting $h\to0$ and getting

$$1\le\lim_{h\to0}\frac{\phi(h)-1}{h}\le1+\frac{1}{k},$$

and since k is arbitrary, the limit is 1. Hence by (17)

$$\phi'(x)=\phi(x). \tag{18}$$

The interested reader may wish to compare this part with the infinitesimal generators of vol. 1, p. 109. The idea here is expressed by equation (17): On account of the functional equation of the exponential function the derivative at x is immediately deducible from the derivative at 0. The same phenomenon occurs at a higher level of abstraction with topological groups: Tangent structure at any place is simply obtained from the tangent structure at the identity of the group.

Finally we re-label $\phi(1)$ as e and we show that

$$\phi(x)=e^x, \qquad 0\le x\le1. \tag{19}$$

First, by repeated use of (16)

$$\left[\phi\!\left(\frac{1}{q}\right)\right]^q=\phi\!\left(\frac{q}{q}\right)=e$$

for any positive integer q, so that $\phi(1/q)=e^{1/q}$. Next, we choose any positive integer p so that $1 \leq p \leq q$ and we apply (16) again, to get

$$\phi\left(\frac{p}{q}\right) = (e^{1/q})^p = e^{p/q}.$$

This proves (19) for all rational x and the rest follows by continuity. Having developed the principal properties of the exponential on the interval $[0, 1]$ we continue them to all real numbers by exploiting the functional equation (16).

Since there are already so many standard and other approaches to the exponential, a new one calls for some motivation and explanation. We emphasize an interesting use of important inequalities, introduce early and rather painlessly the concept of uniformity, and employ a very important natural law of growth $(f_n(x) = (1 + x/n)^n$, cursed with the unfragrant name of "compound interest law"). Above all these, we present on a small but realistic example the constructive approach built into the very definition: an entity (here the exponential $\phi(x)$) is to be regarded as given only if it is computable to arbitrary accuracy by some effective means. This is afforded by the two-sided bracketing (12), together with (14).

In connection with (12) the reader may wish to show that the sequence

$$\left\{\left(1 + \frac{x}{n}\right)^{n+a}\right\}, \qquad 0 \leq x \leq 1, \qquad n = 1, 2, \dots$$

is increasing for $a < 1/2$ and decreasing for $a > 1/2$ [136].

(c) The arithmetic, geometric, and harmonic means can be extended from the discrete to the continuous domain. To be able to define all three of them for functions $f(x)$ on an interval $a \leq x \leq b$ we shall suppose that $f > 0$. The arithmetic mean of f is simply handled:

$$A(f; a, b) = \lim_{n \to \infty} \frac{1}{n} \sum_{j=1}^{n} f\left(a + j\frac{b-a}{n}\right) = \frac{1}{b-a} \int_a^b f(x)\, dx. \qquad (20)$$

To define similarly the geometric and harmonic means $G(f; a, b)$ and $H(f; a, b)$ we use the conjugacy principle:

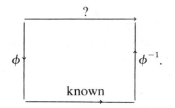

For the geometric mean we observe that the arithmetic and geometric means are conjugate under $L(x) = \log x$:

$$G(x_1, \ldots, x_n) = L^{-1}(A(L(x_1), \ldots, L(x_n)))$$

which leads us to define

$$G(f; a, b) = e^{\frac{1}{b-a}\int_a^b \log f(x)\, dx}. \tag{21}$$

For the harmonic mean we use the inverse $1/x$ in place of the logarithm $L(x)$, getting

$$H(f; a, b) = (b - a)\Big/ \int_a^b \frac{dx}{f(x)}. \tag{22}$$

Putting together (20), (21), and (22) we obtain the continuous version of the arithmetic–geometric–harmonic inequality (8):

$$(b-a)\Big/ \int_a^b \frac{dx}{f(x)} \le e^{\frac{1}{b-a}\int_a^b \log f(x)\, dx} \le \frac{1}{b-a}\int_a^b f(x)\, dx.$$

As an example, we take $f(x) = 1 + x$, $a = 0$, and $b = 1/N$ where N is large, and we get the bracketing

$$\frac{1}{(N+1)\log(1+1/N)} \le \frac{(1+1/N)^N}{e} \le \frac{2N+1}{2N+2}.$$

Expanding the $\log(1+x)$ by the usual alternating series we find the estimate

$$\frac{2N+1}{2N+2} - \frac{1}{12N(N+1)} \le \frac{(1+1/N)^N}{e} \le \frac{2N+1}{2N+2}.$$

With $f(x) = 1 + x^2$ we get under the same conditions

$$\frac{N}{(1+N^2)\arctan 1/N} \le e^{2N(\arctan 1/N - 1/N)} \le \frac{3N^2+1}{3N^2+3}.$$

There is also a continuous analog

$$M_t(f) = \left[(b-a)^{-1}\int_a^b f^t(x)\, dx\right]^{1/t}$$

of $F(t)$ in (8a–c).

Related to the geometric mean $G(f; a, b)$ there is the so-called product integral $P(f)$, of some use in probability, especially with continuous Markov chains [58]. This is defined as

$$P(f) = \lim \prod_{j=1}^n [1 + f(x_j)\, \Delta x_j]$$

where the limit is taken over partitions of $[a, b]$ in the same way as with the Riemann integral. We have then

$$P(f) = e^{\lim \sum_{j=1}^{n} \log[1+f(x_j)\Delta x_j]} = e^{\int_a^b f(x)\,dx}.$$

Recalling the definition (20) of $A(f; a, b)$ we ask: Under suitable conditions, what is the limit of

$$n\left[\frac{1}{n}\sum_{j=1}^{n} f\left(a+j\frac{b-a}{n}\right) - \frac{1}{b-a}\int_a^b f(x)\,dx\right]$$

as $n \to \infty$? The reader may wish to show that it is $1/2(f(b)-f(a))$; higher limits of the same type are suggested by examining the Euler–McLaurin formula (vol. 1, p. 125). An application due to Polya [136] is the following: If

$$P_n = 1^1 \cdot 2^2 \cdot 3^3 \cdots \cdot n^n$$

then for large n

$$P_n^{1/n} \sim e^{-n/4}\, n^{(n+1)/2}. \tag{23}$$

Similarly, the behavior of

$$Q_n = 1^n \cdot 2^{n-1} \cdot \ldots \cdot (n-1)^2 \cdot n = 1!\,2! \ldots n!$$

for large n may be obtained from (23) since

$$P_n Q_n = (n!)^{n+1}. \tag{24}$$

As with the extension of the factorial $n!$ to the Γ-function $\Gamma(1+n)$, we might ask here whether the "iterated" factorials P_n and Q_n extend to functions $P(z)$ and $Q(z)$ defined for nonintegral arguments. Since (24) suggests that

$$P(z)Q(z) = \Gamma^{1+z}(1+z)$$

it is enough to extend one of them, say Q_n, to $Q(z)$. It turns out that such an extension, quite analogous to the Γ-function, is provided by the Alexeievsky function [183]:

$$Q(z) = (2\pi)^{z/2}\, e^{-z(z+1)/2 - \gamma z^2/2} \prod_{m=1}^{\infty}\left[\left(1+\frac{z}{m}\right)^m e^{-z+z^2/2m}\right]$$

and we have then $Q(n) = Q_n$ for positive integers n. This may be compared with the Weierstrass product:

$$\Gamma(z) = z^{-1} e^{-\gamma z} \prod_{m=1}^{\infty}\left[\left(1+\frac{z}{m}\right)^{-1} e^{z/m}\right].$$

(**d**)   We consider here an example of handling together both the discrete and the continuous forms of inequalities. Let us start with the well known identity

$$(1+2+\cdots+n)^2 = 1^3+2^3+\cdots+n^3 \qquad n \geq 1. \tag{25}$$

Denoting the sum $1^k+2^k+\cdots+n^k$ by $S_k(n)$ we show easily that the equality

$$S_k(n) = S_j^p(n) \tag{26}$$

of type (25) occurs only for (25) so that $j=1$, $p=2$, $k=3$. This can be proved, for instance, by equating the degrees and the leading coefficients of the two polynomials (in n) in (26). The reader may wish to explore the possibilities of extended monomial relations

$$S_k(n) = \text{const.} \prod_{i=1}^{s} S_{j_i}^{p_i}(n)$$

and of more general relations. Looking at (25) from the point of view of growth rates suggests the following problem: If $0=a_0<a_1<\cdots<a_n$, what is a sufficient condition on the a_i's, in terms of slowness of growth, for the inequality

$$\left(\sum_{i=0}^{n} a_i\right)^2 \geq \sum_{i=0}^{n} a_i^3 \tag{27}$$

to hold for all n? Using induction on n, or otherwise, the reader may wish to show that

$$a_{i+1}-a_i \leq 1, \qquad i=0,1,\ldots,n-1 \tag{28}$$

is a sufficient condition for (27). Having proved the sufficiency of (28) for (27), we derive the integral counterpart of (27):

$$\left(\int_0^u f(x)\,dx\right)^2 \geq \int_0^u f^3(x)\,dx, \qquad u \geq 0. \tag{29}$$

This holds if $f(0)=0$ and f satisfies the continuous analog of (28): $0<f'(x)\leq 1$. Besides proving (29) by discretizing and referring to (27) we can also prove it directly. Let $F(u)$ be the difference L.H.S.−R.H.S. in (29). Differentiating we get

$$F'(u) = 2f(u)\left[\int_0^u f(y)\,dy - \int_0^u f(y)f'(y)\,dy\right]$$

and so by integrating back

$$\left(\int_0^u f(x)\,dx\right)^2 - \int_0^u f^3(x)\,dx = 2\int_0^u\int_0^y f(y)f(v)[1-f'(v)]\,dv\,dy. \tag{30}$$

The reader may wish to compare this with the questions of isoperimetric deficiency and measures of transcendency, raised in vol. 1, pp. 23, 42. The *precise* form (30) of inequality (29) suggests that there should also be the precise form of the discrete counterpart (27) of inequality (29). Using some simple ideas on telescoping cancellation and so on from Chapter 3 of vol. 1, the reader may wish to show that

$$\left(\sum_{i=0}^{n} a_i\right)^2 - \sum_{i=0}^{n} a_i^3 = 2 \sum_{i=0}^{n} \sum_{j=0}^{i} a_i \frac{a_j + a_{j-1}}{2}[1 - (a_j - a_{j-1})]$$

where we have defined $a_{-1} = a_0 = 0$.

2. DIFFERENTIAL EQUATIONS AND SUMMATION OF SERIES

(a) Certain series involving sequences of solutions of an ordinary differential equation can be evaluated as follows. We set up a partial differential equation for the sum of the series, relate this partial differential equation to the original ordinary one, and verify that a given function is indeed the solution, and so the sum of the series [110]. Consider, for example, the generalized Laguerre polynomial $L_n^a(x)$ (here a is a superscript, not a power). It is a polynomial solution $u = u_n(x)$ of the ordinary differential equation

$$xu'' + (1 + a - x)u' + nu = 0. \tag{1}$$

Laguerre polynomials are a family orthogonal on the interval $[0, \infty)$ with the weight function $x^a e^{-x}$. The case $a = 0$ leads to the (ordinary) Laguerre polynomials $L_n(x)$. An explicit expression is

$$L_n^a(x) = \sum_{k=0}^{n} (-1)^k \binom{n+a}{n-k} \frac{x^k}{k!};$$

there is a formula of Rodriguez type (i.e., giving the function as the nth derivative of an expression associated with the weight-function):

$$L_n^a(x) = \frac{x^{-a} e^x}{n!} \frac{d^n (x^{n+a} e^{-x})}{dx^n}.$$

Also, there is a relation with Hermite polynomials:

$$H_{2n}(x) = (-1)^n \, 2^{2n} \, n! \, L_n^{-1/2} (x^2),$$
$$H_{2n+1}(x) = (-1)^n \, 2^{2n+1} \, n! \, x L_n^{1/2} (x^2).$$

Suppose that we wish to sum the series (or equivalently, evaluate the *generating* function):

$$G(x, y) = \sum_{n=0}^{\infty} L_n^a(x)y^n. \tag{2}$$

We can write (1) as

$$P\left(x, \frac{d}{dx}, n\right)u = 0$$

where P is a polynominal in its three variables:

$$P(x_1, x_2, x_3) = x_1 x_2^2 + (1 + a - x_1)x_2 + x_3.$$

Since

$$\frac{\partial}{\partial x}[u_n(x)y^n] = u_n'(x)y^n,$$

$$y\frac{\partial}{\partial y}[u_n(x)y^n] = nu_n(x)y^n,$$

we find that

$$P\left(x, \frac{\partial}{\partial x}, y\frac{\partial}{\partial y}\right)[u_n(x)y^n] = 0.$$

Consequently by the property of linearity any function $H(x, y)$ of the form

$$H(x, y) = \sum_{n=0}^{\infty} c_n L_n^a(x)y^n \tag{3}$$

satisfies the partial differential equation

$$P\left(x, \frac{\partial}{\partial x}, y\frac{\partial}{\partial y}\right)H = 0$$

or, in explicit form,

$$xH_{xx} + (1 + a - x)H_x + yH_y = 0. \tag{4}$$

In particular, $G(x, y)$ given by (2) is a solution of the above. Now, we verify by an explicit calculation that the function

$$(1 - y)^{-a-1} e^{-xy/(1-y)}$$

satisfies (4) so that

$$(1 - y)^{-a-1} e^{-xy/(1-y)} = \sum_{n=0}^{\infty} c_n L_n^a(x)y^n \tag{5}$$

for some sequence of constants c_n. We assume here that the solutions of the partial differential equation (4) are of the form (3); the method, as we have it, is therefore formal and the end result is subject to verification. To determine c_n we put $x = 0$ and observe that

$$L_n^a(0) = \binom{a+n}{n}$$

and so putting $x = 0$ in (5) we get

$$(1-y)^{-a-1} = \sum_{n=0}^{\infty} c_n \binom{a+n}{n} y^n.$$

Hence by the binomial theorem $c_n = 1$ for all n. In effect, the original series G has been summed:

$$\sum_{n=0}^{\infty} L_n^a(x) y^n = (1-y)^{-a-1} e^{-xy/(1-y)}.$$

Similarly, for the Bessel equation

$$x^2 u'' + x u' + (x^2 - n^2) u = 0$$

with a solution $u_n(x) = J_n(x)$ (the ordinary Bessel function of order n) we find that the function

$$H(x, y) = \sum_{n=-\infty}^{\infty} c_n J_n(x) y^n$$

satisfies the partial differential equation

$$x^2 H_{xx} - y^2 H_{yy} + x H_x - y H_y + x^2 H = 0.$$

The reader may wish to verify that

$$e^{x(y-1/y)/2}$$

is a solution of the above equation. Again it turns out that $c_n = 1$ for all n so that

$$e^{x(y-1/y)/2} = \sum_{n=\infty}^{\infty} J_n(x) y^n.$$

Here some relation with the Hadamard product formula

$$(f \circ g)(x) = \frac{1}{2\pi i} \int_C f(u) g(x/u) \frac{du}{u}$$

suggests itself; for instance, we have

$$J_0(x) = e^{x^2/2} \circ e^{-x^2/2}$$

so that

$$J_0(x) = \frac{1}{2\pi i} \int_C e^{(u^2 - x^2/u^2)/2} \frac{du}{u}.$$

Choosing for the contour C the circle of radius $x^{1/2}$ about the origin, we put $u = x^{1/2} e^{i\theta}$ and transform the above into the standard integral representation

$$J_0(x) = \frac{1}{\pi} \int_0^\pi \cos(x \sin \phi) \, d\phi.$$

With a related method we can sometimes convert ordinary generating functions to exponential generating functions. For instance, let $P_n(x)$ be the nth Legendre polynomial. As is well known, the ordinary generating function is

$$\sum_{n=0}^\infty P_n(x) y^n = (1 - 2xy + y^2)^{-1/2}.$$

We ask: What is the exponential generating function

$$\sum_{n=0}^\infty \frac{P_n(x) y^n}{n!} = G(x, y)?$$

Recalling the Borel transform which expresses the ordinary generating function in terms of the exponential one (Chapter 3, sec. 4) we get

$$\int_0^\infty e^{-t} G(x, yt) \, dt = (1 - 2xy + y^2)^{-1/2}. \tag{6}$$

To use this for the evaluation of G we observe the Bessel function integral

$$\int_0^\infty e^{-at} J_0(bt) \, dt = (a^2 + b^2)^{-1/2} \tag{7}$$

of vol. 1, p. 178. Comparing (6) and (7) we find that the first thing to do is to express the quantity $1 - 2xy + y^2$ of (6) as a sum of two squares. This, of course, can be done in many ways, e.g.,

$$1 - 2xy + y^2 = [(1 - x^2)^{1/2}]^2 + (x - y)^2, \tag{8}$$
$$1 - 2xy + y^2 = (1 - xy)^2 + [y(1 - x^2)^{1/2}]^2, \tag{9}$$

and so on. Using (7) and (9) we express $(1 - 2xy + y^2)^{-1/2}$ as the right type of integral:

$$\int_0^\infty e^{-t} e^{xyt} J_0(yt \sqrt{1 - x^2}) \, dt = (1 - 2xy + y^2)^{-1/2}$$

and by comparing with (6) we get the desired exponential generating function G:

$$\sum_{n=0}^{\infty} \frac{P_n(x)y^n}{n!} = e^{xy} J_0(y\sqrt{1-x^2}).$$

The reader may wish to determine why taking (7) and (8) instead of (7) and (9) would not work.

(b) As another, rather more operatorially oriented, technique of that type, we consider the Truesdell unified approach to special functions [172]. This starts with the Truesdell equation

$$\frac{\partial F(x, a)}{\partial x} = F(x, a+1); \tag{10}$$

we prepare ourselves in advance for things to come by observing that this equation is automatically satisfied by the generalized ath order derivative: $f(x, a) = f^{(a)}(x)$. Thereby we also establish connections with Section 4, p. 59, of vol. 1. Let the shift and differentiation operators be defined by

$$GF(x, a) = F(x, a+1), \qquad DF(x, a) = \frac{\partial F(x, a)}{\partial x}, \qquad E_y F(x, a) = F(x+y, a).$$

Now relation (10) is simply $D = G$ and so the Taylor series expansion can be written symbolically (cf. vol. 1, p. 109) as $E_y = e^{yG}$ or, in the expanded form, as

$$F(x+y, a) = \sum_{n=0}^{\infty} F(x, a+n) \frac{y^n}{n!}. \tag{11}$$

Consider for the purposes of an example the Hermite function $u = H_a(x)$ which satisfies the differential equation

$$u'' - 2xu' + 2au = 0$$

and the recursion

$$H_{a+1}(x) = 2xH_a(x) - H_a'(x). \tag{12}$$

We verify that the function

$$F(x, a) = e^{i\pi a} e^{-x^2} H_a(x)$$

satisfies the Truesdell relation (10). Hence, applying expansion (11) we have

$$e^{2xy-y^2} H_a(x-y) = \sum_{n=0}^{\infty} \frac{H_{a+n}(x)y^n}{n!}. \tag{13}$$

In the special case of $a = 0$ we have $H_0(x) = 1$ and $H_n(x)$ is the nth Hermite polynomial; equation (13) gives us then

$$e^{2xy-y^2} = \sum_{n=0}^{\infty} \frac{H_n(x)y^n}{n!},$$

and on multiplying by e^{-x^2} and using the McLaurin series,

$$e^{-(x-y)^2} = \sum_{n=0}^{\infty} e^{-x^2} \frac{H_n(x)y^n}{n!}$$

and

$$H_n(x) = (-1)^n e^{x^2} \frac{d^n e^{-x^2}}{dx^n}$$

which is the Rodriguez-form representation, yet another example of an orthogonal polynomial given as the nth derivative.

(c) The last item in this section also refers to the summation of series and the use of differential operators to evaluate the sums. It may be prefaced by observing that if $P(x)$ is a polynomial and k a nonnegative integer then

$$P\left(x\frac{d}{dx}\right)x^k = P(k)x^k,$$

which recalls the Heaviside calculus, ordinary differential equations (especially the Euler equations), and so on. Also, we get

$$\sum_{n=0}^{\infty} \frac{P(n)x^n}{n!} = P\left(x\frac{d}{dx}\right)e^x$$

and more generally

$$\sum_{n=0}^{\infty} a_n P(n)x^n = P\left(x\frac{d}{dx}\right)f(x), \qquad \text{if} \qquad f(x) = \sum_{n=0}^{\infty} a_n x^n.$$

In particular, if $P(x)$ takes up integer values for $x = 0, 1, \ldots$ and s is an integer, then

$$\sum_{n=0}^{\infty} \frac{P(n)s^n}{n!} = Ne^s, \qquad N \text{ integer}.$$

The reader may observe here a relation to the set-partition numbers B_n having the generating function (vol. 1, p. 217)

$$e^{e^t-1} = \sum_{n=0}^{\infty} \frac{B_n t^n}{n!}.$$

Writing the left-hand side as

$$\frac{1}{e}\left[1+\frac{e^t}{1!}+\frac{e^{2t}}{2!}+\cdots\right],$$

differentiating n times, and then putting $t=0$ we get

$$B_n=\frac{1}{e}\sum_{k=0}^{\infty}\frac{k^n}{k!}.$$

Let $P(x)$ and $Q(x)$ be polynomials, with $Q(n)\neq 0$ for $n=0, 1, \ldots$, and put

$$f(x)=\sum_{n=0}^{\infty}\frac{P(n)}{Q(n)}x^n;$$

then by a theorem of Abel $y=f(x)$ satisfies the differential equation

$$Q\left(x\frac{d}{dx}\right)y=P\left(x\frac{d}{dx}\right)\frac{1}{1-x}, \tag{14}$$

(recall that $(1-x)^{-1}$ serves as the identity for the Hadamard multiplication, vol. 1, p. 233). This can be proved by an induction on the degree of Q, using the factorization $Q(x)=(x-x_1)Q_1(x)$. An extension due to Polya [136], is the following. Let $P(x)$ and $Q(x)$ be polynomials satisfying the conditions

$$\text{g.c.d.}(P, Q)=1, \qquad Q(0)=0, \qquad Q(n)\neq 0 \qquad \text{for} \qquad n=1, 2, \ldots.$$

If we define

$$y=1+\sum_{n=1}^{\infty}\left[\prod_{k=1}^{n}\frac{P(k)}{Q(k)}\right]x^n$$

then y satisfies the differential equation

$$Q\left(x\frac{d}{dx}\right)y=P\left(x\frac{d}{dx}\right)(xy),$$

which is of the form similar to (14). For instance, if we let

$$y=1+\left(\frac{1}{2}\right)^2 x+\left(\frac{1\cdot 3}{2\cdot 4}\right)^2 x^2+\left(\frac{1\cdot 3\cdot 5}{2\cdot 4\cdot 6}\right)^2 x^3+\cdots,$$

then y satisfies the differential equation

$$x(1-x)y''+(1-2x)y'-\frac{1}{4}y=0. \tag{15}$$

The question arises whether y can be evaluated explicitly. It turns out that it is easier to introduce $x = k^2$ and to treat k as the independent variable. We let

$$h(k) = 1 + \frac{1}{2} k^2 + \frac{1 \cdot 3}{2 \cdot 4} k^4 + \frac{1 \cdot 3 \cdot 5}{2 \cdot 4 \cdot 6} k^6 + \cdots = (1 - k^2)^{-1/2}$$

and we put $y = f(k^2)$. Then the differential equation (15) becomes

$$\frac{d}{dk} \left[k(1 - k^2) \frac{dy}{dk} \right] = ky. \qquad (16)$$

Recalling the properties of Hadamard products (vol. 1, p. 232) we have $y(k) = h(k) \circ h(k)$, and using the integral representation of the Hadamard product we have

$$y(k) = \frac{1}{2\pi i} \int_C \frac{dz}{z\sqrt{(1 - z^2)(1 - k^2/z^2)}}$$

with a suitable contour C. We introduce the new variable ϕ by $z = k^{1/2} e^{i\phi}$ and apply the modular transformation of Landen to the resulting elliptic integral (vol. 1, p. 69). Then it turns out that

$$y(k) = \frac{2}{\pi} K(k) \qquad (17)$$

where $K(k)$ is the complete elliptic integral

$$K(k) = \int_0^{\pi/2} \frac{d\theta}{\sqrt{1 - k^2 \sin^2 \theta}}.$$

The same result may be obtained in a second way by identifying the series

$$1 + \left(\frac{1}{2} \right)^2 k^2 + \left(\frac{1 \cdot 3}{2 \cdot 4} \right)^2 k^4 + \cdots$$

with $(2/\pi) K(k)$, then expanding the integrand in (18) and integrating term by term. Finally, we can evaluate y in the third way: We prove (17) by observing that the elliptic integral $K(k)$, considered as a function of the modulus k, satisfies the differential equation (16)—the so-called Legendre relation.

It may be observed that somewhat similar expressions enter into the computation of the three Watson integrals [181], referring to the probabilities of return to the origin in certain random walks on three-dimensional cubic lattices. For instance, one of the Watson integrals is

$$I = \int_0^\pi \int_0^\pi \int_0^\pi \frac{dx \, dy \, dz}{1 - \cos x \cos y \cos z}$$

which, by straightforward expansion in series gives us

$$I = \sum_{n=0}^{\infty} \left(\int_0^{\pi} \cos^n x \, dx \right)^3 = \pi \left[1 + \left(\frac{1}{2} \right)^3 + \left(\frac{1 \cdot 3}{2 \cdot 4} \right)^3 + \left(\frac{1 \cdot 3 \cdot 5}{2 \cdot 4 \cdot 6} \right)^3 + \cdots \right],$$

so that, employing our previously introduced function $h(k)$, we have

$$I = \pi F(1), \qquad F(k) = h(k) \circ h(k) \circ h(k).$$

3. DISCONTINUOUS FACTORS, RANDOM WALKS, AND HYPERCUBES

(a) A discontinuous factor is a definite integral, usually between infinite limits, of a product involving some oscillatory functions. These are most often solutions of second order ordinary differential equations such as, for instance, the trigonometric or Bessel functions. The whole integral depends on one or more parameters and its chief feature of usefulness is that it is a suitably discontinuous function of those parameters. The simplest discontinuous factor is

$$\text{sgn } x = \frac{2}{\pi} \int_0^{\infty} \frac{\sin xu}{u} \, du = \frac{1}{\pi} \int_{-\infty}^{\infty} \frac{\sin xu}{u} \, du \tag{1}$$

with value 1 for $x > 0$, -1 for $x < 0$, and 0 for $x = 0$. Some applications involving sgn x were given in vol. 1 (pp. 177 and 194) in evaluating integrals

$$P = \int_A f(x) \, dx$$

where x is a vector variable ranging over a complicated subset A of some Euclidean space E^n. We attempt to evaluate P by introducing the characteristic function of A:

$$\chi_A(x) = \int_G F(x, u) \, du$$

and getting then

$$P = \int_{E^n} f(x) \chi_A(x) \, dx = \int_G \int_{E^n} f(x) F(x, u) \, dx \, du$$

in which the inner integral is sometimes more tractable than the original integral. The characteristic function χ is often synthesizable out of several functions of sgn-type (1).

We mention another simple application of discontinuous factors, which arose in functional analysis [56]: Letting n and p be positive integers, with $n > p$, we wish to evaluate

$$S = S(n, p) = \sum_{k=0}^{n} (-1)^k \binom{n}{k}(n - 2k)^{p-1} \operatorname{sgn}(n - 2k). \qquad (2)$$

If we use the representation (1), then express sines by imaginary exponentials, and integrate by parts, we find eventually that

$$S = 0 \qquad n - p \text{ odd},$$

$$S = (-1)^{(n-p)/2} \pi^{-1} \cdot 2^{n+1}(p-1)! \int_0^\infty \frac{\sin^n u}{u^{p+1}} \, du, \qquad n - p \text{ even.} \qquad (3)$$

Another application closely related to our discontinuous factor (1) occurs in answering the following question: If an nth degree polynomial equation has coefficients that are random real numbers distributed according to some simple reasonable probability law, then what is the expected number of real roots? Let $f(t)$ be a sufficiently smooth function, then

$$T(f; a, b) = \int_a^b |f'(t)| \, dt$$

is the total variation of f over $[a, b]$; we could say that this is the total amount of vertical travel done by a point running over the arc $a \le t \le b$ of $y = f(t)$. Hence

$$N(f; a, b) = \frac{1}{2} T(\operatorname{sgn} f(t); a, b) \qquad (4)$$

is the number of zeros of $f(t)$ on $[a, b]$; here the zeros are counted without multiplicity but the end points a and b, if they happen to be zeros of f, are counted only with the weight $1/2$. This may be compared with the much easier formula for complex variables

$$\frac{1}{2\pi i} \int_C g(z) \frac{f'(z)}{f(z)} \, dz = \sum_{z_k \in Z(f)} g(z_k) \qquad (5)$$

giving the sum of $g(z_k)$ taken over the set $Z(f)$ of all zeros of the function f inside a closed contour C (no zeros on C), counted with multiplicity. The reader may wish to attempt to produce a real-variable equivalent of this, which would generalize equation (4).

To continue with the use of (4) we would require the derivative of sgn x which is the so-called delta-function and is, properly speaking, a distribution rather than a function. Proceeding purely formally, we would have

$$N(f; a, b) = \frac{1}{2\pi} \int_a^b \int_{-\infty}^{\infty} \cos[xf(t)] |f'(t)| \, dx \, dt$$

in which the inner integral is undefined. However, if we just interchange the order of integrations, we get

$$N(f; a, b) = \frac{1}{2\pi} \int_{-\infty}^{\infty} \int_a^b \cos[xf(t)] |f'(t)| \, dt \, dx \tag{6}$$

which is fairly easily shown to give the correct expression for the number N of zeros of f in $[a, b]$, at any rate for functions f which are as simple as the polynomials [89]. With N given by (6) we take f to be an nth degree polynomial with the coefficients distributed, for instance, independently in normal fashion, with mean 0 and variance 1. After some calculations it can be shown then (see [89] again) that for this, and for a few other distributions, the expected number of real roots of an nth degree equation is about $(2/\pi) \log n$.

(b)   In the following we establish connections between some types of random walk, numbers of solutions of certain linear Diophantine inequalities, discontinuous factors, and the geometry of the n-dimensional cube. We start by computing the number $N = N(n, s, k)$ of integer solutions $x_1, \ldots, x_k$ of the system of inequalities

$$-s \le \sum_{j=1}^k x_j \le s \tag{7}$$

$$-n \le x_j \le n, \qquad j = 1, \ldots, k, \tag{8}$$

where s, n, k are positive integers; later, we evaluate certain limits relating to N. We use here the characteristic device for certain additive problems: introduce an exponential sum, raise it to the requisite power, and pick out the desired number as the suitable coefficient in that power. We put

$$\left(\sum_{j=-n}^n e^{ijt} \right)^k = \sum_{j=-nk}^{nk} b_j e^{ijt}$$

and conclude that b_p is the number of solutions of (8) which satisfy the condition

$$\sum_{j=1}^k x_j = p.$$

Hence

$$N(n, s, k) = \sum_{p=-s}^s b_p.$$

Using the sum of a geometric series we find

$$\sum_{j=-n}^{n} e^{ijt} = \frac{e^{i[(2n+1)/2]t} - e^{-i[(2n+1)/2]t}}{e^{it/2} - e^{-it/2}} = \frac{\sin[(2n+1)/2]t}{\sin(t/2)}.$$

To pick now the pth coefficient from a trigonometric sum we use the formula (essentially, the residue theorem)

$$\frac{1}{2\pi} \int_{-\pi}^{\pi} e^{-ipt} \sum_{j=N}^{M} b_j e^{ijt}\, dt = b_p, \qquad N \le p \le M.$$

Hence

$$N(n, s, k) = \frac{1}{2\pi} \int_{-\pi}^{\pi} \left(\frac{\sin[(2n+1)/2]t}{\sin(t/2)} \right)^k \frac{\sin[(2s+1)/2]t}{\sin(t/2)}\, dt. \qquad (9)$$

Geometrically expressed, $N(n, s, k)$ is the number of the points of integer lattice in the k-dimensional Euclidean space E^k, common to the hypercube H given by (8) and to the slab of E^k lying between the two parallel planes related to (7):

$$\sum_{j=1}^{k} x_j = s, \qquad \sum_{j=1}^{k} x_j = -s.$$

(However, we note that in (7) and (8) x_j were integers whereas above they are real coordinates). Let us now evaluate the limit

$$\lim (2n)^{-k} N(n, s, k) = \lim \frac{N(n, s, k)}{\text{vol}(H)} \qquad (10)$$

when n and s tend to infinity so that s/n tends to a finite limit b. We use the geometrical interpretation of N as the number of certain lattice points, and we invoke a crude form of that regularity principle which asserts that for a sufficiently regular solid (here, the intersection of the cube and the slab) surface phenomena are in the limit negligible against volume phenomena (alternatively, in such a solid "most" points lie "far" from the boundary). We find formally by the use of (9) that the limit (10) is

$$L(k, b) = \frac{2}{\pi} \int_0^{\infty} \frac{\sin^k u \sin bu}{u^{k+1}}\, du; \qquad (11)$$

but it is also a certain fraction of the volume of a k-dimensional cube:

$$L(k, b) = 2^{-k} \text{Vol}\left\{ (x_1, \ldots, x_k): -1 \le x_j \le 1, -b \le \sum_{j=1}^{k} x_j \le b \right\}. \qquad (12)$$

To connect all this to the foregoing material on discontinuous factors we observe preliminarily that by (11) and (1) $L(0, b) = \text{sgn } b$. Next, take $k = 1$ so that by (11)

$$L(1, b) = \frac{2}{\pi} \int_0^\infty \frac{\sin u \sin bu}{u^2} \, du,$$

but by the geometrical interpretation (12) we get at once

$$L(1, b) = b \quad \text{if} \quad b \le 1$$
$$= 1 \quad \text{if} \quad b \ge 1$$

or $L(1, b) = \min (1, b)$. Equating the two forms, we have the discontinuous factor

$$\frac{2}{\pi} \int_0^\infty \frac{\sin u \sin bu}{u^2} \, du = \min (1, b). \tag{13}$$

With a simple substitution this becomes

$$\frac{2}{\pi} \int_0^\infty \frac{\sin au \sin bu}{u^2} \, du = \min (a, b).$$

Of course, equation (13) could also be integrated directly (as shown in vol. 1, p. 177):

$$\frac{dL(1, b)}{db} = \frac{2}{\pi} \int_0^\infty \frac{\sin u \cos bu}{u} \, du$$
$$= \frac{1}{\pi} \int_0^\infty [\sin u(1+b) + \sin u(1-b)] \frac{du}{u}$$
$$= \tfrac{1}{2}[\text{sgn } (1+b) + \text{sgn } (1-b)]$$

so that, integrating, $L(1, b) = \min (1, b)$. For $k = 2$

$$L(2, b) = \frac{2}{\pi} \int_0^\infty \frac{\sin^2 u \sin bu}{u^3} \, du$$

and by the geometrical interpretation (12)

$$L(2, b) = b - \frac{b^2}{4}, \quad 0 \le b \le 2;$$
$$= 1, \quad 2 \le b.$$

The reader may wish to compare this with vol. 1, p. 251. Computing $L(3, b)$ by the geometrical interpretation (12) is a simple exercise in elementary solid geometry. By reference to Fig. 1a and b, if necessary, we

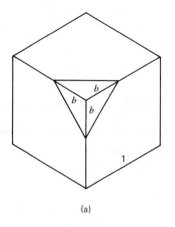

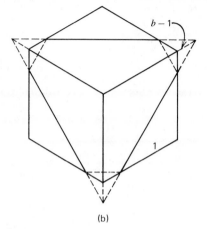

(a)

(b)

Fig. 1. Subsets of cubes.

find that

$$= 1 - \frac{1}{24}(3-b)^3 + \frac{1}{8}(1-b)^3 \quad \text{if} \quad 0 \le b \le 1,$$

$$L(3, b) = \frac{2}{\pi} \int_0^\infty \frac{\sin^3 u \sin bu}{u^4} du = 1 - \frac{1}{24}(3-b)^3 \quad \text{if} \quad 1 \le b \le 3,$$

$$= 1 \quad \text{if} \quad 3 \le b.$$

In general, $L(k, b)$ is piecewise polynomial, each polynomial being of degree k. The reader may wish to verify that $L(k, b)$ is $(k-1)$ times continuously differentiable for $k = 1, 2, 3$, and to prove this in general. From the knowledge of $L(k, b)$ we can compute the $(k-1)$-dimensional content $V_k(b)$ of the section $S_k(b)$ of the k-dimensional cube with the vertices $(\pm 1, \ldots, \pm 1)$ by the plane $x_1 + \cdots + x_k = b$. We note that in changing b to $b + db$ the plane moves by the amount $k^{-1/2} db$, that the slab used between (9) and (10) has *two* faces, and that there is a scale factor 2^k, getting

$$V_k(b) = 2^{k-1} k^{1/2} \frac{dL(k, b)}{db} = \frac{2^k \sqrt{k}}{\pi} \int_0^\infty \left(\frac{\sin u}{u}\right)^k \cos bu \, du. \quad (13a)$$

As a small check we recall that $S_3(0)$ is a regular hexagon of side $\sqrt{2}$.

Actually, the content of equations (9)–(12) can be generalized as in Polya [135]: If

$$R_n = \left\{(x_1, \ldots, x_n): -a_j \le x_j \le a_j, \ -a_{n+1} \le \sum_{j=1}^n x_j \le a_{n+1}\right\}$$

then

$$\text{Vol } R_n = \frac{2^{n+1}}{\pi} \int_0^\infty x^{-n-1} \prod_{j=1}^{n+1} \sin a_j x \, dx$$

$$= \frac{1}{n!} \sum (-1)^m \{\max [0, \, \varepsilon_1 a_1 + \varepsilon_2 a_2 + \cdots + \varepsilon_{n+1} a_{n+1}]\}^n \quad (14)$$

where $\varepsilon_1, \ldots, \varepsilon_{n+1}$ range independently over the 2^{n+1} systems of values $(\pm 1, \ldots, \pm 1)$ and m is the number of -1's in the particular combination of ε's. The above integral is obtained by using the residues in evaluating

$$\int_C z^{-n-1} \sum (-1)^m \, e^{i\Sigma \varepsilon_j a_j} \, dz$$

where C is the standard contour of two semicircles, one large and one small, centered at the origin and joined by parts of the real axis. In particular, putting all a's equal to 1 in (14), we get

$$\int_0^\infty \left(\frac{\sin x}{x} \right)^{n+1} dx = \frac{\pi}{2^{n+1} n!} \sum_{j=0}^{(n+1)/2} (-1)^j \binom{n+1}{j} (n+1-2j)^n.$$

(c) We take up now a seemingly unrelated matter, the problem of simple continuous random walk in three dimensions. This is: given n unit vectors $u_1, \ldots, u_n$ in E^3, independently and uniformly random as to direction, find the distribution of their sum $s = \sum_{j=1}^n u_j$. Questions such as this, and other much more general ones of similar type, can be answered by the Markov method [32], which by using suitable discontinuous factors allows us to write down at once the Fourier transforms of the desired probability distributions. But in our problem of random walk in E^3 there is a simple solution based, somewhat surprisingly, on the theorem of Archimedes (vol. 1, p. 3). According to this the area of a set on a sphere is the same as the area of the axial projection of that set onto the cylinder enveloping the sphere. In particular, if u is a unit vector from the origin O, uniformly at random as to direction, then the tip of u lies on the unit sphere and is uniformly distributed on that sphere. Hence it follows that the projection of u on any diameter of the sphere is *uniformly* distributed over that diameter. Thus our whole random walk problem is reduced to the problem of finding the distribution of the sum of n independent identically distributed random variables $x_1, \ldots, x_n$ (the projections of the vectors $u_1, \ldots, u_n$), each uniform over the interval $[-1, 1]$. If now $x = (x_1, \ldots, x_n)$ is considered as a point in the n-dimensional space E^n then x fills the n-cube

$$H = \{(x_1, \ldots, x_n) : -1 \le x_i \le 1\}$$

and is uniformly distributed in H. Thus, with b positive, say, the probability that $x_1 + \cdots + x_n \le b$ is that fraction of the volume of H that lies on the origin-side of the plane $x_1 + \cdots + x_n = b$. In fact, except for some slight renormalizations, the random walk problem and the problem of the discontinuous factor $L(n, b)$ have coalesced. For other random walks we also get a similar reduction but integrals over subsets of H will replace simple volumes—this is due to nonuniformity of the distribution of the representative point x over H. Feller, writing in 1963, [55, vol. 2, pp. 32–33], stated that "the reduction to one dimension seems to render this famous problem trivial." But Polya's analysis [135] which underlies this trivialization, antedates Feller's words by some 50 years. We remark that random walk with steps of differing lengths $a_1, \ldots, a_n$ is easily handled in the same way: we just use (14) in place of the discontinuous factors $L(n, b)$.

Because of random walks, and for other reasons, we examine briefly what happens to the discontinuous factor $L(n, b)$ (or rather, its derivative) as n gets large. In the expression

$$L(n, b) = \frac{2}{\pi} \int_0^\infty \left(\frac{\sin u}{u} \right)^n \frac{\sin bu}{u} \, du$$

we replace the difficult factor $\sin u/u$ by a possibly good fit which is easier to handle: $e^{-u^2/6}$, based of course on the approximation $\sin u/u = 1 - u^2/6 + 0(u^3)$. We reason, very roughly, that the difficult factor is raised to a high power and so only the contribution near $u = 0$ will matter. Or, more precisely, we use a straightforward piece of asymptotic analysis (vol. 1, p. 241). With this approximation we have

$$L(n, b) \cong \frac{2}{\pi} \int_0^\infty e^{-nu^2/6} \frac{\sin bu}{u} \, du.$$

Using expression (13a) but with reference to *unit* cube (rather than one of edge-length 2) we get from the above

$$V_n(b) \cong \sqrt{n} \int_0^\infty e^{-nu^2/6} \cos bu \, du$$

for the content of a section of the cube by the plane $x_1 + \cdots + x_n = b$. Therefore, evaluating the integral (vol. 1, p. 183)

$$V_n(b) \cong \sqrt{\frac{3\pi}{2}} \, e^{-3b^2/2n}.$$

Finally, we renormalize the units changing b to $\sqrt{n}b$ (because of the reasons given above (13a)), getting

$$V_n(b) \cong \sqrt{\frac{3\pi}{2}} \, e^{-3b^2/2}.$$

Except for a constant factor this is the normal density with the mean 0 and the standard deviation $3^{-1/2}$, giving us what may be considered at first a rather strange appearance of the normal law. We can express all this rather simply: Take a cube of large dimension and symmetrize it with respect to a long diagonal, then after a renormalization the section areas are normally distributed.

According to certain results of J. V. Whittaker [184] the proposition that the cross-section area is asymptotically normal as the dimension $n \rightarrow \infty$, holds for other convex bodies beside the cube, for instance, for any simplex. Moreover, the plane of the section may be arbitrarily oriented (except for a finite number of special positions). This suggests that it might be possible to attempt a novel proof of the central limit theorem. The latter asserts, roughly, that the sum of n independent identically distributed random variables is asymptotically normally distributed as n becomes large. The idea of such a hypothetical proof would be to show by a direct argument that the cross-section areas for cubes or other convex figures of high dimension are approximately normally distributed.

(d)   There is a connection between discontinuous factors, order statistics, and order factors. If $x_1, \ldots, x_n$ are n real numbers then after being rearranged in the order of nondecrease they are $x_{(1)}, \ldots, x_{(n)}$ and we call $x_{(k)}$ the kth order statistic. For instance $x_{(1)} = \min(x_1, \ldots, x_n)$, $x_{(n)} = \max(x_1, \ldots, x_n)$ and if $n = 2m+1$ is odd then $x_{(m+1)}$ is the median of the x's. The relation to discontinuous factors is by way of recursion formulas such as

$$\max(x_1, \ldots, x_n) = g_1 x_1 + (1 - g_1) \max(x_2, \ldots, x_n) \tag{15}$$

where

$$g_1 = 2^{1-n} \prod_{i=2}^{n} [1 - \operatorname{sgn}(x_i - x_1)],$$

or

$$\min(x_1, \ldots, x_n) = h_1 x_1 + (1 - h_1) \min(x_2, \ldots, x_n) \tag{16}$$

where

$$h_1 = 2^{1-n} \prod_{i=2}^{n} [1 + \operatorname{sgn}(x_i - x_1)].$$

Similar though more complicated formulas exist for other order statistics. These can be handled by means of order factors. Given n numbers $x_1, \ldots, x_n$ and a permutation π on $(1, 2, \ldots, n)$ the order factor $F_\pi(x_1, \ldots, x_n)$ is a special discontinuous factor which is 1 if the x_i's are in the order indicated by π, and 0 otherwise. That is, we have $x_{\pi(i)} < x_{\pi(j)}$ whenever $i < j$. We get, therefore,

$$F_\pi(x_1, \ldots, x_n) = 2^{1-n} \prod_{i=1}^{n-1} [1 + \operatorname{sgn}(x_{\pi(i+1)} - x_{\pi(i)})],$$

with the additional property that if the desired inequalities hold, but not sharply, then $F_\pi = 2^{-k}$ where k is the number of equalities. It is now simple to show that the kth order statistic is a pseudolinear combination of its arguments $x_1, \ldots, x_n$, with order factors as coefficients. For instance, we have

$$x_{(1)} = \min(x_1, \ldots, x_n) = 2^{1-n} \sum_{i=1}^{n} \left\{ \prod_{\substack{j=1 \\ j \neq i}}^{n} [1 - \operatorname{sgn}(x_i - x_j)] \right\} x_i \qquad (17)$$

and

$$x_{(n)} = \max(x_1, \ldots, x_n) = 2^{1-n} \sum_{i=1}^{n} \left\{ \prod_{\substack{j=1 \\ j \neq i}}^{n} [1 + \operatorname{sgn}(x_i - x_j)] \right\} x_i. \qquad (18)$$

As a small application, let us consider two random variables x_1, x_2 with the joint density function $f(x_1, x_2)$. By (18)

$$\max(x_1, x_2) = \tfrac{1}{2}[1 - \operatorname{sgn}(x_2 - x_1)]x_1 + \tfrac{1}{2}[1 + \operatorname{sgn}(x_2 - x_1)]x_2.$$

Using the integral representation (1) we find the expectation M of $\max(x_1, x_2)$ to be

$$
\begin{aligned}
M = \frac{1}{2} \iint & (x_1 + x_2)f(x_1, x_2)\, dx_1\, dx_2 \\
& + \frac{1}{\pi} \iint \int_0^\infty \frac{\sin(x_2 - x_1)u}{u}(x_2 - x_1)f(x_1, x_2)\, du\, dx_1\, dx_2.
\end{aligned}
\qquad (19)
$$

Suppose now that x_1 and x_2 are identically and independently distributed with the density $f(x)$. Then (19) simplifies to

$$M = \int xf(x)\, dx + \frac{1}{\pi} \int_0^\infty \frac{1}{u} \left[\iint (x_2 - x_1) \sin(x_2 - x_1)u f(x_1) f(x_2)\, dx_1\, dx_2 \right] du.$$

Specializing to the uniform density $f \equiv 1$ on $[0, 1]$ and 0 elsewhere, we get

$$M = \tfrac{1}{2} + \frac{2}{\pi} \int_0^\infty \frac{2 - 2\cos u - u \sin u}{u^4}\, du. \tag{20}$$

However, in this uniform case M is quite simply computed by a direct argument. Generally, if $x_1, \ldots, x_n$ are identically distributed independent random variables with cumulative function $F(x)$ and $x_{(k)}$ is as before the kth order statistic, then

$$\text{Prob } (x_{(k)} \leq x) = \text{Prob } (k \text{ or more } x_i\text{'s are } \leq x)$$

$$= \sum_{j=k}^{n} \binom{n}{j} F^j(x)[1 - F(x)]^{n-j}.$$

In particular, if the x_i's are uniformly distributed on $[0, 1]$ then $F(x) = x$, $0 \leq x \leq 1$, and so the distribution for $x_{(k)}$ is

$$P_k(x) = \text{Prob } (x_{(k)} \leq x) = \sum_{j=k}^{n} \binom{n}{j} x^j (1 - x)^{n-j},$$

and the expectation of $x_{(k)}$ is

$$\int_0^1 x P_k'(x)\, dx = \frac{k}{n+1}$$

as we might have guessed. For our simple problem we find $M = 2/3$; equating this to equation (2) we find that

$$\int_0^\infty \frac{2 - 2\cos u - u \sin u}{u^4}\, du = \frac{\pi}{12}.$$

This is quite simply obtainable by residues and complex integration, but it is not quite so simple to obtain by real-variable methods alone.

The following physical application may help to illustrate the order statistics. Let n identical perfectly elastic heavy balls be marked with the numbers $1, 2, \ldots, n$; suppose that all of them are moving in the same smooth-fitting groove so that the displacement of the ith one, as function of time, when measured along the groove, would have been $X_i = f_i(t)$. Since on impact any two balls interchange their velocities the reader may wish to show that in the presence of all n balls the correct dependence of the ith displacement on t is $x_i = X_{(i)}$. Several problems can be formulated now, relating to analytical formulas for x_i, estimating the number of impacts when, say, each X_i is a polynomial of degree k in t and so on.

If the variables $n_1, \ldots, n_k$ are nonnegative integers we may observe the following generating function:

$$\sum_{n_1=0}^{\infty} \cdots \sum_{n_k=0}^{\infty} \min (n_1, \ldots, n_k) x_1^{n_1} \cdots x_k^{n_k}$$

$$= \frac{x_1 x_2 \cdots x_k}{(1-x_1)(1-x_2) \cdots (1-x_k)(1-x_1 x_2 \cdots x_k)}.$$

More generally, a generating function for the $(k+r-1)$-th order statistic $M_r(n_1, \ldots, n_k)$ of the numbers $n_1, \ldots, n_k$ can be obtained as follows [28]:

$$S_r = \sum_{n_1=0}^{\infty} \cdots \sum_{n_k=0}^{\infty} M_r(n_1, \ldots, n_k) x_1^{n_1} \cdots x_k^{n_k},$$

and the quantities S_r obey the system of equations

$$\sum_{j=0}^{k-r} \binom{k-j-1}{r-1} S_{j+1} = \left[\prod_{j=1}^{k} (1-x_j) \right]^{-1} \sum \frac{x_1 \cdots x_r}{1-x_1 \cdots x_r}$$

and in the last sum the summation is over the $\binom{k}{r}$ r-tuples of x's.

4. CALCULUS OF VARIATIONS

(a)   Some simple isoperimetric problems in which a quantity

$$Q(C) = \int F(x, f', f)\, dx$$

associated with an arc C of the curve $y = f(x)$ is to be extremized, while the length

$$L(C) = \int \sqrt{1+y'^2}\, dx$$

of C is kept fixed, can be solved by the following direct method. We approximate to C by a polygonal arc C_n which we take to be a chain of n rigid straight links and we determine the angles between these so as to extremize $Q(C_n)$. The length of C_n being L, we justify the passage to the limit and obtain the solution C as $\lim_{n\to\infty} C_n$. One example for this procedure is the isoperimetric problem of the circle: to find the closed plane curve of circumference L, which encloses largest possible area. This was solved, partially at least, by the argument due to Steiner, vol. 1, p. 21. An alternative way is to consider C_n to be a closed n-gon with n sides of lengths L/n, and to maximize the area $Q(C_n)$ which the n-gon encloses. Here we use the minimum perturbation principle—in extremizing a function of many variables apply first those perturbations that change possibly few variables. This was used in vol. 1, p. 149, to show that of all n-gons inscribed into a circle the regular one has largest area. In that case we could perturb just one vertex at a time. Now we must perturb two consecutive vertices, say v_{i+1} and v_{i+2}, rather than just one. Hence we are led to maximizing the area A of the quadrilateral $v_i v_{i+1} v_{i+2} v_{i+3}$ of fixed side-lengths, say, a, b, c, d (actually three of them are L/n). By using some simple calculus, either with or without Lagrange multipliers, we show that A is largest when the four vertices lie on a circle. The same follows, without any calculus, from a generalization of Brahmagupta's formula

$$A^2 = (s-a)(s-b)(s-c)(s-d), \qquad 2s = a+b+c+d,$$

for the area of a cyclic quadrilateral; this itself generalizes Heron's formula

$$A^2 = s(s-a)(s-b)(s-c)$$

for the area of a triangle (we let $d = 0$). The generalization we need holds for any quadrilateral, cyclic or otherwise, in which the sum of a pair of opposite angles is 2α:

$$A^2 = (s-a)(s-b)(s-c)(s-d) - abcd \cos^2 \alpha.$$

Or, we can express A in terms of the diagonals x and y, and the angle β they make

$$A = \tfrac{1}{2}xy \sin \beta.$$

Now it follows that $Q(C_n)$ is largest when every four consecutive vertices lie on a circle. Hence C_n is a regular n-gon and $C = \lim_{n\to\infty} C_n$ is the circle of radius $L/2\pi$.

Another problem of the same type is that of determining the equation of the catenary: the curve C formed by a heavy flexible chain of uniform density, suspended from two points at equal height (the reader may wish to show how to reduce the case of unequal to that of equal height). This is indeed a variational isoperimetric problem, for C will be such that its center of mass will be lowest possible while the length stays fixed. With reference to Fig. 1a the problem is:

$$\text{minimize} \quad 2\int_0^a y\sqrt{1+y'^2}\, dx \quad \text{while} \quad 2\int_0^a \sqrt{1+y'^2}\, dx = \text{const.}$$

We consider here the approximating polygonal chain consisting of n

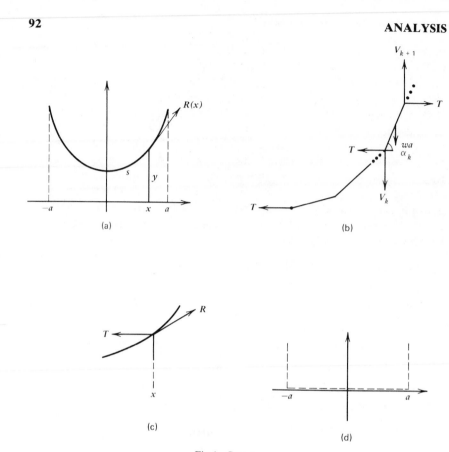

Fig.1. Catenary.

equal links of length a and weight wa, taking n even for convenience. Consider the lowest link, shown in Fig. 1b, in which T is the horizontal tension; if the kth link is examined for static equilibrium we find that the horizontal force T is constant throughout the chain, and $V_{k+1} - V_k = wa$, with $V_1 = wa$. Hence $V_k = kwa$ and so

$$\tan \alpha_k = \frac{wa}{T} k. \tag{1}$$

If we pass to the limit on n in C_n, equation (1) informs us that y', which replaces $\tan \alpha_k$, is proportional to the arc length s, which replaces ka, counted from the lowest point (Fig. 1a). We get then the analog of equation (1): $y' = (w/T)s$, where w and T have the same meaning as before: weight per unit length, and the horizontal tension (which is constant throughout). Differentiating the last equation we get the differential equation of C:

$$y'' = \frac{w}{T}\sqrt{1+y'^2}.$$

This is solved and we get

$$y = \frac{T}{w}\cosh\frac{wx}{T} + C_1, \qquad (2)$$

where C_1 is an inessential constant positioning the chain up or down. The total tension is $R(x)$, as in Fig. 1a and c, and $R = T \sec \alpha$ so that $R = T \cosh(wx/T)$. Therefore the material of the chain is *not* used optimally, for the chain is uniform and so its lower parts are unnecessarily heavy not having as much weight to support as the upper parts. The reader may wish to verify that with the optimum use of the material of the chain we have a variable density p, constant total tension T, and the equation of this catenary of equal tension is

$$y = \frac{T}{p}\log\sec\frac{px}{T};$$

the reader may also wish to find p. Further, the reader may attempt solving the problem of optimal design by generalizing our $2n$-links approach to the case of variable link-weights. A strange piece of duality has been noted in the section on natural equations of the geometry chapter: The *natural* equation of the catenary of constant tension has the same form as the *Cartesian* equation of the ordinary catenary.

It may be noted that the catenary (2) also solves another problem: that of determining the arc C so that the surface generated by rotating C about the x-axis has least area (while the length $L(C)$ is constant). For, by the Pappus–Guldin theorems (vol. 1, p. 202) the area is $L(C)$ times the path traversed in rotation by the center of mass of C. However, we note that in this formulation the problem is location-dependent so that the vertical positioning constant C_1 of equation (2) becomes of importance. For suitable values of C_1, a, and $f(a)$, there may be solutions which are not smooth, such as the polygonal line of Fig. 1d.

(b) The variational problem of the shortest line joining two points p, q in the plane is so simple as to be considered obvious: The solution is the straight segment $C = pq$. One can justify it to oneself in several ways, two of which are the mechanical and the geometrical. The mechanical argument is that of the equilibrium position of a smooth elastic rubber band stretched from p to q; hence this rests ultimately on the minimal potential energy as the requirement for equilibrium. So, unless the rubber band is straight some small part of it has nonzero resultant force arising from tensions—hence no equilibrium. The geometrical argument is that $C = pq$ is shorter than any other arc $S = \widehat{pq}$ because we project S onto C: Such projection never increases the length and since $C \neq S$, the length gets actually decreased. Those two simple arguments are in themselves quite trivial but their chief point of usefulness is that they can be elaborated to determine shortest lines of surfaces other than the plane.

For instance, we show that the shortest curve C on a sphere K joining two points p and q is the (shortest) great-circle arc. For if S is another arc $\widehat{pq}$ on K, we can use the above projection argument, and project S on C by a suitable system of circles (circles of latitude corresponding to a north and a south pole which lie on the same great circle as p and q).

Suppose next that we take a convex and sufficiently smooth surface S and we ask for the shortest arc C in S, which joins two points p and q of S. Here we can use the argument of the equilibrium of the smooth elastic band pq. Consider a small arc $A = \widehat{rs}$ of C; it is acted upon by three forces: the two tensions $T(r)$ and $T(s)$, and the reaction N of S on C. N is really distributed along A, and we are interested in the limiting case when r and s approach the same limit point, say, the midpoint m of A. Since three forces in equilibrium must be coplanar and since the reaction N is normal to S at m, we find that this normal lies in the osculating plane to A at m (for the osculating plane is the limiting position of the plane rms). Also, the vector sum of the tensions $T(r)$ and $T(s)$ has the limiting direction of the principal normal to A at m because by the Frenet–Serret formulas the principal normal is the direction of the derivative of the tangent vector with respect to the arc length. Thus our heuristic reasoning leads us to the celebrated theorem of Johann Bernoulli: The shortest curve joining two points in S has the property of having principal normal coincide at each point with the surface normal.

(c)   When light moves through the interface of two media such as air and water where its velocities are c and u, we find the familiar phenomenon of refraction: the path AB of a light ray is bent and we find that $\cos \alpha / \cos \beta = c/u = n$ – the refractive index (see Fig. 2a).

This is easily derived from Fermat's principle: The path AB is such as to minimize the total time of passage $T = \int_A^B ds/v$. For, as in Fig. 2a, we have

$$T = \frac{\sqrt{a^2 + x^2}}{c} + \frac{\sqrt{b^2 + (d - x)^2}}{u},$$

and setting dT/dx equal to 0 we get

$$\frac{x}{c\sqrt{a^2 + x^2}} = \frac{d - x}{u\sqrt{b^2 + (d - x)^2}}$$

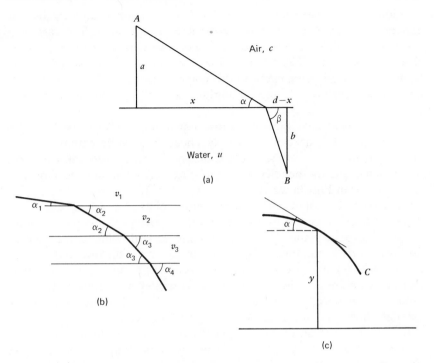

Fig. 2. Snell's law.

which is what we want: $\cos \alpha/c = \cos \beta/u$. Similarly, with several layers, as in Fig. 2b, we have

$$\frac{\cos \alpha_1}{v_1} = \frac{\cos \alpha_2}{v_2} = \cdots = \text{const.},$$

and in the limit of a continuous variation of velocity with the vertical distance y, as in Fig. 2c, we get a trajectory C for which

$$\frac{\cos \alpha}{v} = \text{const.} \tag{3}$$

all along C. We may consider the general problem of light-ray trajectories when light moves in a plane medium with variable index of refraction $n(x, y)$. Taking $c = 1$ we wish then to minimize the time of passage

$$T = \int \frac{ds}{v} = \int n(x, y) \sqrt{1 + y'^2} \, dx. \tag{4}$$

A case of special interest arises when $n(x, y) = y^a$; the Euler–Lagrange equation for (4) is then $a(1 + y'^2) = yy''$. With $a = 1$ we have a catenary as solution, as shown earlier. With $a = 1/2$ the solution is any parabola of equation $y = Ax^2 + Bx + C$ provided that $B^2 - 4AC = -1$. The cases $a = -1/2$ and $a = -1$ lead to a cycloid and a (semi)circle, respectively, and we shall take them up now. We note that for the general a the trajectories through the origin are tangent to the x-axis if $a > 0$, to the y-axis if $a < 0$, and are straight lines through the origin (with all possible inclinations) if $a = 0$.

If $a = -1$ we have the differential equation $1 + y'^2 + yy'' = 0$ so that $(x - A)^2 + y^2 = B^2$, as can be simply checked by differentiating twice. Neglecting the case $y < 0$ we get as solutions semicircles of radius B in the upper half-plane, perpendicular to the x-axis. This can also be shown by geometry: as in Fig. 3a we verify that $\cos \alpha / y = 1/B$ so that (3) holds. We observe that the upper half-plane, with points defined as usual and "straight lines" taken to be light-ray paths, provides the Poincaré model of the hyperbolic plane. The distance between two points is the "optical distance," that is, the time it takes light to pass from the one to the other; points on the x-axis are the "infinitely remote" points, and two "lines" that have a point at infinity ($= x$-axis) in common, are parallel. As shown in the Fig. 3b, for a "straight line" L and a point P off L there will be two "straight lines" L_1 and L_2 through P, "parallel" to L.

If $a = -1/2$ we can prove analytically that the trajectory is a cycloid. Again, the same thing can be proved by some simple geometry. We suppose that the cycloid C is described by a point A on the rim of a rolling wheel of radius r. Taking the geometry of Fig. 4 we have then $y = r(1 - \cos \omega) = 2r \sin^2 \omega/2$. The instantaneous motion of the wheel is a rotation about the point B of contact with the x-axis and so the tangent T to the curve C at A is perpendicular to AB, hence $\alpha + \gamma - \pi/2 = \pi/2 - \omega$. Also $\omega + 2\gamma = \pi$ so that $\alpha = (\pi - \omega)/2$ and $\cos \alpha = \sin \omega/2$. Therefore $y = 2r \cos^2 \alpha$, or $(\cos \alpha)/y^{1/2}$ is a constant, verifying equation (3).

We observe that if a heavy body slides down a smooth curve C in the

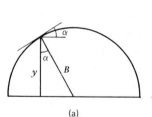

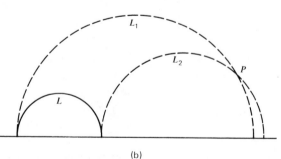

(a)   (b)

Fig. 3. Poincaré's model.

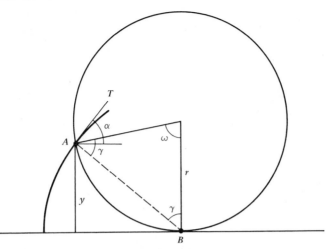

Fig. 4. Cycloid.

vertical plane under no forces other than gravity, then its velocity v after sliding down C through the vertical distance y is $\sqrt{2gy}$. It follows that the (upside down) cycloid is the curve of fastest descent—the so-called brachistochrone (from Greek: *brachys*—short, *brachystos*—shortest, *chronos*—time). This property of the cycloid—being the curve of fastest fall—was proved by Johann Bernoulli in 1696. The problem was also solved within one year by Leibniz, Newton, and Jacob Bernoulli (the younger brother of Johann; the Bernoullis, through brothers, were on bad terms—a not uncommon phenomenon in mathematical families. So, for instance, Gauss drove both his sons into exile (to U.S.A.) and Hilbert expelled from his home his somewhat simpleminded son Franz).

(**d**)   Consider a geodesic C on a circular cone as in Fig. 5a. The simplest way to study C is to slit the cone along a generator and to unroll it onto a plane. Geodesics are preserved by the unrolling operation and so C becomes a straight line, as shown in Fig. 5b. If h is the closest approach to the vertex of the cone then we find that $h = d \sin \alpha$. If r is the distance of the point of the cone which became P on unrolling, from the cone axis, and γ is the semivertical angle of the cone, then from $h = d \sin \alpha$ and $r = d \sin \gamma$ we get

$$r \sin \alpha = h \sin \gamma = \text{const.} \tag{5}$$

We can express it thus: If the geodesic of the cone cuts the meridians (i.e., the generators) at the (variable) angle α, and r is the distance from the cone axis, then $r \sin \alpha$ is constant along the geodesic. It follows in

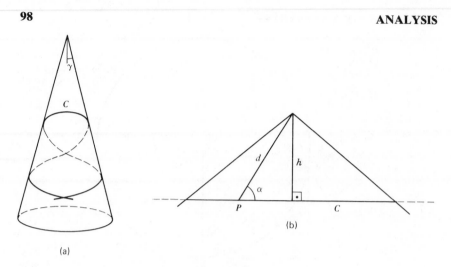

Fig. 5. Geodesic on a cone.

particular that the geodesic, which is horizontal (cone being vertical) at its closest approach to the vertex, gets steeper and steeper as it recedes from the axis. Using the constancy of $r \sin \alpha$ we can compute the number N of double points of the complete geodesic, and obtain a theorem given in [105]: if $1/2 \operatorname{cosec} \gamma$ is not an integer then $N = [1/2 \operatorname{cosec} \gamma]$, otherwise $N = [1/2 \operatorname{cosec} \gamma] - 1$. The reader may wish to verify this, and also to show that using the staircase functions of Fig. 6, we have the single formula: $N = -[1 - 1/2 \operatorname{cosec} \gamma]$. Here $[x]$ is, as usual, the greatest integer $\leq x$. By using some elementary spherical trigonometry we can also show that (5) holds for the geodesics on a sphere: If the great circle arc G joins a to b, if r is the distance of a point to the axis of the sphere, and if G cuts the parallel circles at the (variable) angle α, then $r \sin \alpha = \text{const.}$ along G.

This arc G, joining a to b, has the property of being shortest for the

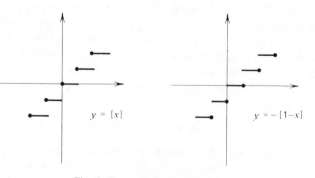

Fig. 6. Two staircase functions.

purpose of sailing (or flying) from a to b on the sphere. In this context the path G is called an orthodrome (from Greek: *orthos*—straight, and *dromos*—course). As the constancy of $r \sin \alpha$ along G shows, the orthodrome G has the disadvantage of cutting the meridians at different angles. This made classical navigation rather difficult and for that reason it was sometimes preferred to move from a to b along a loxodrome or rhumbline (from Greek: *loxos*—oblique), which cuts all meridians at the same angle and so moves up toward the pole winding itself round it infinitely often, like the logarithmic spiral.

Actually the constancy of $r \sin \alpha$ along the geodesics is true for all surfaces of S of revolution, not just for cones and spheres. This is the theorem of Clairaut. The reader may wish to construct a proof by supposing that S is generated by rotating a plane curve C, then approximating to C by a polygonal line which when rotated together with S, generates a polyconical approximation to S. On such a polyconical surface Clairaut's theorem follows from (5) and from the fact of continuity of α along the circular edges of the polycone. The reader may also wish to use Clairaut's theorem to find the geodesics on certain surfaces of revolution (e.g., torus, paraboloid).

Clairaut's theorem has an extension to Liouville surfaces (those are the surfaces where the element of arc ds^2 is given by the curvilinear coordinates u,v as $ds^2 = [F(u) + G(v)](du^2 + dv^2)$—together with surfaces of revolution, these exhaust the class of surfaces on which geodesics can be found by integration alone, without having to solve differential equations); for this see [165].

(e) In vol. 1, pp. 20–26, we have considered two related isoperimetric problems: that of the circle, and that of the helix. Their common formulation (altered slightly to bring out the similarities) is this: find a rectifiable curve C (closed or open) of fixed length, in the Euclidean n-space (n even or odd) so as to maximize the volume of its convex hull $\bar{C}$. We get the circle problem if $n = 2$ and C is closed, and the helix problem if $n = 3$ and C is an open arc. This, together with some further evidence, suggests that we face here four (classes of) problems:

(a) C closed, n even,
(b) C open, n even,
(c) C open, n odd,
(d) C closed, n odd.

Under certain mild restrictions the problem (a) has been solved by Schoenberg [152]. Just as the solution of the special case $n = 2$ is (similar to) the circle given parametrically by

$$x(t) = \sin t, \quad y(t) = \cos t, \quad 0 \le t < 2\pi, \tag{6}$$

so here the solution is (similar to) the "hypercircle" generalizing (6):

$$x_1(t) = \sin t, \quad x_3(t) = \tfrac{1}{2}\sin 2t, \ldots, x_{2j-1}(t) = \frac{1}{j}\sin jt,$$

$$x_2(t) = \cos t, \quad x_4(t) = \tfrac{1}{2}\cos 2t, \ldots, x_{2j}(t) = \frac{1}{j}\cos jt,$$

where $0 \le t < 2\pi$. It is not very difficult to show, using symmetries, mirror images, and so on, that the problem (b) has the same solution (7) but with $0 \le t \le \pi$, what we might call a semihypercircle. In vol. 1, p. 24, it was indicated that the solution of the special case $n = 3$ of problem (c) is, under some further restrictions, one turn of the circular helix of pitch $2^{-1/2}$. For general odd n the reader may wish to verify whether the solution is analogous: one "turn" of the hypercircular helix.

The last problem, (d), turns out to be rather different from the other three. Under certain rather stringent assumptions the special case $n = 3$ can be attacked by ordinary variational techniques. Let us suppose that the maximizing closed curve C has the xz-plane and the yz-plane as planes of symmetry, projecting as a convex arc on each, and that $x, y,$ and z are expressible in terms of the arc length s as sufficiently smooth functions so that $x(0) > 0$, $x'(0) = 0$, $y(0) = 0$, $y'(0) = 1$, $z(0) = 0$, $z'(0) = 0$, $z(s) \ge 0$ (there are some grounds to believe that these requirements are not as stringent as they may seem at first, and the reader may wish to relax some of them, showing that they follow from the assumption of the maximality of the volume of $\bar{C}$). The approximate appearance of C in the first octant is shown in Fig. 7. Choosing for the volume element the slab shown in the figure, we express the volume V of $\bar{C}$ as

$$V = \int_0^L xyz' \, ds.$$

This is to be maximized subject to a side-condition: a differential equation, $x'^2 + y'^2 + z'^2 - 1 = 0$, stating that s is the arc length, and subject to certain further initial and periodicity conditions (which need not concern us now). Using the Lagrange multiplier technique we form the integral

$$F = \int_0^L [xyz' - \lambda(x'^2 + y'^2 + z'^2 - 1)] \, ds$$

and maximize it getting as the Euler–Lagrange equations

$$(2\lambda x')' = yz', \quad (2\lambda y')' = -xz', \quad (xy - 2\lambda z')' = 0.$$

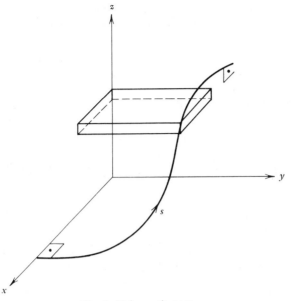

Fig. 7. Volume element.

With the special additional conditions we've assumed, it is not hard to show that λ is constant which (by scaling, if needed) we take to be 1/2, getting the differential equations for the curve C

$$x'' = -xy^2, \quad y'' = -yx^2, \quad z' = xy \tag{8}$$

with the initial conditions

$$x(0) = c > 0, \quad x'(0) = 0, \quad y(0) = 0, \quad y'(0) = 1, \quad z(0) = 0. \tag{8a}$$

Equations (8), though with different initial conditions, also occur in a completely different context: They describe the motion of charged particles in certain high-energy accelerators.

We notice at once some important differences between the problems (a), (b), and (c), and the (special case of) problem (d). In the first three the coordinate functions are expressible as elementary functions of the arc length s, in the problem (d) they appear not to be. Further, in the first three cases we deal with circles, circular helices, hypercircles and hypercircular helices, all of which are simply characterized as curves for which *all* curvatures are constant (this is deduced from the n-dimensional Frenet–Serret equations, vol. 1, p. 50). We also recall that the isoperimetric problem of the sphere also has as its solution a surface with both principal curvatures constant. So, it might appear that isotropic and homogeneous isoperimetric problems maximizing the volumes of convex hulls have as their solutions structures of constant curvature. On the other hand, in the solution C of the case $n = 3$ of (d), given by (8) and (8a) it is not clear whether there are any curvatures which are constant (the reader may wish to express C in *natural* coordinates and look for curvature relations).

Some further problems may be raised here. Using the Liouville theory of functional complexity (vol. 1, p. 205) the reader may wish to show that C is not expressible by elementary functions of s. Generalizing the number π from two to three dimensions, the reader may wish to devise a numerical procedure to show that for any closed curve K in E^3

$$\frac{\text{Vol}(\bar{K})}{\text{Length}^3(K)} \le B = 0.0031816877\ldots$$

and the inequality is strict except when K is (similar to) C. A closed strictly convex (i.e., without straight segments in the boundary) solid S cannot be the convex hull of an arc of finite length; the reader may wish to estimate the length of the shortest arc whose convex hull approximates S sufficiently well.

In our problem we have started with a maximization for a closed space-arc C, and from there we have reached by way of Euler–Lagrange equations the differential system (8). The reader may wish to reverse this and prove that equations (8) (and some others like them) have periodic solutions by showing directly that a maximizing arc C exists. Here one invokes an equivalent of the Bolzano–Weierstrass theorem, the Blaschke principle, of vol. 1, p. 46.

It may be observed that for the special case $n = 2$ the solutions of problems (a) and (b), namely the circle and the semicircle, are convex curves. Recalling the n-convexity generalization for curves in n-space (vol. 1, p. 50), we notice that n-convexity appears to enter, globally not just locally, into all four problems (a)–(d).

5. VECTOR, MATRIX, AND OTHER GENERALIZATIONS

In this section we consider a few simple classical formulas involving numbers, which become less simple and classical when the numerical parameters are replaced by vectors, matrices, or other generalized entities. The prototype, of course, is the passage from $ax = b$ implying $x = b/a$ for numbers, to $Ax = b$ implying $x = A^{-1}b$ for vectors and matrices. The reader may detect a partial relation with what had been elsewhere (vol. 1, p. 106) called the reification principle: extending to operators the properties of numbers (quantities).

(a)   For $a > 0$ we have the well-known formula

$$I(a) = \int_{-\infty}^{\infty} e^{-ax^2} \, dx = \left(\frac{\pi}{a}\right)^{1/2} \tag{1}$$

provable, for instance, by writing down the Cartesian double integral for $I^2(a)$ and evaluating it by transforming to polar coordinates:

$$I^2(a) = \int_{-\infty}^{\infty} e^{-ax^2} \, dx \int_{-\infty}^{\infty} e^{-ay^2} \, dy = \int_{-\infty}^{\infty} \int_{-\infty}^{\infty} e^{-a(x^2 + y^2)} \, dx \, dy$$

$$= \int_{0}^{2\pi} \int_{0}^{\infty} e^{-ar^2} \, r \, dr \, d\theta = \frac{\pi}{a}.$$

For n dimensions we have a simple extension: Positive number a gets replaced by a positive definite $n \times n$ matrix A, and the scalar variable x gets replaced by the vector variable $x = (x_1, \ldots, x_n)$. We have then the analog of equation (1):

$$\int_{-\infty}^{\infty} \cdots \int_{-\infty}^{\infty} e^{-xAx^T} \, dx_1 \cdots dx_n = \pi^{n/2} |A|^{-1/2}. \tag{2}$$

Here and in the sequel the vector x^T is the transpose of x. Equation (2) can be proved in (at least) two different ways:

(A)   We can rotate the coordinate system thereby diagonalizing A and replacing the quadratic form xAx^T by the sum of squares $\sum_{j=1}^{n} \lambda_j \, y_j^2$, λ_j's being the (necessarily positive) eigenvalues of A. Then we take advantage of the "separated" form of the exponent and the functional equation $e^x e^y = e^{x+y}$. Finally, we calculate the Jacobian and use the fact that the determinant $|A|$ is the product of all eigenvalues.

(B)   We can use a similar procedure but "separate" the exponent term-by-term, inductively rather than at once, using Lagrange's formula

$$xAx^T = \sum_{k=1}^{n} \frac{|A_k|}{|A_{k-1}|} \left(x_k + \sum_{j=k+1}^{n} b_{kj} x_j \right)^2;$$

here the quantities $|A_k|$ are the leading minors, all of them ≥ 0, $|A_0| = 1$. The determinant $|A|$ appears in (2) as a result of multiplicative cancellation

$$|A| = |A_n| = \prod_{k=1}^{n} \frac{|A_k|}{|A_{k-1}|}.$$

We also use the translational invariance

$$\int_{-\infty}^{\infty} f(x)\, dx = \int_{-\infty}^{\infty} f(x+c)\, dx.$$

Method (B) is longer than (A) and not as neat perhaps, but it is capable of considerable generalizations some of which we shall mention later.

A simple extension of (1), used "in reverse," together with the Cartesian-to-polar transformation, is one of the simplest means of computing the volume $v_n a^n$ of the n-dimensional ball B_n of radius a. Since the normal to the sphere S_{n-1} bounding B_n is everywhere in the radial direction, we find the area of S_{n-1} to be $d(v_n a^n)/da = n v_n a^{n-1}$. From (1) we have

$$\pi^{n/2} = \left(\int_{-\infty}^{\infty} e^{-x^2}\, dx \right)^n = \int_{-\infty}^{\infty} \cdots \int_{-\infty}^{\infty} e^{\sum\limits_{i=1}^{n} x_i^2}\, dx_1 \cdots dx_n.$$

Letting $r^2 = \sum\limits_{j=1}^{n} x_j^2$ and observing the previous remark on the area of S_{n-1} we have

$$\pi^{n/2} = n v_n \int_0^{\infty} e^{-r^2} r^{n-1}\, dr.$$

In the above $u = r^2$ is taken as the new variable and we get the integral in terms of Γ-functions thus getting v_n:

$$v_n = \frac{\pi^{n/2}}{\Gamma(1+n/2)}.$$

We note a generalization of (1)

$$\int_{-\infty}^{\infty} e^{-ax^2} \cos 2tx\, dx = \sqrt{\frac{\pi}{a}}\, e^{-t^2/a} \tag{3}$$

valid for $a > 0$ and arbitrary t (see vol. 1, p. 183). Since $\sin x$ is an odd function this can also be written as

$$\int_{-\infty}^{\infty} e^{-ax^2+2itx}\, dx = \sqrt{\frac{\pi}{a}}\, e^{-t^2/a}. \tag{4}$$

Suppose now that we wished to generalize this to n dimensions in the same way that (2) generalizes (1); we might perhaps guess that the generalization is

$$\int_{-\infty}^{\infty} \cdots \int_{-\infty}^{\infty} e^{-xAx^T+2it\cdot x}\, dx_1 \cdots dx_n = \pi^{n/2} |A|^{-1/2} e^{-tA^{-1}t^T}, \tag{5}$$

where A is as before a positive definite matrix and t is an arbitrary n-vector $(t_1, \ldots, t_n)$. Using the notion of completing the square, and either of the methods (A) and (B), the reader may wish to show that equation (5) is correct.

It might be asked: Why replace (3) with the equivalent form (4)? A part of the answer concerns the use of characteristic functions in probability theory. Given a probability density $f(x)$, its Fourier transform

$$F(t) = \int_{-\infty}^{\infty} e^{ixt} f(x) \, dx \tag{6}$$

is called the characteristic function. Assuming sufficient differentiability, and the existence of integrals, we can generate the moments: differentiate (6) k times and put $t = 0$ getting

$$M_k(f) = \int_{-\infty}^{\infty} x^k f(x) \, dx = (-i)^k F^{(k)}(0).$$

In the so-called moment problem we ask whether $f(x)$ is determined if all its moments $M_k(f)$, $k = 0, 1, \ldots$, are known. For sufficiently well-behaved functions we could reason thus: if $F^{(k)}(0)$ is known for all k then the McLaurin series will determine $F(t)$, and then the inverse Fourier transform

$$f(x) = \frac{1}{2\pi} \int_{-\infty}^{\infty} e^{-ixt} F(t) \, dt$$

gives us the density $f(x)$. To show that some sort of condition is necessary for the unique determination of $f(x)$, we note the exceptional cases, such as $f(x) = Ae^{-x^{1/4}} \sin x^{1/4}$. Here, using the substitution $x = y^4$ and then the formulas of vol. 1, p. 185, we verify that *all* moments vanish: $M_k(f) = 0$ for $k = 0, 1, \ldots$. But it must be added that this function could not be a probability density being negative sometimes.

Next, we ask what is the characteristic function for the normal density. Except for a constant multiplier, the normal density is $f(x) = e^{-ax^2}$, $a > 0$; hence its characteristic functions is

$$\int_{-\infty}^{\infty} e^{-ax^2 + itx} \, dx.$$

Replacing, for reasons of mere convenience, t by $2t$ we get just the right-hand side of (4). We can appreciate (2) and (5) better now. Consider an n-variate normal distribution with the vector $x = (x_1, \ldots, x_n)$ of variates; except for a constant normalizing multiplier the density is

$$f(x_1, \ldots, x_n) = e^{-xAx^T}$$

with the positive definite matrix A replacing $a > 0$ of (1). The characteristic function F is the multidimensional Fourier transform

$$F(t_1, \ldots, t_n) = \int_{-\infty}^{\infty} \cdots \int_{-\infty}^{\infty} e^{i(t_1 x_1 \cdots + t_n x_n)} f(x_1, \ldots, x_n) \, dx_1 \cdots dx_n. \quad (7)$$

As before, we can produce any mixed moment in terms of a partial derivative of F at $(0, \ldots, 0)$. Letting $m = m_1 + \cdots + m_n$ and applying the operator

$$\left(\frac{\partial}{\partial t_1}\right)^{m_1} \left(\frac{\partial}{\partial t_2}\right)^{m_2} \cdots \left(\frac{\partial}{\partial t_n}\right)^{m_n}$$

to (7), then putting all t_i's $= 0$, we get

$$\int_{-\infty}^{\infty} \cdots \int_{-\infty}^{\infty} x_1^{m_1} \cdots x_n^{m_n} f(x_1, \ldots, x_n) \, dx_1 \cdots dx_n$$

$$= (-i)^m \frac{\partial^m F(t_1, \ldots, t_n)}{\partial t_1^{m_1} \cdots \partial t_n^{m_n}} \Big|_{t_1 = \cdots = t_n = 0}.$$

Introducing the vector $t = (t_1, \ldots, t_n)$ and replacing t by $2t$, we find that equation (7) becomes equation (5).

(b) The starting point for our next topic is the Euler integral for the Γ-function

$$\int_0^{\infty} e^{-xy} x^{t-1} \, dx = \Gamma(t) y^{-t}, \qquad \mathrm{Re}(t) > 0, \qquad \mathrm{Re}\,(y) > 0. \quad (8)$$

Next, we write down two formulas due respectively to A. E. Ingham [82], and to C. L. Siegel [157]. Ingham's form is

$$(2\pi)^{-n(n+1)/2} \int_{-\infty}^{\infty} \cdots \int_{-\infty}^{\infty} e^{-i\mathrm{tr}(CT)} |A - iT|^{-k} \prod_{j \leq k} dt_{jk} =$$

$$= e^{-\mathrm{tr}(CA)} |C|^{k-(n+1)/2} \Big/ \left[(2\pi^{1/2})^{n(n-1)/2} \Gamma(k) \Gamma\left(k - \tfrac{1}{2}\right) \ldots \Gamma\left(k - \frac{n-1}{2}\right) \right], \quad (9)$$

where A and C are real symmetric $n \times n$ matrices, A is positive definite, $T = (t_{jk})$ is also an $n \times n$ matrix, and tr is the usual trace function. Siegel's form is

$$\int \cdots \int_{D_n} e^{-\mathrm{tr}XY} |X|^{t-(n+1)/2} \prod_{j \leq k} dx_{jk} = \pi^{n(n-1)/4} |Y|^{-t} \prod_{j=0}^{n-1} \Gamma\left(t - \frac{j}{2}\right), \quad (10)$$

where X and Y are $n \times n$ matrices, Y is symmetric with positive definite real part, $t - (n+1)/2 \geq 0$, and the integration is over that part of n^2-dimensional space which makes $X = (x_{jk})$ positive definite. For a clue to the sense in which (9) and (10) generalize (8) we examine what happens when $n = 1$. From (10) we get (8) itself exactly. But (9) reduces for $n = 1$ to

$$\frac{1}{2\pi} \int_{-\infty}^{\infty} e^{-ict}(a - it)^{-k} \, dt = \frac{e^{-ca}c^{k-1}}{\Gamma(k)}. \tag{11}$$

While some relation to (8) is clear, a slight re-working of (11) will make it much clearer. Taking the factor e^{-ca} to the other side and introducing the new variable $u = a - it$, we can write (11) as

$$\frac{1}{2\pi i} \int_{a-\infty i}^{a+\infty i} e^{cu}u^{-k} \, du = \frac{c^{k-1}}{\Gamma(k)}. \tag{12}$$

This, except for the symbols used, is almost the same thing as (8) once we notice that (8) asserts that $\Gamma(t)y^{-t}$ is the Laplace transform of x^{t-1} while (12) asserts that, conversely, $c^{k-1}/\Gamma(k)$ is the inverse Laplace transform of u^{-k}. Further, on account of the very special form of (8) and (12), the duality can be taken with respect to the Mellin transform as well: Equation (8) states that $\Gamma(t)y^{-t}$ is the Mellin transform of e^{-xy} while (12) states that $c^{k-1}/\Gamma(k)$ is the inverse Mellin transform of e^{cu}.

Ingham produced (9) in 1933 for some colleagues who came across this integral in a problem arising from statistics. Siegel came across (10) in 1934 in his number-theoretical investigations on the analytic theory of quadratic forms. We sketch now, very briefly, the origin of the formulas (9) and (10).

Starting with the n-variate density

$$f(x_1, \ldots, x_n) = \pi^{-n/2}|A|^{1/2}e^{-xAx^T},$$

where $x = (x_1, \ldots, x_n)$ and A is positive definite, we consider the joint distribution of the quantities x_j, $x_{jk}(= x_j x_k)$, for $j, k = 1, \ldots, n$. The characteristic function, that is, the Fourier transform $M = M(t_1, \ldots, t_n; t_{11}, \ldots, t_{nn})$ of this distribution, is

$$M = \pi^{-n/2}|A|^{1/2} \int_{-\infty}^{\infty} \cdots \int_{-\infty}^{\infty} e^{-x(A-iT)x^T}e^{2it \cdot x} \, dx_1 \cdots dx_n.$$

To integrate this we extend the validity of (5) to the case when A is a complex matrix with positive definite real part (compare this with the condition next to equation (8)) and we get

$$M = |A|^{1/2}|A - iT|^{-1/2}e^{-t(A-iT)^{-1}t^T}.$$

Next, let x_j^s, with $j = 1, \ldots, n$, and the superscript (not power) s ranging over $1, \ldots, N$, be N independent values of the random variable x_j, and put

$$c_j = \sum_{s=1}^{N} x_j^s, \qquad c_{jk} = \sum_{s=1}^{N} x_j^s x_k^s.$$

We use the theorem that if a random variable has the characteristic function F then the sum of N independent values of that variable has characteristic function F^N. With this we get for the characteristic function G of the joint distribution $f(c)$ of the variables c_j and c_{jk}

$$G = |A|^{N/2}|A - iT|^{-N/2} e^{-Nt(A-iT)^{-1}t^T}.$$

Our aim is to find $f(c)$ and so we just take the inverse Fourier transform of G. After some integrations using (5), we find that $f(c)$ is a product of two factors, one of which (the one corresponding to products c_{jk}, with $C = (c_{jk})$) is the left-hand side of (9).

Siegel's form arises as follows. We start with

$$\sum_{k=1}^{\infty} k^{t-1} e^{-kx} = \Gamma(t) \sum_{n=-\infty}^{\infty} (x + 2\pi in)^{-t}, \qquad t > 1, \qquad \mathrm{Re}\ x > 0; \qquad (13)$$

this is simply proved by applying Poisson's summation formula (vol. 1, p. 122)

$$\sum_{n=-\infty}^{\infty} g(n) = \sum_{n=-\infty}^{\infty} \int_{-\infty}^{\infty} e^{-2\pi i n u} g(u)\, du$$

to the function

$$\begin{aligned} g(u) &= u^{t-1} e^{-ux}, & u &\geq 0; \\ &= 0, & u &< 0. \end{aligned} \qquad (14)$$

Here we note another appearance of the Laplace-and-Mellin kernel $u^{t-1}e^{-ux}$. In the process of verifying (13) we would use (8). Siegel, for certain number-theoretic purposes, needed a matrix generalization of (13):

$$\sum_{T} |T|^{t-(n+1)/2} e^{-\mathrm{tr}(TY)} = \pi^{n(n-1)/4} \prod_{j=0}^{n-1} \Gamma\left(t - \frac{j}{2}\right) \sum_{F} |Y + 2\pi iF|^{-t} \qquad (15)$$

where T runs over all positive definite $n \times n$ matrices with integer elements, and xFx^T runs over all quadratic forms in n variables $x_1, \ldots, x_n$, with integer coefficients. We note here how the left-hand side of (13) with its summation over positive integers generalizes to a summation in (15) over positive definite matrices with integer elements, while the right-hand side of (13) in which the summation is over all integers corresponds to a summation in (15) which also extends over integral, but not necessarily positive, elements. One can prove (15) much like (13): a (multidimensional) Poisson summation formula is applied to a matrix equivalent of (14), and in the process we use Siegel's vector form (10) in place of the scalar form (8). It may be also added here that Siegel observes a generalization of the Euler B-function (just as (10) might be said to generalize (8)): we have

$$\int_0^\infty x^{t-1}(x+y)^{-s} \, dx = y^{t-s} B(t, s)$$

and the generalization is

$$\int \cdots \int |X|^{t-(n+1)/2} |X+Y|^{-s} \prod_{j \leq k} dx_{jk}$$

$$= \pi^{n(n-1)/4} |Y|^{t-s} \prod_{k=0}^{n-1} \frac{\Gamma(t-k/2)\Gamma(s-t-k/2)}{\Gamma(s-k/2)}$$

where Y has a positive definite real part, and X is symmetric and positive definite. Siegel in [157] remarks that the above generalization of the B-function integral is deduced by the method used by Euler in the proof of the connection between the B- and the Γ-functions. In this respect the reader may recall the Dirichlet transform D_s (vol. 1, p. 110) and the fact that it provided there an almost instantaneous proof of the connection

$$B(s_1, s_2) = \frac{\Gamma(s_1)\Gamma(s_2)}{\Gamma(s_1+s_2)}$$

mentioned by Siegel; this might tempt the reader to provide an n-dimensional generalization of the D_s-transform (possibly on the following lines: where D_s transforms power series (in one variable) into ordinary zeta functions, and L-series generally, so the hypothetical n-dimensional transform would change power series (in several variables) into generalizations of Dirichlet series, perhaps the Epstein-type zeta functions).

The reader may attempt to prove the Ingham and Siegel forms (9) and (10) using a slight generalization of the method (B) of proving (2).

(c)   For the next topic we return to the Mahler theorem (vol. 1, p. 206): If $f(z) = \sum_{n=0}^{\infty} a_n z^n$ is a rational function, regular at $z = 0$, and if infinitely many a_n's are 0, then n_0 exists, such that the set

$$\{n: n \geq n_0, \ a_n = 0\}$$

consists of one or more arithmetic progressions with the same difference. That is, there exists a positive integer L and integers $L_1, \ldots, L_k$ such that for $n \geq n_0$ $a_n = 0$ if and only if $n \equiv L_i \pmod{L}$. An alternative equivalent form of Mahler's theorem was mentioned in vol. 1, p. 206; if we put

$$\hat{f}(z) = \sum_{n=0}^{\infty} \text{sgn} \, |a_n| z^n$$

then Mahler's theorem is equivalent to the following seemingly simple statement: $\hat{f}$ is rational. That is, the property of being a rational function is inherited under the mapping $I: f \to \hat{f}$. This property can also be considered in a much more general setting. Consider the set S of all power series regular around the origin, with coefficientwise addition and multiplication; the coefficient field is the set of all complex numbers, and so we can speak about algebras in S. For instance, S itself is an algebra, and so is the subset of S consisting of power series with, say, constant term 0. An element $f \in S$ is idempotent if $f \circ f = f$ which means, of course, that all coefficients in f are 0 or 1. Given an element $g \in S$, we define the set I_g of idempotents corresponding to g:

$$I_g = \{f: f \in S, f \circ f = f, f \circ g = g\}.$$

That is, the gap-set ($=$ set of all n such that the nth coefficient $= 0$) of each idempotent of I_g is contained in the gap-set of g. There is a natural order on idempotents: $f_1 \le f_2$ if the gap-set of f_2 is contained in that of f_1; this allows us to distinguish *the* minimal idempotent $\hat{g}$ in I_g. Let us call a subalgebra of S an MI-algebra if, together with each element g, it also contains the associated minimal idempotent $\hat{g}$. Mahler's theorem reads now: the rational functions in S form an MI-algebra. The reader might wish to find other MI-algebras and try to investigate (the possibility of) representation and structure theorems for MI-algebras.

It is also possible to put Mahler's theorem in yet another, quite different, guise. First, we show that we can limit ourselves without loss of generality to the case of real coefficients. Now, let V be the (real) m-dimensional vector space, u a vector in V, T a linear transformation, and P a plane. The reader may wish to show that Mahler's theorem is equivalent to the following: If the vector $T^n u$ ($n = 0, 1, 2, \ldots$) lies in P infinitely often, then it visits the plane periodically. Here it is clear that we can renormalize u and T without loss of generality so that both are of norm 1. Let us consider the special case of $m = 2$ (with the renormalization), then we have this: If the powers $e^{in\alpha}$ of a complex number $e^{i\alpha}$ are equal to $e^{i\beta}$ infinitely often, then α and β are rational multiples of 2π and the equality occurs periodically. We see now Mahler's theorem as a far-reaching "dimensional" generalization of the preceding simple proposition on complex numbers.

Actually, rather more can be added to the above generalization. Let us have, as before,

$$f(z) = \sum_{n=0}^{\infty} a_n z^n, \qquad \hat{f}(z) = \sum_{n=0}^{\infty} \operatorname{sgn} |a_n| z^n$$

but this time without requiring that f be rational, only that $a_n = 0$ for infinitely many n, and let us examine the possibilities for $\hat{f}$. There are only two, without any middle ground whatever. Either a_n vanishes in the most regular manner possible, that is, periodically, and then $\hat{f}$ is itself of the "most regular" kind possible, namely, a rational function. Or a_n vanishes in a nonperiodic, irregular, manner and then $\hat{f}$ is of the "most irregular" kind possible, namely a function with the natural boundary. This is a special case of Szegö's theorem [42] which states that if the coefficients of a power series assume only a finite number of distinct values, then either these values are assumed periodically from some place onward and the series represents a rational function, or the assumption of values is not periodic and then the series represents a function with the unit circle as a natural boundary.

In the other approach to Mahler's theorem this correlates with the phenomenon of equidistribution: if $\alpha/2\pi$ is irrational then the powers 1, $e^{i\alpha}$, $e^{2i\alpha}, \ldots$ are equidistributed on the unit circle (cf. vol. 1, p. 32). Taking the vector u and the linear transformation to be both of norm 1, and supposing that the sequence u, Tu, $T^2 u, \ldots$ visits no plane P periodically, the reader may wish to formulate, and attempt to prove, some sort of equidistribution conjecture for the sequence u, Tu, $T^2 u, \ldots$. This may take the form of equidistribution on some unit-sphere (of suitable dimension); a further generalization allows us to introduce irrationality concept into certain topological groups G ($x \in G$ is fully irrational if the group it generates is dense in G, etc.).

The phenomenon of the completely regular ($=$ rational) versus completely irregular ($=$ with natural boundary) dichotomy for power series whose coefficients are only 0 or 1, suggests some further developments. These concern the two basic approaches to the classical theory of functions of a complex variable. In the "holistic" or global approach, due to Riemann, one studies the analytic function as a whole, for example, classifying complete functions by their associated Riemann surfaces. In the "elementalistic" Weierstrassian approach one studies the functional elements, that is, the coefficients of a local Taylor series of the function (the two approaches are well characterized by these two objects: the Riemann surface—a quintessence of globality, on the one hand, and the functional element for $f(z)$—as local an object as one could ask). Since any single element determines the whole function f it also determines all of its properties. Therefore we should be able, in principle at least, to read off all the properties of f from any single sequence of its Taylor coefficients. But this is usually exceedingly difficult. For instance, very little is known about the sequence $a_0, a_1, \ldots$, which would guarantee that

$$f(z) = \sum_{n=0}^{\infty} a_n (z - z_0)^n$$

should be single-valued. We may therefore try to convert into an easier form the difficult Weierstrassian program: from the sequence of coefficients of a power series to obtain the properties of the function it represents. Such a sub-Weierstrassian program replaces the *sequence* of coefficients by the *set* of coefficients, together with their *multiplicity*. To make this precise we use the permutations $\pi(n)$ of the set $\{0, 1, 2, \ldots\}$, and we call the power series

$$\sum_{n=0}^{\infty} a_n z^n, \qquad \sum_{n=0}^{\infty} a_{\pi(n)} z^n$$

permutations of each other. Returning now to our completely regular–completely irregular dichotomy, we ask: What conditions on the sequence $a_0, a_1, \ldots$ will guarantee that every permutation

$$f_\pi(z) = \sum_{n=0}^{\infty} a_{\pi(n)} z^n$$

has the unit circle as the natural boundary (and, say, is regular inside)?

Several sufficient conditions might be conjectured:

(1) $|a_n| = 1$ for all n; among the a_n's there are infinitely many distinct ones, each one repeated infinitely often.

(2) $|a_n| = 1$ for all n; the distinct ones among the coefficients are a_0, $a_1 = a_2$, $a_3 = a_4 = a_5, \ldots$ (and these assume suitable values).

(3) The closure of the set $\{a_0, a_1, \ldots\}$ is, say, the set

$$\{z : 1 \le |z| \le 2\}.$$

(4) What we wish occurs for "almost all" sequences $a_0, a_1, \ldots$ If we impose the condition $|a_n| = 1$ for all n then the phrase "almost all" $\ldots$, can be made rigorous: we can use the Jessen infinite-torus measure [86].

Instead of the property of singularity (= having a natural boundary) we could choose other properties, and instead of the invariance under the full symmetric group S_∞ of all permutations we might require the invariance under a subset or a subgroup of S_∞. We can also modify this in various ways. Sample questions the reader might wish to answer are:

(A) What is the exact subset (subgroup) of S_∞ which preserves the radius of convergence of any function $f(z) = \sum\limits_{n=0}^{\infty} a_n z^n$? (we may assume that the radius for f is positive);

(B) What is the exact subset (or subgroup) of permutations which preserve the property of being rational? algebraic? entire?

Somewhat more generally, the reader may wish to compare the dual lattices, that of "properties" and that of, say, subgroups of S_∞ which preserve these "properties." One may attempt to discover new subgroups of S_∞ corresponding to known properties, and vice versa.

(d) Finally, we mention very briefly two generalizations, not exactly similar to the vector and matrix generalizations of Sections (a)–(c), but more distantly related. The starting point for both is the binomial theorem for positive integer exponents

$$(a+b)^n = \sum_{k=0}^{n} \binom{n}{k} a^{n-k} b^k. \tag{16}$$

The first generalization is concerned with a rigorization of the symbolic calculus (see vol. 1, pp. 106 and 118) in the direction of the polynomial sequences $p_n(x)$ satisfying the functional equation

$$p_n(x+y) = \sum_{k=0}^{n} \binom{n}{k} p_{n-k}(x) p_k(y).$$

We observe that with the umbral convention ($p_k \leftrightarrow p^k$) the above may be written as

$$[p(x+y)]^n = \sum_{k=0}^{n} \binom{n}{k} [p(x)]^{n-k} [p(y)]^k$$

bringing out even more the relation to (16). Such a sequence of polynomials $p_n(x)$ is called a binomial sequence. Examples are: powers of x, falling factorials $x(x-1)\cdots(x-n+1)$, rising factorials $x(x+1)\cdots(x+n-1)$, Abel polynomials

$$A_n(x) = x(x-na)^{n-1},$$

Laguerre polynomials

$$L_n(x) = \sum_{k=1}^{n} \frac{n!}{k!} \binom{n-1}{k-1} (-x)^k,$$

and so on; to these definitions we add the conditions that $p_n(x)$ is of degree n and $p_0(x) = 1$. The binomial sequences $p_n(x)$ give rise to Sheffer sequences $s_n(x)$ where the defining property of the polynomial $s_n(x)$ is the functional equation

$$s_n(x+y) = \sum_{k=0}^{n} \binom{n}{k} p_{n-k}(x) s_n(y).$$

There are also doubly indexed polynomial sequences $p_n^{(j)}(x)$ with

$$p_n^{(i+j)}(x+y) = \sum_{k=0}^{n} \binom{n}{k} p_{n-k}^{(i)}(x) p_k^{(j)}(y)$$

relating, for instance, to the associated Laguerre and Hermite polynomials.

A basic result is that of Mullin–Rota [130, 150]: A polynomial sequence is binomial if and only if it is basic for some delta operator. Here the operators map polynomials onto polynomials, the sequence $p_n(x)$ is basic relative to the operator

$$g(D) = \sum_{n=1}^{\infty} g_n D^n \qquad \left(g_1 \neq 0, \; D = \frac{d}{dx} \right)$$

if $p_0 = 1$, $p_n(0) = 0$ for $n \geq 1$, and also

$$g(D) p_n(x) = n p_{n-1}(x), \qquad n \geq 1,$$

and an operator Q is a delta operator if $Q(1) = 0$, $Q(x) \neq 0$, and Q commutes with a shift.

As an application we note the Schur–Carlitz expansion [150]

$$\sin \pi x = \sum_{n=1}^{\infty} \frac{a_n}{n!} x^n (1-x)^n. \tag{17}$$

Here the delta operator $D(I-D)$ is basic for the polynomials

$$p_n(x) = x(I-D)^{-n} x^{n-1};$$

recalling how to handle differentiation and integration of nonpositive-integer order we have

$$p_n(x) = \frac{x}{(n-1)!} \int_0^{\infty} e^{-t} t^{n-1} (x+t)^{n-1} \, dt$$

and then

$$e^{ax} = \sum_{n=0}^{\infty} \frac{p_n(a)}{n!} x^n (1-x)^n.$$

Hence

$$\sin \pi x = \frac{1}{2i} \sum_{n=1}^{\infty} \frac{p_n(\pi i) - p_n(-\pi i)}{n!} x^n (1-x)^n.$$

The reader may wish to verify now that the coefficients a_n in (17) are positive.

Our second generalization of the binomial (16) is entirely different. Instead of the sum of two numbers, $a + b$, we consider the vector sum of two n-dimensional convex sets, $A + B$:

$$A + B = \{z: z = x + y, x \in A, y \in B\},$$

with the usual identification of points in the n-dimensional Euclidean space E^n, and the corresponding vectors from the origin. Since clearly

$$A + B = \bigcup_{x \in A} (B + x)$$

we can describe the vector sum $A + B$ as follows. We fix an arbitrary point in B and we move it, *by translations alone*, so that the fixed point rides over the whole of A. Then the various positions of B will sweep out just the vector sum $A + B$. For instance, if B is a ball of radius r then $A + B$ is the outer parallel body A_r to A. A special case was mentioned in vol. 1, p. 18, for a three-dimensional case K_r; we had then

$$V(K_r) = V(K) + rA(K) + r^2 M(K) + r^3 \cdot \frac{4\pi}{3}; \tag{18}$$

here V and A refer to the volume and surface area, and $M(K)$ is the integral of the mean curvature (we add: 4π here happens to be the integral of the Gaussian curvature).

For certain technical reasons it is preferable to deal with the linear combination $\lambda A + (1 - \lambda)B$ rather than with the vector sum $A + B$; we have $0 \le \lambda \le 1$ and there is no difficulty with scaling A and B. Instead of the nth power, as in (16), we consider the function "volume of" and so we are led to the problem of expressing the volume $V[\lambda A + (1 - \lambda)B]$. It turns out that

$$V[\lambda A + (1 - \lambda)B] = \sum_{k=0}^{n} \binom{n}{k} \lambda^{n-k} (1 - \lambda)^k V_k(A, B) \tag{19}$$

where $V_k(A, B)$ are the so-called mixed volumes of A and B. As (18) indicates, the mixed volumes of a convex body K with a ball are of special importance, and include the surface area of K, integrals of curvatures over the boundary of K, and so on. When A and B are plane convex bodies, $\lambda A + (1 - \lambda)B$ can be simply visualized as follows. We place A in a plane P_0 and B in a plane P_1 parallel to P_0 and one unit away from it. Let X be the convex hull of $A \cup B$ and let P_λ be a plane between P_0 and P_1 at the distance λ from P_0. Then the section of X by P_λ is precisely $\lambda A + (1 - \lambda)B$. For the theory and applications of mixed volumes see [18].

6. CONTINUITY AND SMOOTHNESS

(a) We define the continuity and differentiability of a function $f(x)$ in the usual way and we say that $f \in C^k$ if f belongs to the class C^k of functions possessing continuous derivatives of order up to and including k. Similarly, $f \in C^\infty$ if the preceding holds for every positive integer k. Finally, we define the collection A of all analytic functions and we say that $f \in A$ if f is analytic. Here it might seem necessary to distinguish between the usual analytic functions of a complex variable, and the real-analytic functions. We say that $f(x)$ is real-analytic on an open interval if every point of the interval is the center of an expansion of f into a power series (with radius of convergence > 0). However, the restriction of an analytic function to an open interval $I = (a, b)$ is real-analytic, and conversely, a function which is real-analytic on I is continuable to a function which is analytic in a domain including I.

Informally, we say that f is smooth, or nice, or well behaved, if it belongs sufficiently far in the sequence

$$C^0, C^1, C^2, \ldots, C^n, C^{n+1}, \ldots, C^\infty, A. \tag{1}$$

Rigorous definitions of the foregoing, and various consequences and theorems on the subject, form of course a considerable part of mathematics. Being a companion, we do not propose to develop the standard theory. Rather, we shall consider some uses of continuity, certain simple aspects of pathology (i.e., misbehavior) of functions, different alternative approaches to the concept and the operation of differentiation, and finally a sort of concept interpolation and extrapolation. These last will hinge on a question raised before (vol. 1, p. 59): how to extend to a larger domain a quantity or operation defined initially for integers alone, as for instance $n! \to \Gamma(x)$ or integer iteration $f(f(\cdots f(x) \cdots))$ to real-order iteration $f_a(x)$. With respect to the series (1) our interest will be not in the technique of fractional-order differentiation. Rather, we shall be interested in the possibility of classifying, from the point of view of smoothness, the region between C^∞ and A; this will lead us to the definition of quasi-analyticity. We shall also consider the possibility of continuing the sequence (1) beyond A. That is, we shall consider smoothness conditions even stronger than analyticity. Good candidates for such "hypersmoothness" are the sequentially monotonic functions ($= C^\infty$- functions with derivatives of constant sign); these will be suggested by a somewhat detailed examination of just how such C^∞ functions as e^{-1/x^2} fail to be analytic. We note that these hypersmooth classes are reminiscent of G. Cantor's transfinite continuation of closures of linear sets [26]. Starting from some questions in trigonometric series, and continuing the successive closures of subsets of the line, Cantor was led to transfinite ordinals and eventually to set theory.

Besides considering the degrees of smoothness by ranking functions in (1) according to their goodness, the reverse is also possible. The idea is to classify the discontinuous functions according to their badness by continuing (1) to the left of C^0. One form that this might take is the Baire classification, [8]. Baire class B^0 is just C^0 of (1)—the continuous functions. A discontinuous function belongs to B^{n+1} if it is not in B^m for $m \leq n$, but is a limit of a sequence of functions from B^n. So, for instance, sgn x belongs to B^1. The function $f(x)$ defined to be 1 for rational x and 0 for irrational x can be expressed analytically by the formula

$$f(x) = \lim_{n \to \infty} \lim_{m \to \infty} (\cos n! \pi x)^{2m}. \tag{1a}$$

The proof runs schematically like this. Let x be rational, since we want $\lim_{n \to \infty} \cdots$ we can without loss of generality take $n \geq n_0$ where n_0 is as large as necessary; take $n_0 = $ denominator of x. Now $\cos^2 n! \pi x = 1$ for all $n \geq n_0$ and so $f(x) = 1$. Next, let x be irrational; then for arbitrary fixed n we have $|\cos n! \pi x| < 1$ so that $\lim_{m \to \infty} \cdots = 0$ and hence $f(x) = 0$. The reader may note that we go to work, so to say, on the double limit from one end for rational x, and from the other end for irrational x.

From the above representation of $f(x)$ it follows that it must be in B^1 or in B^2. As it turns out, $f \in B^2$ (and the reader may wish to prove it by showing that $f(x)$ is not a limit of continuous functions). Similarly, it can be shown that each B^n is not empty. The "bad" scale $B^0, B^1, B^2, \ldots$ can be continued into the transfinite domain rather more easily and naturally than the "good" scale (1). The reader may wish to generalize (1a) to the following: let $F(x, y)$ be a function of two variables, such that $F \to 0$ as $x^2 + y^2 \to \infty$; find an explicit formula like (1a) for the function $f_F(x)$ defined as follows:

$$f_F(x) = F(p, q), \quad x \text{ rational and} = p/q \text{ where}$$
$$\text{g.c.d. } (p, q) = 1;$$
$$= 0, \quad x \text{ irrational.}$$

As a further exercise in obtaining formulas for pathological curves the reader may wish to consider the von Kooch curve K obtained as the limiting case of a certain process similar to the one that yields the Cantor set. The first four approximations K_0, K_1, K_2, K_3 are shown in Fig. 1 a–d. The curve K has a number of interesting properties, the principal one being that every subarc contains a subarc similar to K (the reader may wish to find other curves with this property, and to modify the situation so

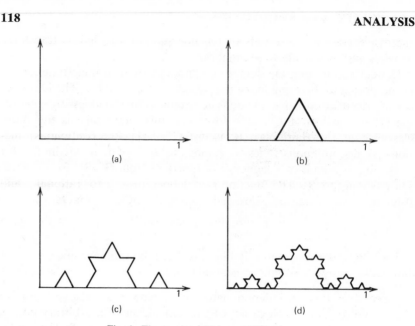

Fig. 1. The curve of Helge von Kooch.

that the property applies to surfaces). We parametrize K by a real parameter t, $0 \le t \le 1$, writing t in the quaternary scale as $t = 0.t_1 t_2 t_3 \ldots$, where $t_i = 0$, 1, 2, or 3. Then the four segments making up K_1 determine in the obvious fashion the first digit t_1 of t, similarly the sixteen segments of K_2 determine t_1 and t_2, and so on, so that in the limit a unique value of t corresponds to each point of K (with some simple proviso for the numbers with two quaternary expansions, e.g., $0.133 \cdots = 0.200 \cdots$). We observe that t_n is simply expressed as a function of t:

$$t_n = [4^n t] - 4[4^{n-1} t]$$

and we note that the cubic

$$h(x) = x^3 - \frac{9x(x-1)}{2}$$

has values $h(0) = 0$, $h(1) = 1$, $h(2) = -1$, $h(3) = 0$. Using these the reader may wish to determine the complex number $z = z(t)$ that gives us the point on K corresponding to the parameter value t.

(b) As to the uses of continuity, these are of course innumerable. Here we mention a few that are perhaps not quite typical. We start with two examples in which continuity is used to show that something important is constant: Rouché's theorem, and Liouville's theorem. For Rouché's theorem we take two functions $f(z)$ and $g(z)$ of a complex variable z, both of them analytic inside and on a simple closed contour C in the complex plane. We suppose that f dominates g on C: for $z \in C$ we have $|f(z)| > |g(z)|$ (as a consequence f does not vanish on C). It is required to show that inside C the equations $f = 0$ and $f + g = 0$ have the same number of roots (counting these with multiplicity). One way of proving this is to recall that

$$N_F = \frac{1}{2\pi i} \int_C \frac{F'}{F}\, dz$$

is the number of zeros an analytic function has inside the closed contour C (with $F \neq 0$ on C). We form the expression

$$N(\lambda) = \frac{1}{2\pi i} \int_C \frac{f' + \lambda g'}{f + \lambda g}\, dz, \qquad 0 \le |\lambda| \le 1,$$

and we show that under our hypotheses this is continuous in λ. But it is also an integer, being the number of something (namely, of zeros of $f + \lambda g$ inside C). Since a continuous integer-valued function is constant, we get our theorem in a slightly more general form than requested: For every complex λ such that $0 \le |\lambda| \le 1$ the equations $f = 0$ and $f + \lambda g = 0$ have equally many roots inside C.

For the following proof of Liouville's theorem the intervention of continuity is indirect and rather minimal though not absent. The theorem we want is that if $f(z)$ is analytic and bounded in the whole complex plane then it is a constant. We use the mean-value theorem for f, which is easily deduced from Cauchy's formula

$$f(z_0) = \frac{1}{2\pi i} \int_C \frac{f(z)}{z - z_0}\, dz. \tag{2}$$

Taking for C the circle which is the boundary of the disk $D(z_0, R)$, centered at z_0 and of radius R, we let $z - z_0 = R^{i\theta}$, etc., eventually getting our mean-value theorem: $f(z_0)$ is the mean value of $f(z)$ over the disk. To show that $f(z)$ is constant we show that for arbitrary z_1 and z_2 $f(z_1) = f(z_2)$. Now, by the mean-value theorem

$$\begin{aligned} f(z_1) &= \text{mean of } f \text{ over } D(z_1, R), \\ f(z_2) &= \text{mean of } f \text{ over } D(z_2, R). \end{aligned} \tag{3}$$

If R is sufficiently big relative to $|z_1 - z_2|$ then it follows from the hypothesis of *boundedness* that most of the contribution to both means in (3) comes from the common part

$$D(z_1, R) \cap D(z_2, R).$$

Hence, with R sufficiently large, $f(z_1)$ is arbitrarily close to $f(z_2)$, and so $f(z_1) = f(z_2)$. In effect, we have used a sort of continuity argument thus: At two sufficiently near points a continuous function assumes arbitrarily close values, and by a sort of scale change we have brought the points z_1 and z_2 into what is effectively a sufficiently close neighborhood.

On the other hand, it must be allowed that the more usual proof of Liouville's theorem is no more complicated, and can be considerably extended. We start again from Cauchy's formula (2) and we form the difference

$$f(z_2) - f(z_1) = \frac{z_2 - z_1}{2\pi i} \int_C \frac{f(z)}{(z - z_1)(z - z_2)} \, dz. \tag{4}$$

Using the basic estimate

$$\left| \int_C F(z) \, dz \right| \leq \max_{z \in C} |F(z)| \, \text{Length} \, (C)$$

we show that $f(z_2) - f(z_1)$ is 0, much as before. The extension arises from the applicability of the above to show that if $f(z)$, instead of being bounded, obeys

$$|f(z)| \leq O(R^n) \quad \text{on} \quad D(z, R), \, R \text{ large},$$

then the conclusion is that $f(z)$ is a polynomial of degree $\leq n$. The proof is modelled on (4): Instead of showing that something is constant by showing that the *first* difference vanishes, we show that it is a polynomial of degree $\leq n$ by showing that its $(n+1) - st$ difference vanishes.

However, the proof of Liouville's theorem, as based on the mean-value property, also has its extensions. We observe first that Liouville's theorem applies to real-valued functions $u = u(x, y)$ which are harmonic and bounded in the whole plane: $u(x, y) \leq M$ for all x, y. To prove this we apply the useful device of exponential map: There exists an analytic function $f(z)$ such that $f(x + iy) = u(x, y) + iv(x, y)$, and now we apply Liouville's theorem (for analytic functions) *not to f directly* but to $g = e^f$. This g is also analytic and bounded: $|g| \leq e^M$ since $|e^{it}| = 1$. Hence g, and therefore u, must be constant. The reader may wish to try applying the theorem to f itself.

But harmonic functions are not restricted to be of two real variables. We can define $u(x_1, \ldots, x_n)$ to be a real-valued harmonic function of n real variables using as the harmonicity definition precisely the mean-value property, either over spheres or over balls. Here the spheres and balls correspond to the plane case of circles and circular disks. Of course, Liouville's theorem for these n-variables harmonic functions holds since our first proof extends at once from 2 to n dimensions. This is so even though there are no "analytic functions in n dimensions." The reader may wish to recall (cf. vol. 1, p. 14) that, accidentally or otherwise, this too was first proved by Liouville.

Both Rouché's theorem and Liouville's theorem can be used to prove the fundamental theorem of algebra: If $P(z)$ is an nth degree polynomial then the equation $P(z) = 0$ has n roots (counting multiplicity). In using Rouché's theorem here we compare the principal term $a_0 z^n$ of $P(z)$ with the sum of all others, on a sufficiently big circle; in using Liouville's theorem we argue indirectly with $1/P(z)$. A frequent use of Liouville's theorem occurs in the elementary theory of elliptic functions. For example, to show that an elliptic function satisfies a differential equation, an algebraic addition theorem, or a similar relation, we form the difference of the two sides of the (presumed) equation. This must also be an elliptic function, and by an explicit calculation we show that it has no poles. Hence it is bounded in a period parallelogram, and consequently everywhere. Therefore it is a constant by Liouville's theorem. We now choose a suitable value z_1 and show by computing the value at z_1 that the constant is 0.

(c) A frequent use of continuity occurs in the form of the I.V.P.— intermediate value property. One of its simpler forms is this: If $f(x)$ is a continuous real-valued function defined on $[a, b]$ and $f(a)f(b) < 0$, then $f(x_1) = 0$ for some $x_1 \in (a, b)$. This form of I.V.P. can be generalized into some basic theorems of topology. To see this we observe that if our I.V.P. were false then we could deform $f(x)$ to a continuous function $g(x)$ on $[a, b]$ which takes up two values only: $f(a)$ and $f(b)$. The domain of this hypothetical continuous g would be a closed interval and the range would be a two-point set that can be regarded as the boundary of the domain. Applying dimensional generalization, we change the closed interval to the closed n-dimensional ball or, somewhat more generally, a closed n-cell. The result is a topological theorem: No continuous map f of an n-cell into its boundary can keep all boundary points fixed. An easy consequence is the Brouwer fixed-point theorem [81]: For a continuous map f of an n-cell into itself there is always x such that $x = f(x)$. Indeed, if this were false we would define $F(x)$ as follows: Supposing the domain B to be an n-ball, let $F(x)$ be the point in which the ray from x through $f(x)$ cuts the boundary of B. This F clearly contradicts the previous theorem.

It may be noticed that the original I.V.P. has been changed, when we spoke of $g(x)$ above, into the form used before to prove Rouché's theorem. This was done by reasoning that a continuous integer-valued function is constant. The original form of I.V.P. is the beginning of numerical procedures for finding (or localizing) the (or a) zero x_1 of $f(x)$ on (a, b). Sometimes, even, the same is claimed for some of the topological theorems which are the dimensional generalizations of the simple

I.V.P. We shall now consider a somewhat different dimensional generalization of I.V.P., intended to serve as a starting point for solving systems of simultaneous equations

$$f_i(x_1, \ldots, x_n) = 0, \qquad i = 1, \ldots, n. \tag{4a}$$

Each f_i is a continuous real-valued function of its n real variables $x_1, \ldots, x_n$ and we suppose for convenience that each f_i is defined everywhere. In the original I.V.P. there were two ($= 2^1$) values of x, namely a and b, such that the different signs, namely $+$ and $-$, were achieved by f at those values of x. We now generalize the single equation $f(x) = 0$ to the system (4a), and in n dimensions we first have 2^n possible sign sequences of n signs, each $+$ or $-$. A sign sequence will be said to be realized, at the n-tuple $\xi_1, \ldots, \xi_n$, if

$$\text{sgn } f_i(\xi_1, \ldots, \xi_n), \qquad i = 1, \ldots, n,$$

is the sign of the ith member of the sequence. Now we have the 2^n rule as a generalization of the I.V.P.: If every one of the 2^n sign sequences is realized then (4a) has a solution. It is not hard to show that this conclusion is not always true if only some $2^n - 1$ out of the 2^n sign sequences are realized. A counterexample can even be given in which each f_i of (4a) is a linear function. The reader may wish to prove this, as well as attempt a proof of the 2^n rule itself. Here one might perhaps proceed either by induction, or by making use of the following topological proposition [81], which is itself a yet another n-dimensional generalization of the I.V.P.: If in the closed n-cube H the ith pair of $(n-1)$-dimensional opposite faces is (strictly) separated by a closed subset of H, then the n closed separating subsets have a nonempty intersection.

We note that with the use of the original I.V.P. the zero x_1 is restricted to lie between a and b, and now it is easy to develop procedures (e.g., the sequential localization type of vol. 1, p. 159) which localize x_1 to smaller and smaller intervals. On the other hand, the 2^n rule, while guaranteeing the existence alone of a solution of (4a), does not a priori limit this solution to lie in some subset of E^n. The reader may wish to show first that without some further conditions on f_i's in (4a) no such limitation is possible: No matter how close to each other the 2^n n-tuples $(\xi_1, \ldots, \xi_n)$ may lie, the solution may be arbitrarily far from them. Then the reader may wish to put suitable conditions on f_i's, for instance in terms of bounds on partial derivatives, gradients, or Lipschitz constants, so as to produce a priori limitation of the solution(s) of (4a). Also, the reader may wish to attempt the development of complete, optimal or suboptimal, deterministic or stochastic, sequential localizing algorithms, based on the 2^n rule. Finally, there is another generalization we might attempt, in some ways the most important one. Using the equivalance of functional ($=$ differential, integral, etc.) equations to systems of infinitely many linear, or other, equations in infinitely many unknowns, we may try to generalize the 2^n rule to an infinite-dimensional space and thereby obtain criteria for existence of solutions of linear, or especially nonlinear, functional equations.

(**d**) We shall now consider a standard example of question in mathematical pathology: Given a succession of properties of increasing stringency, do there always exist objects having any given property but failing to have the next one in some conspicuously spectacular fashion? We ask: Are there functions $f(x)$ in the class C^∞ (for all real x, throughout) which fail to be analytic on any subinterval (a, b)?

We start by showing that the function

$$f(x) = e^{-1/x}, \qquad x > 0,$$
$$= 0, \qquad x \le 0,$$

is a C^∞-function but not analytic on any interval including $x = 0$. This, of course, is common mathematical knowledge. What may be not quite so well known is how badly f fails to be analytic at the origin, and how relatively easy it is to prove this failure. First, by simple induction on n we show that

$$f^{(n)}(x) = x^{-2n} Q_n(x) e^{-1/x} \tag{4b}$$

where $Q_0(x) = Q_1(x) = 1$ and for $n \ge 1$

$$Q_{n+1}(x) = x^2 Q_n'(x) + (1 - 2nx) Q_n(x). \tag{5}$$

Hence $Q_n(x)$ for $n \ge 1$ is a polynomial of degree $n - 1$; the substitution $1/x = u$ in (4b) and the fact that e^u tends to infinity with u faster than any polynomial in u, show that

$$\lim_{x \to 0} f^{(n)}(x) = 0, \qquad n = 0, 1, 2, \ldots.$$

Hence, if we define $f^{(n)}(0) = 0$ for all n, we see that $f(x) \in C^\infty$ since clearly $x = 0$ is the only place where difficulty might arise. One can finish now the argument in the standard fashion, short but rather unilluminating: $f(x)$ is not analytic at or near $x = 0$ because otherwise its McLaurin series would have had all coefficients identically 0, hence by the identity theorem for power series $f(x)$ would be $\equiv 0$ in some neighbourhood of $x = 0$ and so everywhere $\equiv 0$. What we are after is not just knowing that $f^{(n)}(x)$ exists and approaches 0 as $x \to 0$, for every n, but we wish also to know how $f^{(n)}(x)$ behaves near $x = 0$. It turns out that with increasing n $f^{(n)}(x)$ starts oscillating near $x = 0$ rather wildly, with both frequency and amplitude growing, somewhat after the manner of the function $(1/x) \sin (1/x)$. To show this, we first examine the graphs of f and its first few derivatives. These are drawn one under another in Fig. 2a–d. The function f itself is illustrated in Fig. 2a, there is one point of inflection, at the root of $f'' = 0$ or $Q_2(x) = 0$ so that $x = 1/2$; this is also the maximum of f' in Fig. 2b and the first positive zero of f'' in Fig. 1c. The graph of f' in Fig. 2b has two points of inflection; in Fig. 2c these function as the extrema, and in Fig. 2d as zeros. Generally, imagining the graphs of Fig. 2 continued downward we would need vertical dotted lines joining triples of graphs. The corresponding abscissa functions as a point of inflection on the top graph, as an extremum on the middle one, and as a zero on the bottom graph of the three. This simple graphical representation, together with (4b) and (5), shows that $Q_n(x)$ has $n-1$ distinct positive zeros for $n \geq 1$. Also, these, the zeros of $f^{(n)}(x)$, tend to crowd more and more toward $x = 0$ as n grows, and the values at the extrema grow bigger and bigger. For instance, let us examine the first positive zero x_n of $f^{(n)}(x)$. Then, from Fig. 2, $x_1 = 1/2$ and $f'(x_1) = 4e^{-2}$. From the graph of Fig. 2c $f''(x_2)$ is the slope of the first inflection tangent of Fig. 2b. Since the highest point on the curve of Fig. 2b is clearly below that tangent, we have $f''(x_2) \geq 2f'(x_1)$. In the same way $f'''(x_3) \geq 2^2 f'(x_1)$ since $x_3 \leq 1/2$. Similar argument applied to the first inflection tangent of any successive graph shows that if $M_n = f^{(n)}(x_n)$ then $M_n \geq A_i 2^n$ ($A_1, A_2, \ldots$ are positive constants). Actually rather more can be shown. The reader may wish to show that

$$Q_n(x) = 1 - \cdots + (-1)^{n+1} n! \, x^{n-1}, \qquad n \geq 1$$

so that the product of the $n-1$ roots, which we know to be distinct positive numbers, is $1/n!$. Therefore the smallest root, which is of course x_n, satisfies the inequality

$$x_n \leq (n!)^{-1/(n-1)}.$$

Using Stirling's approximation we have for large n $x_n \leq e/n$. With this, and with a more careful repetition of the previous argument concerning inflection tangents, we find that for large n

$$M_n \geq A_2 \left(\frac{n}{e} \right)^n$$

or, what amounts to the same thing,

$$M_{n+1} \geq A_3 n!$$

A precise evaluation of M_n could be attempted from (4b) and (5) since x_n is the smallest positive root of $Q_n(x) = 0$. A number of other problems can be formulated concerning $f(x)$, $f^{(n)}(x)$, and especially the polynomials

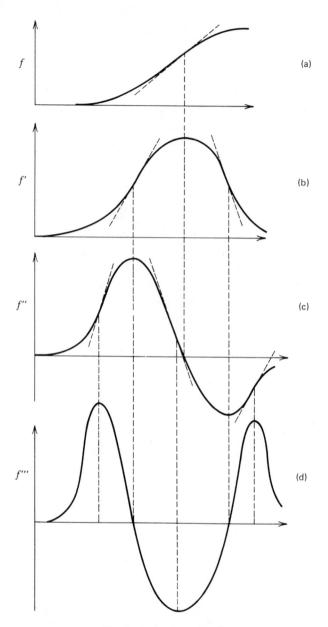

Fig. 2. Derivatives of $e^{-1/x}$.

$Q_n(x)$. The reader may wish to recall the use of Faa di Bruno's formula in vol. 1, p. 215, to calculate the nth derivative of e^{-1/x^2}, and use the same method to get an explicit expression for $Q_n(x)$ (and perhaps investigate some combinatorial properties of its coefficients). However, the problems of estimating quantities like M_n are probably of greater importance than the other problems since some tentative classifications of the functions in C^∞ that are not analytic can be attempted on that basis. In this connection we recall the definition of quasi-analyticity: functions of a class $S \subset C^\infty$ are called quasi-analytic if $f \in S$ is identically 0 whenever it and all of its derivatives vanish at a point. This, of course, is very strongly reminiscent of the uniqueness theorem for power series: An analytic function cannot have two different power series about the same point as center. For analytic functions f the knowledge of f and its derivatives at a point gives us the whole of f throughout its domain of definition—we just use power series and standard analytic continuation. It has been shown by Carleman [27] that a different process will yield us the whole of a quasi-analytic function as soon as we know it and all of its derivatives at a single point. With M_n defined to be $\sup |f^{(n)}(x)|$ we have the Denjoy–Carleman theorem [41]: If

$$Q(x) = \sum_{n=0}^{\infty} \frac{x^n}{M_n}, \qquad q(x) = \sup_{n \geq 0} \frac{x^n}{M_n}, \qquad x > 0,$$

then the following are equivalent:

(a) f is not quasi-analytic,

(b) $\displaystyle\int_0^\infty (1+x^2)^{-1} \log Q(x)\, dx < \infty,$

(c) $\displaystyle\int_0^\infty (1+x^2)^{-1} \log q(x)\, dx < \infty$

(d) $\displaystyle\sum_{n=1}^{\infty} M_n^{-1/n} < \infty,$

(e) $\displaystyle\sum_{n=1}^{\infty} \frac{M_{n-1}}{M_n} < \infty.$

The idea here is that $f(x)$ is quasi-analytic provided that its derivatives do not oscillate too badly (with respect to amplitude). Superficially at least, we can reason that (d) (which is the original Denjoy form) makes sense on the following grounds. For an *analytic* f the power series converges, this places a bound of the form

$$|f^{(n)}(x)| \leq AK^n n!$$

on its derivatives; a significantly faster increase of derivatives would make the series in (d) convergent. Here we may note a result of Holmgren [79]: f is not quasi-analytic if $M_n \geq [(1+\varepsilon)n]!$ for some $\varepsilon > 0$; this follows simply from (e) above and from the Γ-function relation: For large x we have $\Gamma(x)/\Gamma(x+a) \sim x^{-a}$. Carleman's contribution to the theorem above [91] was based on the following elementary lemma which is of independent interest: Let $\sum_{n=1}^{\infty} a_n$ be a convergent series of positive terms, then

$$\sum_{n=1}^{\infty} \left(\prod_{j=1}^{n} a_j \right)^{1/n} < e \sum_{n=1}^{\infty} a_n$$

and the constant is best possible. This is proved by maximizing the L.H.S. subject to the side condition $\sum_{n=1}^{\infty} a_n = 1$, using the Lagrange multipliers, and so on. For further information on quasi-analytic functions and their use in Fourier series see references [91, 151].

We return now to our function $f(x)$ which is $e^{-1/x}$ for $x > 0$, and 0 for $x \leq 0$. Having produced this f that has the desired pathology (namely being in C^∞ but not in A) at one point, $x = 0$, we next manufacture a function which has the same pathology everywhere, not just at one point. Let $\{r_1, r_2, \ldots\}$ be any set dense in the real line; for definiteness, let us say, the set of all rational numbers in some enumeration. Let also $\sum_{n=0}^{\infty} a_n$ be a convergent series of positive constants, and put

$$F(x) = \sum_{n=1}^{\infty} a_n f(x - r_n). \tag{6}$$

Then $F(x)$ can be proved to be the desired nowhere analytic C^∞-function. This is so *not* because the partial sums of the series in (6) have the C^∞/A pathology at bigger and bigger sets of points but by the following reasoning. If, on the contrary, F were somewhere analytic, then we would have had the Taylor expansion

$$F(x) = \sum_{n=0}^{\infty} \frac{F^{(n)}(r_N)}{n!} (x - r_N)^n \tag{7}$$

valid for $|x - r_N| < \varepsilon$, $\varepsilon > 0$, since the rationals $\{r_n\}$ are dense. But (6) shows that none of the derivatives $F^{(n)}(r_N)$ involves a_N. Hence $F(x)$, as given by (7), is independent of a_N which is a contradiction since a_N can be found from (6) and (7).

Under the same circumstances we could have used instead of (6)

$$F(x) = \sum_{n=1}^{\infty} \left[a_n \prod_{j=1}^{n} f(x - r_j) \right]; \tag{8}$$

if $\sum a_n$ converges sufficiently fast F belongs to C^∞ and the conclusion then would have been that $F(x)$ in (7) depends only on $a_1, a_2, \ldots, a_{N-1}$. The condensation-of-singularities methods (6) and (8) enable us to produce numerous other pathologies. For instance, taking suitable $f(x)$, we can produce continuous nowhere differentiable functions $F(x)$. Again, with suitable sets $\{r_n\}$ and $f(x) = x$ we can produce with (8) functions that at certain prescribed countable dense point-sets take up values which lie in similar sets (e.g., whose values are never algebraic for any algebraic real x).

(e)   Starting with our function $f(x)$ which is $e^{-1/x}$ for $x > 0$ and 0 for $x \le 0$, we can synthesize other C^∞ functions with useful interpolatory properties. Two simple examples of C^∞ functions which vanish outside an interval are given in Fig. 3a and b. Their equations are as follows:

$$y = cf(x-a)f(b-x)/f^2\left(\frac{b-a}{2}\right) \qquad \text{for Fig. 3a,}$$

$$y = \frac{f(1-x^2)}{f^2(1)} \qquad \text{for Fig. 3b.}$$

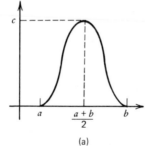

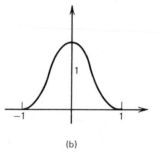

(a)   (b)

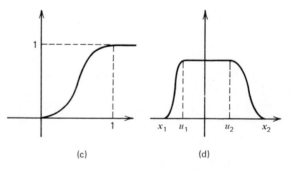

(c)   (d)

Fig. 3. Examples of C^∞ functions.

To produce a smoothly bridging C^∞ function $h(x)$ which is 0 for $x \leq 0$ and 1 for $x \geq 1$ we define

$$h(x) = \begin{cases} 0 & x \leq 0, \\ f\left(\dfrac{x}{1-x}\right) & 0 < x < 1, \\ 1 & 1 \leq x, \end{cases}$$

with the graph given in Fig. 3c. There is no difficulty in matching the derivatives at $x = 0$ and at $x = 1$; at $x = 1$, for instance, we can either verify directly that $\lim_{x \to \infty} f^{(n)}(x) = 0$ for $n \geq 1$ or we can exploit the functional relation

$$f\left(\frac{x}{1-x}\right) = ef(x).$$

The construction of $h(x)$ amounts to compressing, by means of the bilinear function $x/(1-x)$, the whole infinite interval $(0, \infty)$ for $f(x)$ to the finite interval $(0, 1)$ for $h(x)$. With $h(x)$ it is not hard to synthesize in turn the flat-roof C^∞ function which is 0 outside an interval I and 1 on a subinterval T of I. We could call I the base and T the roof of the flat-roof function. An example is

$$h\left(\frac{x - x_1}{u_1 - x_1}\right)\left[1 - h\left(\frac{x - u_2}{x_2 - u_2}\right)\right]$$

with graph shown in Fig. 3d. Here we observe that the change $h \to 1 - h$ reverses the 0–1 bridging by h; then the multiplication of two shifted bridging functions, one direct and the other reversed, accomplishes our purpose. Let us consider a special case of the above with $x_1 = -1$, $u_1 = -1/2$, $u_2 = 1/2$, $x_2 = 1$:

$$\phi(x) = h(2x + 2)[1 - h(2x - 1)]. \tag{9}$$

Using this function we shall produce an example of a C^∞ function f on $(-\infty, \infty)$ all of whose derivatives are arbitrarily prescribed at a point, say at $x = 0$:

$$f^{(n)}(0) = c_n, \qquad n = 0, 1, \ldots,$$

where $c_0, c_1, \ldots$ are arbitrary constants. Note first that an elementary use of Taylor series suggests the function

$$F(x) = \sum_{n=0}^{\infty} \frac{c_n x^n}{n!} \tag{10}$$

as an answer to our problem. Indeed, this would have been so if the series in (10) always converged. Since it may not, we proceed differently, though still starting with (10) and modifying it to

$$G(x) = \sum_{n=0}^{\infty} \frac{c_n x^n \, \phi_n(x)}{n!} \tag{11}$$

The multipliers $\phi_n(x)$ will be flat-roof C^∞ functions of the type shown in Fig. 3d, whose bases get shorter and shorter. The use of C^∞ multipliers ϕ_n assures that G is a C^∞ function, the use of flat-roof functions guarantees that (11) has the same derivatives at $x = 0$ as (10), and the contraction of bases is arranged to make (11) convergent. The reader may wish to verify that an adequate set of multipliers is

$$\phi_n(x) = \phi(C_n x), \quad C_n = \max(|c_0|, |c_1|, \ldots, |c_n|),$$

with $\phi(x)$ given by (9). Our example (11) is not only a C^∞ function but it is even real-analytic for all x except, possibly, near $x = 0$. This is so because for $x \neq 0$ either (10) already converges, or the number of nonzero terms in the summation of (11) is finite. In the latter case this finiteness is, of course, nonuniform: There will be more and more terms as x gets closer to 0. However, the reader may wish to modify the construction and provide a C^∞ function with prescribed derivatives at a point, which fails to be analytic on any subinterval.

We consider now a very different method of generating C^∞ functions with certain prescribed interpolatory properties. This method depends on starting with the right initial function f_0, and iterating (to infinity) the running-average operation on f_0. Suppose that $f(x)$ is a continuous function defined for all x, then

$$S_a f(x) = \frac{1}{2a} \int_{x-a}^{x+a} f(u) \, du \tag{12}$$

is called the running average of f (with the interval $2a$). This is an analog of the method used sometimes to smoothe out discrete sequences of data. Here the doubly infinite sequence $\ldots x_{-1}, x_0, x_1, \ldots$ would be replaced by

$$\tilde{x}_n = \frac{1}{2k+1} \sum_{j=-k}^{k} x_{n+j}, \quad n = \ldots, -1, 0, 1, \ldots.$$

In smoothing singly infinite or finite sequences in this way we lose something at one or both ends; one may attempt to make up this loss by extrapolation from the smoothed-out data. It is not hard to show that $S_a f(x)$ is differentiable, with

$$\frac{d}{dx} S_a f(x) = \frac{1}{2a} [f(x+a) - f(x-a)].$$

It follows that the n times iterated running average

$$\prod_{i=1}^{n} S_{a_i} f(x) = S_{a_n} S_{a_{n-1}} \cdots S_{a_1} f(x)$$

is a function with continuous first n derivatives. It may be proved that the averaging preserves total area:

$$\int_{-\infty}^{\infty} f(x) \, dx = \int_{-\infty}^{\infty} [S_a f(x)] \, dx$$

if the L.H.S. exists. Also, if $f(x)$ is monotone then so is $S_a f(x)$.

Suppose now that we start with the function

$$\begin{aligned} &= 0 && x \leq \varepsilon \\ f_0(x) = j(x) \quad && \varepsilon \leq x \leq 1 - \varepsilon \\ &= 1 && 1 - \varepsilon \leq x, \end{aligned}$$

where $0 < \varepsilon < 1/2$ and $j(x)$ is any monotone increasing function, such that $f_0(x)$ is continuous. Let $\sum_{n=1}^{\infty} a_n$ be a convergent series of positive constants with sum $< \varepsilon/2$. Now, let us consider the sequence of iterated running averages:

$$f_0(x), \quad S_{a_1} f_0(x), \quad S_{a_2}[S_{a_1}(f_0(x))], \ldots$$

with the limit

$$F(x) = \prod_{n=1}^{\infty} S_{a_n} f_0(x). \tag{13}$$

Then $F(x)$ is a well-defined monotone C^{∞} function and $F(x) = 0$ for $x \leq 0$, while $F(x) = 1$ for $x \geq 1$. Hence we have constructed another type of the C^{∞} bridging function of Fig. 3c.

We observe finally the weighted running average

$$S_a f(x) = \int_{x-a}^{x+a} w(x) f(x) \, dx \Big/ \int_{x-a}^{x+a} w(x) \, dx.$$

This reduces to the ordinary running average when we take $w = \text{const.}$ $(\neq 0)$.

(f) Our next topic we could approach in several ways. First, it could be asked whether the series (1) can be continued beyond A—are there functions even "smoother" or "nicer" than analytic ones? Second, one might ask what are the relevant mathematical properties that either imply or require more than analyticity—this is perhaps just a reformulation of the preceding. Third, it may have occurred to the reader, on looking through part (c) of this section, to ask whether the oscillatory behavior of derivatives is really necessary in C^∞-functions which are not analytic; what follows if $f(x)$ has all derivatives at and near $x = 0$ and if these are of constant signs? Could such $f(x)$ still fail to be analytic? These questions turn out to be the key to the definitions of significant classes of functions which go beyond analyticity: the absolutely monotonic, the completely monotonic, the completely convex, and generically, the sequentially monotonic functions. It may be remarked that this continuation of the series (1) beyond A is different from the smoothness classes before A: Up to and including A, even though each new class is a proper subset of the preceding one, it is nevertheless dense (in a reasonable sense) in this preceding bigger class. On the other hand, this no longer holds for the sequentially monotonic classes. The question arises whether the series (1) can be significantly continued to some hypersmooth classes which are nevertheless still dense among the analytic functions.

The sequentially monotonic functions are defined as follows. Let $\sigma(n)$, $n = 0, 1, \ldots$, be a function on nonnegative integers taking up the values ± 1 only. Then $f(x)$ which is of the class C^∞ on an interval I is sequentially monotonic there, with the signature σ, if for $x \in I$

$$\sigma(n)f^{(n)}(x) \geq 0, \qquad n = 0, 1, \ldots.$$

When $\sigma(n) = 1$ for all n, f is an absolutely monotonic function; if $\sigma(n) = (-1)^n$, f is completely monotonic, and if $\sigma(n) = (-1)^{n/2}$ for n even (the values for n odd being irrelevant) then f is called completely convex. The reader may wish to show on the example of

$$f(x) = \sum_{n=0}^{\infty} \frac{\sigma(n)\lambda^n x^n}{n!},$$

where $0 < \lambda < \log 2$, that a function sequentially monotonic on $[0, 1]$ may have its signature prescribed arbitrarily. The bound $\log 2$ occurs here on account of the inequality $e^{\lambda x} - 1 < 1$, $0 \leq x \leq 1$.

Suppose next that $f(x)$ is absolutely monotonic for $a \leq x < b$. Then it must be analytic: The Taylor series

$$\sum_{n=0}^{\infty} \frac{f^{(n)}(a)}{n!} (z - a)^n$$

converges for $|z - a| < b - a$. This is simply proved by working on the real axis and examining the remainder in Taylor's theorem. It turns out that by using our strong monotonicity properties the remainder can be shown to converge to 0 [185]. We have

$$f(x) = \sum_{k=0}^{n} \frac{f^{(k)}(a)}{k!} (x-a)^k + R_n(x, a),$$

where

$$R_n(x, a) = \frac{1}{n!} \int_a^x (x-t)^n f^{(n+1)}(t) \, dt$$

$$= \frac{(x-a)^{n+1}}{n!} \int_0^1 (1-t)^n f^{(n+1)}[a+(x-a)t] \, dt.$$

Since $f^{(n+2)}(x) \geq 0$, $f^{(n+1)}(x)$ is not decreasing; for $x < c < b$ we have, integrating by parts,

$$0 \leq R_n(x, a) \leq \frac{(x-a)^{n+1}}{n!} \int_0^1 (1-t)^n f^{(n+1)}[a+(c-a)t] \, dt$$

$$= \left(\frac{x-a}{c-a}\right)^{n+1} \left[f(c) - \sum_{j=1}^{n} \frac{f^{(j)}(a)}{j!} (c-a)^j \right]$$

$$\leq \left(\frac{x-a}{c-a}\right)^{n+1} f(c)$$

so that $R_n(x, a) \to 0$ as $n \to \infty$. The reader may wish to compare this with the proof that $R_n \to 0$ for the complex-variable case. Further, with the above as model, the reader may wish to prove that the same conclusion holds for every sequentially monotonic function; it is analytic. On the other hand, the question may be asked: In how big a region is a sequentially monotonic function with the signature σ necessarily analytic? Here we quote the following: A completely convex function is necessarily entire [185]. The reader may wish to consider which other signatures σ force the same conclusion.

Perhaps the principal importance of sequentially monotonic functions derives from the appearance of complete monotonicity on $0 \leq x < \infty$ in Bernstein's theorem [185]: A necessary and sufficient condition for $f(x)$ to be completely monotonic on the interval $0 \leq x < \infty$ is that it be a generalized Laplace transform

$$f(t) = \int_0^\infty e^{-xt} \, dF(t), \qquad x \geq 0, \qquad (14)$$

where the integral is in the sense of Stjeltjes, with F bounded and nondecreasing. We sketch a bare outline of the proof based on the use of Krein–Milman theorem: Working in a suitable function space, we prove first that not only are the negative exponentials e^{-ax} completely monotonic on our interval but, moreover, they are the extreme ones. It follows that all others are their convex combinations (linear combinations with nonnegative coefficients adding up to 1). However, a convex combination of negative exponentials is just the above generalized Laplace transform (14).

(g) We come now to the last topic of this section: derivatives and different approaches to derivatives. Most mathematicians would probably be inclined to say that derivative is, at once, two things: a limit, and the result of applying a certain linear operation to a function. We have here the two distinct ideas behind a derivative: (a) limit, (b) linear operation. Under (a) we note ordinary and partial derivatives (of functions of real or complex variables), directional derivatives of several kinds, and eventually Frechet and Gâteaux derivatives. These are, essentially, limits of difference quotients for vector-valued functions; the type of derivative depends on the topology employed in taking the limits. With respect to (b) we have the tangent structures for manifolds (tangent vectors, tangent spaces), and derivations (linear maps on transcendental extensions of a ground field, which obey the Leibniz rule, and vanish on the ground field). Both the derivative-as-limit and the derivative-as-linear-operator are well-known standard mathematical topics. We mention next some minor variants of these.

(A) Following van der Waerden [178] we can define derivatives for polynomials $P(x)$ by congruences: $P(x + h) - P(x) \equiv hP'(x)$ (mod h^2), this can be extended to rational functions and eventually to algebraic functions.

(B) In pattern differentiation it is possible to introduce certain limits as generalized derivatives which, roughly, bear the same relation to the usual derivatives as the Stjeltjes integrals do to the usual Riemann integrals. We explain this on the example of a real-valued "smooth" function $y = f(x)$ of a real variable; let C be its graph. A movable frame $F(p)$ is, to begin with, the set of all lines in the plane through the point p. The collection of lines is parametrized: Each line L is $L(m)$ where m is its slope. To find the derivative of f at the point $q = q(x_1, y_1)$ of C, we consider $F(q)$ and observe that small neighborhoods of q on C lie between some $L(m_1)$ and $L(m_2)$. If these neighborhoods shrink to q then m_1 and m_2 will have a common limit. This is, of course, the ordinary derivative $f'(x_1)$. However, we can take frames other than of straight lines, use other parametrizations and generally vary the procedure (e.g., we can use a several-parameter frame, say, a collection of parabolas, and try to define several consecutive derivatives at once, instead of one after another).

(C)   Under certain conditions the derivative, however defined, has a type of invariance. For instance, we have the Schwarzian derivative $S_x y$ of a function $y = f(x)$:

$$S_x y = \frac{y'''}{y'} - \frac{3}{2}\left(\frac{y''}{y'}\right)^2 \tag{15}$$

where primes denote ordinary differentiation. If a, b, c, d are constants with the determinant $ad - bc \neq 0$ we verify easily the following invariance property of S:

$$\text{if} \quad u = \frac{ay+b}{cy+d} \quad \text{then} \quad S_x y = S_x u. \tag{16}$$

This immediately suggests that the Schwarzian derivative has some possible projective significance. We note that since the general bilinear transformation involves three independent constants, anything like the Schwarzian S must involve derivatives of order up to at least three. The reader may try to find out whether (15) is given by the invariance requirement (16), and perhaps also to construct derivatives similar to the Schwarzian that are invariant under other transformations.

We finish this section with some concluding remarks on the possibility of a third approach to derivatives. This centers neither on limits nor on linear operators. Consequently, it may come as something of a surprise to those who do not *use* mathematics. It happens that the derivatives are used to solve equations numerically by means of procedures that are fast-converging. This speed of the convergence depends so essentially on the use of the derivative that, under certain conditions, we can invert the procedure and obtain a definition of the derivative almost by the sole requirement of fast convergence. We outline this on a very simple example. Let $f(x)$ be a real-valued function of the real variable x, defined where and as needed, with an isolated zero at $x = a$: $f(a) = 0$. Suppose that the zero a is found by our being given an initial approximation x_0 and then producing the successive approximations $x_1, x_2, \ldots$ from a correcting recursion

$$x_{n+1} = x_n - E, \qquad n = 0, 1, 2, \ldots. \tag{17}$$

The error $E = E(f, x_n)$ is a functional depending on f and on x_n. The whole procedure depends on E and we place now one condition: The procedure is stable in the sense that $E = 0$ if x_n is the true zero, a. The simplest way of achieving it is to build f as a factor into E. We have then $E = f(x_n)\phi(f, x_n)$ and (17) becomes

$$x_{n+1} = x_n - f(x_n)\phi(f, x_n). \tag{18}$$

Next we call the convergent sequence $x_0, x_1, x_2, \ldots \to a$, fast-convergent if

$$\frac{x_{n+1} - a}{x_n - a} \to 0 \quad \text{as} \quad n \to \infty.$$

This means essentially that the event $x_n \to a$ occurs just faster than exponentially; for with $x_n = b^n$ (where $|b| < 1$) we have $a = 0$ and

$$\frac{x_{n+1} - a}{x_n - a} = b.$$

Suppose now that the sequence $x_0, x_1, x_2, \ldots$ of successive approximations, produced by (18), is fast-convergent to a. Then, if f happens to be differentiable at a with $f'(a) \neq 0$, the functional ϕ is $1/f'$:

$$\lim_{n \to \infty} \frac{f(x_n)}{x_n - x_{n+1}} = f'(a). \tag{19}$$

We have

$$\frac{f(x_n)}{x_n - x_{n+1}} = \frac{f(x_n) - f(a)}{x_n - a} \frac{x_n - a}{x_n - x_{n+1}} \tag{20}$$

and, under the assumption of fast convergence, difference quotients are equivalent to quotients:

$$\frac{x_n - x_{n+1}}{x_n - a} = 1 - \frac{x_{n+1} - a}{x_n - a} \to 1$$

by the fast convergence. Therefore (19) follows from (20).

This is a mere outline of the subject. We should like to: (1) be assured that no fast convergence occurs without derivatives, (2) place perhaps some stronger conditions on the subject under which the *existence* of f' itself would follow from a sufficiently fast convergence, and (3) place the subject in a more general setting (e.g., in a Banach space, with Gâteaux derivatives instead of Newton's). For these matters see the work of R. S. Booth [19].

7. GENERALIZED FUNCTIONS

With sufficient degrees of smoothness we can set up differentiation and integration, getting the basic apparatus of analysis. But these two operations break down sometimes and what is distressing is that they break down in situations where they seem needed. It certainly would be useful to be able to differentiate a continuous function any number of times.

Note that this would require differentiation of some discontinuous functions as well since continuous functions may have discontinuous derivatives.

However, because of the scale (1) of the last section this cannot be, and the best we can hope for is an extension of the concept of a function to that of a generalized function. For the latter ones there is to be no such thing as the smoothness scale (1)—they are to be all of the class C^{∞}. We shall indicate first how the need for such an extension arose, not in mathematics out of some inner necessity, but in physics and electrical engineering. Then we shall sketch three alternative ways of introducing generalized functions.

(**a**) Our considerations will revolve around the so-called Dirac δ-function $\delta(x)$. This was apparently first introduced by the German physicist G. Kirchoff in 1882 [95, 96]. Briefly put, Kirchoff's interest was in the distinction between sharply localized phenomena and the diffuse ones. So, he introduced the function

$$K(x) = m\pi^{-1/2} e^{-m^2 x^2} \tag{1}$$

with the proviso that "m denotes a very great positive constant." The function $K(x)$ and its first two derivatives are plotted in Fig. 1a–c. We note that the curve of Fig. 1a is just the Gaussian curve with small bandwidth or standard deviation. Technically, bandwidth is the spacing from x_2 to x_1 for two values referring to the curve $y = f(x)$ which enclose

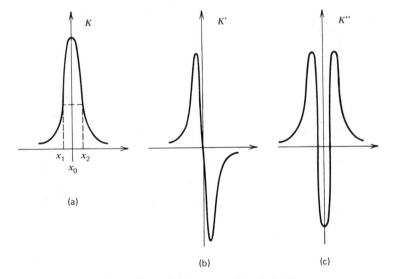

Fig. 1. Kirchoff's function and its derivatives.

a maximum x_0 between them, as in Fig. 1a, and are such that $f(x_1) = f(x_2) =$ fixed given fraction of $f(x_0)$. The term bandwidth is used in communication theory. The probabilistic counterpart is any measure of the tendency to cluster about a point, for example, the standard deviation.

As $m \to \infty$ in (1), $K(x) \to 0$ for every $x \neq 0$, but since $K(0) = m/\sqrt{\pi}$ we have $\lim\limits_{m \to \infty} K(0) = \infty$. Some ten years after Kirchoff's formulation, Oliver Heaviside, an English electrical engineer, was interested in a seemingly different sort of thing. He wanted to calculate the voltage and current surges occurring when a button was pressed or a relay closed. To describe the functioning of the button or the relay he introduced the unit impulse function

$$H(x) = 0 \qquad \text{if} \qquad x < 0,$$
$$= 1 \qquad \text{if} \qquad x \geq 0.$$

Of course, it may be objected that this $H(x)$ is not a realistic description and that a truer description of the state of affairs is some $H_1(x)$ that is also 0 for $x < 0$ and which rises continuously and very steeply to the value 1 over a small range $(0, \varepsilon)$ of x. But empirically such $H_1(x)$ would not be easy to determine, as compared with the brief and elegant $H(x)$. We note that the functions $H(x)$ and $H_1(x)$ concern the same sort of thing as Kirchoff's $K(x)$: sharpness versus diffusion. And attempting to differentiate $H(x)$ we run straight into the Dirac $\delta(x)$.

Considerably after Heaviside's time, P.A.M. Dirac, an English–Belgian physicist, was interested for various quantum-theoretical reasons in functions such as

$$\sum_{n=1}^{\infty} \phi_n(x)\phi_n(y) = D(x, y), \tag{2}$$

where $\phi_1, \phi_2, \ldots$ is a complete orthonormal function system, that is, an orthonormal basis in some function-space. Supposing that the domain of the functions ϕ_n is the whole real line $(-\infty, \infty)$, let $f(y)$ be sufficiently smooth, and let us attempt to evaluate

$$\int_{-\infty}^{\infty} f(y)D(x, y) \, dy. \tag{3}$$

We expand $f(y)$ into the series

$$f(y) = \sum_{k=1}^{\infty} a_k \phi_k(y)$$

and use this, together with (2), in (3). With some elements of orthonormality and justifying the necessary operations (or not) we get

$$\int_{-\infty}^{\infty} f(y)D(x, y)\,dy = \sum_{k=1}^{\infty} \left[a_k\phi_k(x) \sum_{n=1}^{\infty} \int_{-\infty}^{\infty} \phi_n(y)\phi_k(y)\,dy \right]$$

$$= \sum_{k=1}^{\infty} a_k\phi_k(x) = f(x).$$

Thus $D(x, y)$ may be dubbed "the operator of sharp localization," with the defining property of the Dirac δ-function:

$$\int_{-\infty}^{\infty} f(y)D(x, y)\,dy = f(x) \tag{4a}$$

so that $D(x, y) = \delta(x - y)$ and so the problem of Dirac is connected with those of Kirchoff and Heaviside. It is a sample of something often met in theoretical physics or applied mathematics: how to expand a "square" entity in terms of a basis of "round" objects. Examples are expansion of a plane wave in a series of spherical waves and the series for e^{ix} in terms of Bessel functions. Here of course the problem is how to represent the sharp localization operator in terms of a prescribed basis.

With formal manipulations which involve integration by parts, we find

$$\int_{-\infty}^{\infty} f(y)\frac{dD(y, a)}{dy}\,dy = f(y)D(y, a)\big|_{-\infty}^{\infty} - \int_{-\infty}^{\infty} f'(y)D(y, a)\,dy$$

$$= -f'(a)$$

and generally

$$\int_{-\infty}^{\infty} f(y)D^{(n)}(y, a)\,dy = (-1)^n f^{(n)}(a). \tag{4b}$$

One can get some rough idea of the "behavior" of those derivatives $D^{(n)}(x, a)$, or $\delta^{(n)}(x - a)$, from Fig. 1b and c where a is 0. The reader may wish to compare this with the behavior of $e^{-1/x}$, discussed in the last section.

It will be observed that (4) may require many derivatives of f as well as a suitable behavior "at infinity," hence the idea of using for f a C^{∞} function, to ensure all required derivatives, and one that vanishes outside a bounded interval. In technical parlance the last property is called being of compact support. Since it may not always be true that what replaces $D^{(n)}(y, a)$ in (4b) vanishes at infinity, we ensure the correct functioning of the integration by parts, by requiring that f should be a C^{∞}-function of compact support.

The graphs of Fig. 1, when idealized to the limit, explain perhaps why the δ-function is sometimes called the measure induced by the unit mass at the origin, while δ' is compared to the physical notions of a doublet or a dipole. We recall that in hydrodynamical terms a doublet is a limiting case of a source and a sink, of equal capacities, at a distance d. We let the capacity approach ∞ while $d \to 0$, in such a way that the product has a finite limit. The same thing can be done in electromagnetic theory with positive and negative charges, and we get then a dipole. Since, by its definition, a measure is not allowed to assume both values $\pm\infty$, it follows that while δ may be a measure, $\delta^{(n)}$ for $n \geq 1$ is not.

Next, we indicate very briefly the connection of δ-like functions with the operatorial or symbolic methods. Consider a very simple functional equation which is perhaps the simplest possible initial value problem:

$$y' = F(x), \qquad y(0) = a. \tag{5}$$

If we wish to solve it operatorially we write $y' = F$ as $Dy = F$ and we get $y = D^{-1}F$. The obvious interpretation of D^{-1} as

$$\int_0^x \cdots dx$$

would give us

$$y(x) = \int_0^x F(x) \, dx.$$

This is the correct solution of (5) but only if $a = 0$. The reader may wish to check the consequences of taking

$$D^{-1} \equiv \int_{x_0}^x \cdots dx.$$

The reason behind our difficulty is that only for functions vanishing at 0 do the operators D and D^{-1} commute: $DD^{-1} = D^{-1}D$ or written out at length,

$$\frac{d}{dx} \int_0^x \phi(x) \, dx = \int_0^x \phi'(x) \, dx.$$

If we still want to represent the solution y as D^{-1} applied to something, we must somehow provide for the building of the initial value right into the equation itself. This is done precisely by employing δ-functions: we replace (5) with

$$Dy = F(x) + a \, \delta(x) \tag{6}$$

and now we have the correct solution

$$y(x) = D^{-1}(F + a\,\delta(x)) = a + \int_0^x F(x)\,dx, \tag{7}$$

provided of course that $\int_0^x \delta(x)\,dx$ is taken as 1.

The solution (7) of (5) was based on the building of the initial conditions "directly" into the equation, by means of the δ-function. However, there is a very well known technique that achieves this "indirectly," the Laplace transform. Like other integral transforms, this depends principally on the conjugacy principle, illustrated in Fig. 2, as it is applied to solving functional equations. The idea is that under the transformation T certain "difficult" operations entering the functional equation F.E. are mapped onto simpler ones. While the F.E. may be difficult to solve directly, it gives rise under T to a simpler transformed equation T.E. This is solved and the remaining problem is how to go back and reconstruct the solution S.F.E. of the F.E. out of the solution S.T.E. of the T.E. Figure 2 shows graphically one of the reasons why transform methods are called "indirect."

The usefulness of the paradigm of Fig. 2 will be clear when we recall the use of conjugacy in groups and other algebraic systems, for matrices (under the name "similarity"), with iteration, and so on.

In Laplace transform the mapping T is

$$f(x) \rightarrow F(s) = \int_0^\infty e^{-sx} f(x)\,dx \tag{8}$$

and the inverse mapping T^{-1} is

$$F(s) \rightarrow f(x) = \frac{1}{2\pi i} \int_C e^{sx} F(s)\,ds. \tag{9}$$

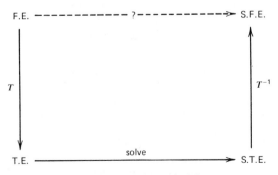

Fig. 2. Conjugacy principle.

Here C is a suitable path of integration in the complex plane, for instance, the straight line from "$c - i\infty$" to "$c + i\infty$" where c is real and sufficiently large. As is usual with any inversion technique, one comes to recognize certain inverse transforms from the direct ones. This is just as in calculus where we recognize the integral

$$\int \frac{dx}{1+x^2} = \text{arc tan } x$$

from the derivative

$$\frac{d \text{ arc tan } x}{dx} = \frac{1}{1+x^2}.$$

So, for instance, we have the inverse Laplace transform

$$T^{-1}s^{-k} = \frac{x^{k-1}}{T(k)} \qquad \text{since} \qquad \int_0^\infty e^{-sx}x^{k-1}\, dx = \Gamma(k)s^{-k}$$

and

$$T^{-1}\frac{s-a}{(s-a)^2+b^2} = e^{ax}\cos bx, \quad T^{-1}\frac{b}{(s-a)^2+b^2} = e^{ax}\sin bx$$

because

$$\int_0^\infty e^{-sx}e^{(a+ib)x}\, dx = \int_0^\infty e^{-x(s-a-ib)}\, dx = (s-a-ib)^{-1}$$

$$= \frac{s-a+ib}{(s-a)^2+b^2}, \text{ etc.}$$

A few such inverse formulas, together with the use of linearity, partial fractions, and the shift-theorem

$$\int_0^\infty e^{-sx}[e^{ax}f(x)]\, dx = F(s-a) \quad \text{if} \quad \int_0^\infty e^{-sx}f(x)\, dx = F(s),$$

will enable us to perform many simple inversions T^{-1} without the use of (9). If

$$\int_0^\infty e^{-sx}f(x)\, dx = F(s)$$

then integration by parts yields

$$\int_0^\infty e^{-sx}f'(x)\, dx = e^{-sx}f(x)\big|_0^\infty + s\int_0^\infty e^{-sx}f(x)\, dx$$

$$= -f(0) + sF(s). \tag{10}$$

Thereby we see how the use of Laplace transforms leads again to the absorption of the initial conditions into the (transformed) equation, just as it was done with the δ-functions before. From (10) we can recognize the following objection to the use of integral transforms: the simple initial value problem of the form (5)

$$y' = xe^{x^2}, \quad y(0) = \tfrac{1}{2},$$

with the obvious solution $y(x) = e^{x^2}/2$, cannot be attacked by Laplace transform since in (8) we have tacitly assumed the "right" behavior of $f(x)$ as $x \to \infty$ whereas

$$\int_0^\infty e^{-sx} x e^{x^2}\, dx$$

is a divergent integral.

As was remarked, a considerable part of the usefulness of integral transforms comes from their mapping of difficult operations, such as differentiation, onto simpler ones, such as multiplication by s. This is particularly well exemplified by the rather natural question of "multiplication": If f and g have Laplace (or other) transforms F and G so that (for example)

$$F(s) = \int_0^\infty e^{-sx} f(x)\, dx, \quad G(s) = \int_0^\infty e^{-sx} g(x)\, dx,$$

then what combination of f and g gets mapped onto the product $F(s)G(s)$? It is well known and easily verified that the combination in question is the (Laplace) convolution

$$\int_0^x f(y)g(x-y)\, dy$$

(and analogous ones for other integral transforms). We write this in the obvious terminology as

$$T(f * g) = Tf \cdot Tg \tag{11}$$

and we verify it by introducing new coordinates. For, with $u = y$ and $v = x - y$, we have

$$\int_0^\infty e^{-sx}\left(\int_0^x f(y)g(x-y)\, dy\right) dx = \int_0^\infty \int_0^x e^{-sx} f(y)g(x-y)\, dy\, dx$$

$$= \int_0^\infty \int_0^\infty e^{-su-sv} f(u)g(v)\, du\, dv = \int_0^\infty e^{-su} f(u)\, du \cdot \int_0^\infty e^{-sv} g(v)\, dv.$$

(b) We have now assembled the rudiments of the apparatus needed for our next task. This deals with the justification of the preceding. Mathematicians have long been bothered, to the point of boggling, by the successes of what seemed an irritating empty formalism of electrical engineering or physics. Due to much zeal, the subject of generalized, δ-like, functions has now been justified in decent rigor, and this in many ways, too. We shall sketch the outlines of three of those ways:

(A) distributions,
(B) equivalent sequences,
(C) fractional operatorial calculus.

It is obvious that any rigorous definition of $\delta(x)$ must rest on an extension of the function concept. Since there are many types of extensions in mathematics we shall collect, for purposes of later use as illustrations, a few examples. First, we have a number of what might be termed binary algebraic extension problems. Here a set X is to be extended to a wider set Y so that an operation D, which could only sometimes be carried out in X, can always be carried out in Y. Briefly, X is extended to a closure under D. We shall not be concerned with the question of the existence of unique, up to an isomorphism perhaps, minimal extension. When we speak of "binary" extensions we mean that Y is obtained from the set of ordered *pairs* of elements of X. However, y is usually not the Cartesian product of X with itself, but arises from it by taking quotients. This means that we have to prescribe which ordered pairs are to be regarded as equal, that is, can be identified; also, we must give definitions for operating on pairs.

For instance, X could be the natural numbers $0, 1, 2, \ldots$ and D could be subtraction. Y is obtained as the integers, in the form of the set of ordered pairs (n, m) of natural numbers, if we define

$$(n, m) = (p, s) \quad \text{if and only if} \quad n + s = m + p, \tag{12}$$

$$(n, m) + (p, s) = (n + p, m + s). \tag{13}$$

If multiplication is also required we add the recipe

$$(n, m)(p, s) = (np + ms, mp + ns). \tag{14}$$

Of course, all along we *mean* $n - m$ when we *speak* of (n, m) but we are not allowed to speak of negatives and minuses, and the extended system Y must be described within the subsystem X. This is entirely analogous to the bookkeeping methods: One is in debt when the positive amount of all debit entries exceeds the positive amount of all credit entries. As is well known, suitable changes in equations (12)–(14) will result in extending the integers X to rational numbers Y, and the real numbers X to the complex numbers Y.

On the other hand, the extension of rational numbers to the real numbers is different. Instead of being "binary" or "ternary" or ..., it is "infinitary," for clearly no n-tuples of rationals will reach to reals, with finite n. Next, this extension appears to be not so much algebraic as analytic, for we are concerned with limits rather than with operations such as sums or products. As is well known, real numbers can be defined as sequences of rational numbers in several ways: by means of Dedekind cuts or similar approximative techniques, as decimal or other expansions, as continued fractions, infinite products or series of certain types, and so on.

As a somewhat generalized version of this type of extension we have the standard theorem that a uniformly continuous function f on a set X of a metric space M has a unique continuous extension to the closure $\bar{X}$ of X in M. For it is only necessary to define

$$f\left(\lim_{n \to \infty} m_n\right) = \lim_{n \to \infty} f(m_n). \tag{15}$$

The interchange of f and lim is of course intimately connected with the uniform continuity as similar interchanges often are. Another way of describing the extension is to say that we consider first the fundamental, or Cauchy, or convergent, sequences $m_1, m_2, \ldots$, of elements in X. Then, we extend the domain of f by defining it at all limits of such sequences by means of (15).

As an application we can take X to be the set of all step-functions on $[0, 1]$. That is, we take all, and only those, functions that assume only a finite number of different values, each one on an interval or a finite union of intervals. As the distance we take the sup-norm:

$$\text{dist}(f, g) = \max_{0 \le x \le 1} |f(x) - g(x)|.$$

For each $f \in X$ we can define the obvious "area under the curve" which is a finite sum $A(f)$. Now we extend the function A to the closure of X and we get in this way the Riemann integral for all continuous functions on $[0, 1]$ and even for some others, too.

With one exception all the extensions we have mentioned have the common property of not interfering with any important structure on the set that is being extended. The sole exception is the concept of order: the reals connot be extended to complex numbers (or for that matter, extended at all) if we require the preservation of the linear order. For it is well known that the properties of order, together with the field axioms, will contradict all three possibilities: $1 = i$, $1 < i$, $i < 1$. We run into exactly the same trouble with structure when we extend the real numbers R either to quaternions or to $n \times n$ matrices over R. In both cases we must give up a part of the algebraic structure of R, namely the commutative law.

It will be similar with our extension of functions to generalized functions. We shall gain the important closure properties: Unlike the ordinary functions, the generalized ones are closed under differentiation, under certain types of limits, and also have certain useful completeness properties with respect to Fourier series. But we pay for this and we lose some essential structure of (ordinary) functions. In the words of G. Temple [167, 168] we lose "the numerical character of functionality." That is, *some* generalized functions have ordinary numerical values for all x just as *some* matrices commute and *some* differences of natural integers are natural integers. But other generalized functions, like $\delta(x)$, do not have this property of having a definite numerical value at each x. This may be something of a shock at first, just as the noncommutativity may be something of a shock, at first. It will be noticed that this "numerical character of functionality" is related to the sharp versus diffuse localization, as with Kirchoff's function, which opened our subject.

(A)   The theory of distributions starts with things like (4): evaluating the improper integrals

$$\int_{-\infty}^{\infty} f(y)\phi(y) \, dy. \tag{16}$$

The aim is to exploit this evaluation in reverse: $\phi(y)$ will be defined as the δ-function $\delta(y)$ by postulating that *no matter what f is*, the value of (16) is to be $f(0)$. This may perhaps explain the usual name "testing function" given to f's which are allowed to appear in (16). As was seen earlier, integration by parts in (4) or in (16) suggests the use of C^∞-functions of compact support as testing functions. How are we to formalize all this? First, we note that (16) is an inner product and we write it as $\langle f, \phi \rangle$. Next, we recall some elements of vector space theory. In particular, if V is a vector space then associated to V there is another vector space, called its dual, whose elements are the linear functionals on V. Here, for V we take the space of all C^∞-functions on $(-\infty, \infty)$ that vanish outside a bounded interval, with real or complex numbers as the scalars. The inner product $\langle f, \phi \rangle$ could also be written as $\phi(f)$. This exhibits somewhat more clearly the fact that ϕ's may be taken as linear functionals on V, and as such, they are the elements of the vector space dual to V.

Hence we have arrived at the starting point of the theory of generalized functions as distributions:

Generalized functions are members of vector spaces dual to the vector spaces of C^∞-functions with compact support.

(B) With the approach via equivalent sequences the definition of a generalized function is more nearly in line with other infinitary extensions, in particular, with the extension from the rational numbers to the reals. First, we observe that the entities we wish to have as generalized functions, $\delta(x)$ for instance, are better and better approximable by sequences of ordinary functions. For $\delta(x)$ we can exhibit many such sequences, with $m = 1, 2, \ldots$:

$$\{\pi^{-1/2} m e^{-m^2 x^2}\} \qquad \text{(Kirchoff functions (1)),}$$

$$\{r_m(x)\} \quad \text{where} \quad r_m(x) = m \quad \text{for } -1/2m \le x \le 1/2m$$
$$= 0 \quad \text{otherwise,}$$

(this may be called the symmetric rectangular approximation),

$$\{s_m(x)\} \quad \text{where} \quad s_m(x) = m \quad \text{for } 0 \le x \le 1/m$$
$$= 0 \quad \text{otherwise}$$

(the shifted rectangular approximation), corresponding triangular approximations, and so on. It is not necessary that each function of an approximating sequence should give 1 as the total area under the curve; we could modify the first sequence to $\pi^{-1/2}(m+1)e^{-m^2 x^2}$ or to $\pi^{-1/2}(m+\sqrt{m})e^{-m^2 x^2}$ but not to $\pi^{-1/2}(2me^{-m^2 x^2})$.

Second, from approximation by sequences we pass over to definition by approximating sequences: Generalized functions *are* certain sequences of ordinary functions. Note that it is entirely superfluous to insist that the generalized functions are *limits*.

Third, it is necessary to declare which sequences are to be regarded as equivalent, in the sense of defining the same generalized function. This last point is just as necessary here as equation (12) was necessary to tell us that, for instance, both $(1, 3)$ and $(4, 6)$ represent the same negative number, viz. -2.

Thus, three things need to be known for our definition: The collection L of ordinary functions which are admissible as members of the approximating–defining sequences, the condition C telling us which sequences of functions in L are to be taken, and the equivalence relation for these sequences identifying the sequences that define the same generalized function. We note that the condition C corresponds to convergence (or Cauchy property) for sequences.

The details can be arranged in a number of slightly different ways. For instance, Lighthill [101] proceeds as follows. First, out of the collection of C^∞-functions on $(-\infty, \infty)$ he singles out a subcollection that is close to all functions of compact support but bigger. Namely, he takes the functions all of whose derivatives fall off rapidly at infinity: All derivatives must decrease faster than any positive power of $|x|^{-1}$ for $|x|$ large. Call such functions rapidly falling off, or just rapid. Next, we prescribe the type of "convergence"—our condition C: A sequence $\{f_n(x)\}$ of rapid functions is called regular if

$$\lim_{n \to \infty} \int_{-\infty}^{\infty} f_n(x)F(x)\, dx \tag{17}$$

exists whenever F itself is rapid. Finally, two regular sequences, $\{f_n(x)\}$ and $\{g_n(x)\}$, are equivalent if

$$\lim_{n \to \infty} \int_{-\infty}^{\infty} f_n(x)F(x)\, dx = \lim_{n \to \infty} \int_{-\infty}^{\infty} g_n(x)F(x)\, dx$$

for any rapid function F. Now:

> *A generalized function is an equivalence class*
> *of regular sequences of rapid functions.*

The use of (17) builds an obvious bridge with the approach via theory of distributions which started from (16). (Another minor variant of approach to generalized functions uses what is called weak convergence. This relies on (17) again, though in a different guise and terminology. It may be recalled that $f_n \to f$ weakly if for every g $\langle f_n, g \rangle \to \langle f, g \rangle$—hence the connection with (17)).

(C)   The development of generalized functions by way of fractional operatorial calculus is due to J. G. Mikusinski [123]. Since the previous theory, that of equivalent sequences, goes through by an infinitary extension of ordinary functions, it may be surprising that one can also get at the generalized functions by a binary algebraic extension. The process parallels closely the extension of integers $\ldots, -1, 0, 1, 2, \ldots$ to rationals, or more generally, imbedding an integral domain in its field of quotients. Instead of integers we take the set T of continuous functions, with the usual addition, but with the convolution

$$f * g = \int_0^x f(y)g(x-y)\, dy$$

as multiplication. T is then a commutative ring. By the theorem of Titchmarsh [123] it is also free of zero-divisors, that is, it is an integral domain:

$$\text{if} \quad f * g = 0 \quad \text{then} \quad f = 0 \quad \text{or} \quad g = 0. \tag{18}$$

It is this crucial result that affords the possibility of embedding T in its field of quotients. Once we have divisions we may ask: Letting g be the function identically equal to 1, what f satisfies $f * g = g$? That is, when is

$$\int_0^x f(y) \, dy = 1$$

for all x? The answer is that f must be the generalized function $f(y) = \delta(y)$. Thus in this theory:

> *Generalized functions are fractions, that is,*
> *members of the quotient field of T.*

On account of its central position we give an outline of a proof of Titchmarsh's theorem (18). First, we observe the theorem of Lerch [123]: If f is continuous and

$$\int_0^a x^n f(x) \, dx = 0, \qquad n = 0, 1, \ldots, \tag{19}$$

then $f \equiv 0$ on $[0, a]$. This is a simple consequence of the Weierstrass approximation theorem: We approximate to f by a polynomial and then conclude from (19) that

$$\int_0^a f^2(x) \, dx = 0$$

so that $f = 0$.

Next, Titchmarsh's theorem (18) is proved for the special case of $f = g$ by a sequence of somewhat lengthy though quite elementary estimates, exploiting the functional equation

$$e^{nx} = e^{ny} e^{n(x-y)}$$

using double integrations, transformations of variables, and so on [123]. In the final step of the proof (18) with f, g is deduced from the special case of f, f. This technique is fairly often met. We suppose now that $f * g = 0$. With $f_1 = xf$, $g_1 = xg$ we check that

$$f_1 * g + f * g_1$$

is a multiple of $f * g$ so that it also vanishes. Hence, too,

$$(f * g_1) * (f_1 * g + f * g_1) = 0.$$

Applying the commutativity, associativity and distributivity of the convolution $*$, we rewrite the last equation as

$$(f * g) * (f_1 * g_1) + (f * g_1) * (f * g_1) = 0.$$

Since $f * g = 0$ by the hypothesis, it follows that the other term is 0. Since this other term is a square (under convolution $*$) by the special case of the theorem, which is already proved, $f * g_1 = 0$. The same reasoning, together with a simple induction, will show that

$$f * g = f * (xg) = f * (x^2 g) = \cdots = 0$$

or

$$\int_0^x y^n f(y) g(x-y)\, dy = 0, \qquad n = 0, 1, \ldots.$$

By the theorem of Lerch $f(y)g(x-y) = 0$ for all x, y. Hence f and g cannot both assume a value $\neq 0$ and so (18) is proved.

In concluding this section we remark briefly on the possibility of a quite different approach to generalized functions. This hinges on the connection of the subject with sharp and diffuse localization and in particular, on the running average in its role of a smoothing and so diffusing operation. We demand the invertibility of running average: At least for a certain class of functions f there must exist functions g such that

$$g = (S_a)^{-1} f, \quad \text{that is,} \quad f = S_a g. \tag{20}$$

For instance, we ask: To what function g must we apply the running average S_a in order to obtain the step-function

$$f(x) = (2a)^{-1} \qquad -a < x < a,$$
$$= 0 \qquad\qquad |x| > a, ?$$

The answer is: No such function g exists but a *generalized* function $\delta(x) = g(x)$ is thereby defined. However, we meet some difficulties when it comes to representing the derivatives of $\delta(x)$ as running averages. These can be managed if we use weighted running averages instead of the ordinary ones, but the matters are then neither so simple as with (20), nor so different from (16) and from the definition of generalized functions as distributions.

3

TOPICS IN COMBINATORICS, NUMBER THEORY, AND ALGEBRA

1. HILBERT MATRICES AND CAUCHY DETERMINANTS

One of the plagues that pursue numerical analysts are the ill-conditioned problems. In the special case of linear systems this comes down to ill-conditioned matrices. These are $n \times n$ nonsingular matrices M with very small determinants; it is therefore difficult to invert M numerically. One of the simplest and most often quoted examples is the Hilbert matrix

$$H_n = \left(\frac{1}{i+j-1} \right), \qquad i, j = 1, \ldots, n. \tag{1}$$

These matrices turn up in numerical work with equispaced data. Equivalently, we meet them in approximation work when the powers 1, x, $x^2, \ldots, x^n$ appear as a basis of a vector space. We show this on the classical simple example. Let $P(x)$ be a kth degree polynomial $a_0 + a_1 x + \cdots + a_k x^k$ and let the numerical data $(x_1, y_1), \ldots, (x_N, y_N)$ be fitted with a kth degree polynomial by the least squares fit ($N > k$). Then the polynomial $P(x)$ of best fit is easily shown to satisfy the equations

$$\overline{x^j y} = \overline{x^j P(x)}, \qquad j = 0, \ldots, k, \tag{2}$$

151

where, as usual, the bar denotes averaging:

$$\overline{f(x, y)} = \frac{1}{N} \sum_{i=1}^{N} f(x_i, y_i).$$

The matrix of the system (2) of linear equations for the coefficients of P is

$$\begin{pmatrix} 1 & \bar{x} & \cdots & \overline{x^k} \\ \bar{x} & \overline{x^2} & \cdots & \overline{x^{k+1}} \\ \cdots & \cdots & \cdots & \cdots \\ \overline{x^k} & \overline{x^{k+1}} & \cdots & \overline{x^{2k}} \end{pmatrix}.$$

If the abscissas $x_1, \ldots, x_N$ are nearly uniformly distributed on $[0, 1]$ then we can replace in the above the average $\overline{x^j}$ by the corresponding integral

$$\int_0^1 x^j \, dx = \frac{1}{j+1}$$

and we have then the Hilbert matrix H_{k+1}. The following simple example indicates some dangers of ill-conditioning: Consider the system

$$x + 0.50y + 0.33z = 1$$
$$0.50x + 0.33y + 0.25z = 0$$
$$0.33x + 0.25y + 0.20z = 0$$

corresponding to the Hilbert matrix H_3 with elements rounded off to two significant places. The exact system has the solution $z = 30$, $y = -36$, $x = 9$. Working with the truncated system and keeping two significant places throughout, we eliminate x getting

$$0.08y + 0.08z = -0.50$$
$$0.08y + 0.09z = -0.33$$

so that $0.01z = 0.17$ and $z = 17$, $y = -23$, $x = 7.0$. This exhibits the catastrophic loss of accuracy due to roundoffs in H_3.

We can evaluate $|H_n|$ as a special case of the Cauchy determinant

$$C_n = \left| \frac{1}{a_i + b_j} \right|, \qquad a_i + b_j \neq 0, \qquad i, j = 1, \ldots, n.$$

This is evaluated by the following sequence of elementary operations. Subtract the last row from each preceding one and remove two series of common factors, one from the rows and one from the columns. The remaining determinant is C_n but with the last row replaced by $1, 1, \ldots, 1$. On this determinant we repeat the same procedure but using this time the last column to subtract from each preceding one. After the removal of row and column factors we remain with the next smaller Cauchy determinant C_{n-1}. Hence we get by induction

$$C_n = \frac{\prod\limits_{i>j} [(a_i - a_j)(b_i - b_j)]}{\prod\limits_{i,j} (a_i + b_j)}. \tag{3}$$

As a special case we take $a_i = i - 1$, $i = 1, \ldots, n$, $b_j = j$, $j = 1, \ldots, n$, and we get

$$|H_n| = \frac{\left(\prod\limits_{j=1}^{n-1} j! \right)^4}{\prod\limits_{j=1}^{2n-1} j!}. \tag{4}$$

To find how fast $|H_n|$ falls off with growing n we form the quotient

$$\frac{|H_{n+1}|}{|H_n|} = \frac{(n!)^4}{(2n)!\,(2n+1)!}. \tag{5}$$

Using Stirling's formula we have

$$\frac{|H_{n+1}|}{|H_n|} \cong \frac{\pi}{2^{1+4n}}, \qquad n \text{ large}. \tag{6}$$

The approximation (6) turns out to be reasonably good, even for small n. For instance, the exact and the approximate ratios for $n = 1, 2, 3$ are

n	exact	approximate
1	$\dfrac{1}{12}$	$\dfrac{\pi}{32}$
2	$\dfrac{1}{180}$	$\dfrac{\pi}{512}$
3	$\dfrac{1}{2800}$	$\dfrac{\pi}{8192}$.

The ratio of the right-hand sides of (5) and (6) tends to 1 with n increasing, by Wallis' infinite product for π. Using (6) for $n = 1, 2, \ldots, m - 1$, and multiplying, we find the approximate relation

$$|H_m| \cong \frac{\pi^{m-1}}{2^{2m^2 - m + 1}},$$

which gives us an idea of the size of $|H_m|$. The reader may wish to develop a better estimate based on the quantities P_n and Q_n from the section on inequalities, in Chapter 2, where

$$Q_n = \prod_{j=1}^{n} j!, \qquad P_n Q_n = (n!)^{n+1}.$$

Hilbert matrix suggests the further problem of finding the worst of all nonsingular ill-conditioned matrices. We ask: Given an $n \times n$ matrix

$$D = \left(\frac{p_{ij}}{q_{ij}}\right), \qquad i, j = 1, 2, \ldots n,$$

in which p_{ij} and q_{ij} are integers subject to the conditions

$$1 \le p_{ij} \le p, \qquad 1 \le q_{ij} \le q \tag{7}$$

(or to similar ones), what is the minimum $d(n, p, q)$ of the absolute value of $|D|$, subject to the side-condition $|D| \ne 0$? In particular, we might wish to compare $d(n, 1, 2n-1)$ with $|H_n|$. Further, we may loosely frame the following sort of problem concerning unrealistic error-control: A computing machine evaluates an $n \times n$ determinant by some specific method; the entries are given rational numbers, before the computation these are converted into decimal form with N significant places; also, all computations are carried on with the same accuracy and the answer is presented in the same way; what is the smallest value of N which will allow us to recover the *exact* value, given only the *approximate* rounded-off one as got by the computer? By what sort of algorithm could such exact value be recovered? Presumably, the answers would use some bounds on numerators and denominators of the entries in the determinant of the same type as (7) above. The problem obviously extends to evaluations other than those of determinants.

2. MULTIPLICATIVE DIOPHANTINE EQUATIONS

By multiplicative Diophantine system we shall understand, loosely speaking, a system of equations to be solved in integers (usually rational ones but sometimes in integers of other number fields) in which no sums occur but only products. Methods for solving such systems have been given by E. T. Bell [12], [13], and M. Ward [180]. They are based on various extensions of Bell's principal lemma: The general solution of

$$xy = zw \tag{1}$$

is given by

$$
\begin{array}{lll}
x = ab & z = ac & \text{g.c.d. } (b, d) = 1. \\
y = cd & w = bd.
\end{array}
\tag{2}
$$

This is obtained by simple divisibility and g.c.d. arguments and holds in algebraic systems more general than rational (or other) integers. As an application we consider the problem of finding all the Pythagorean number-triples x, y, z; these are integers satisfying $x^2 + y^2 = z^2$. We write this in the multiplicative form

$$x \cdot x = (z + y)(z - y)$$

as in (1), and use (2) to get

$$
\begin{aligned}
x &= ab & z + y &= ac \\
x &= cd & z - y &= bd.
\end{aligned}
\tag{3}
$$

This leaves us with the intermediate consistency relation $ab = cd$ to be solved, again by the same method:

$$
\begin{aligned}
a &= mn & c &= mp \\
b &= pq & d &= nq.
\end{aligned}
$$

On substituting this into (3) one gets

$$x = mnpq, \qquad z + y = m^2 np, \qquad z - y = q^2 np.$$

Since n and p occur only as the combination np, we absorb p into n; that is, we put in effect $p = 1$ and have

$$x = mnq, \qquad z + y = m^2 n, \qquad z - y = q^2 n,$$

so that the solution is

$$x = mnq, \qquad z = \frac{n(m^2 + q^2)}{2}, \qquad y = \frac{n(m^2 - q^2)}{2}. \tag{3a}$$

This is not yet a satisfactory form on account of the denominator 2; we get rid of it by simple parity arguments. Either n must be even, or n is odd but then m, q are of like parity. In the second case we write $a + b = m$, $a - b = q$ so that

$$a = \frac{(m + q)}{2}, \qquad b = \frac{(m - q)}{2}$$

and a, b are integers. Replacing m and q by $a + b$ and $a - b$, we get the well-known solution

$$x = n(a^2 - b^2), \qquad y = 2nab, \qquad z = n(a^2 + b^2). \tag{4}$$

As n, a, b run independently over integers we get all the Pythagorean triples; the primitive ones are those obtained from (4) by putting $n = 1$ and observing the condition g.c.d. $(a, b) = 1$. It is easily checked that the other alternative, namely n in (3a) is even, leads to the same solution (4) (even though x and y have interchanged the roles).

Diophantus [45] was an Alexandrian Greek of third century A.D. who, unlike most Greek mathematicians, specialized in algebra and number theory rather than in geometry. P. Fermat (1601–1665), the French jurist and mathematician, wrote his number-theoretic theorems and conjectures in the margin of his copy of Bachet's edition of Diophantus (hence the famous Fermat conjecture that the Diophantine equation $x^n + y^n = z^n$ has no solution in integers x, y, z, none of them 0, if $n \geq 3$, and the accompanying note: "I have found a truly marvellous proof of this but the margin is too small to contain it"). E. T. Bell (1882–1954) was an American mathematician who taught at the California Institute of Technology; he worked principally in number theory and combinatorics and, under the pseudonym John Taine, wrote several science fiction novels and stories (devoted almost entirely to biological rather than mathematical topics).

Before going on we shall solve the equation $x^2 + y^2 = z^2$ again but in a different way. We rewrite it as

$$X\bar{X} = Z\bar{Z} \tag{5}$$

where the capitals denote Gaussian integers $a + bi$ and $\bar{X} = x - iy$ is the conjugate of $X = x + iy$. Solving (5) by the method of (1) and (2) we have

$$\begin{aligned} X &= AB & Z &= AC \\ \bar{X} &= CD & \bar{Z} &= \overline{BD}. \end{aligned}$$

This leads us with two intermediate consistency relations to satisfy:

$$AC = BD \qquad AB = \overline{CD}. \tag{6}$$

The first of these we solve and get

$$\begin{aligned} A &= MN & B &= MP \\ C &= PQ & D &= NQ \end{aligned}$$

so that

$$X = M^2 NP, \qquad \bar{X} = Q^2 NP, \qquad Z = MNPQ.$$

We absorb P into N as before by putting $P = 1$. The second consistency relation in (6) becomes

$$M^2 N = \bar{Q}^2 \bar{N};$$

this is easily shown to lead to $N = \bar{N}$, $M = \bar{Q}$ so that

$$X = nM^2, \qquad Z = nM\bar{M}.$$

Here the use of small n instead of N shows that the Gaussian integer N has imaginary part 0. Putting $M = a + ib$ and separating $X = x + iy$ into real and imaginary parts, we get (4) again.

It will be seen now that the multiplicative form might turn out to be rather more general than it appears at first. This is due to: (a) the fact that we are not restricted to rational integers, and (b) a suitable exploitation of norms as products of conjugates. This is, of course, quite commonplace in number theory. For instance, in the problems of repres⌐ntations of numbers by sums of two or four squares we meet the identities

$$(x^2 + y^2)(u^2 + v^2) = (xu - yv)^2 + (xv + yu)^2,$$
$$(x^2 + y^2 + z^2 + v^2)(a^2 + b^2 + c^2 + d^2)$$
$$= (ax - by - cz - dv)^2 + (ay + bx + cv - dz)^2$$
$$+ (az - bv + cx + dy)^2 + (av + bz - cy + dx)^2.$$

These contract to one simple statement:

$$N(AB) = N(A)N(B);$$

that is, the norm of a product is equal to the product of the norms, both for complex numbers and for quaternions. We conclude from the identities that an integer is a sum of two, or of four, squares of integers as soon as every prime dividing it is such a sum. The problem of representing a number as a sum of three squares is considerably harder; the algebraic part of the reason for it is that the sum of three squares is not a norm.

As a further application of the multiplicative method let us find the numbers which are simultaneously sums of two squares and also of a square and twice a square. That is, we consider the Diophantine equation

$$x^2 + y^2 = u^2 + 2v^2$$

which may be written as

$$y^2 - 2v^2 = u^2 - x^2.$$

Working in the ring I of integers of the quadratic field $R(\sqrt{2})$ we write the above equation as

$$X\bar{X} = UV \tag{7}$$

where

$$X = y + \sqrt{2}\, v, \qquad \bar{X} = y - \sqrt{2}\, v, \qquad U = u - x, \qquad V = u + x,$$

and the capitals stand for integers in I while small letters stand for rational integers. Solving (7) by the multiplicative method we get

$$X = AB, \qquad \bar{X} = CD, \qquad U = AC, \qquad V = BD$$

with three intermediate consistency relations

$$AB = \overline{CD}, \qquad AC = \overline{AC}, \qquad BD = \overline{BD}. \qquad (8)$$

The first one is solved and we get

$$A = MN, \qquad B = PQ, \qquad \bar{C} = MP, \qquad \bar{D} = NQ$$

and now both the second and the third equations of (8) become just

$$N\bar{P} = \bar{N}P. \qquad (9)$$

This is the "second-order" intermediate consistency relation, still of the multiplicative form. We solve it and get

$$N = aT, \qquad P = dT, \qquad (10)$$

for instance, by writing (9) out at length:

$$(n_1 + \sqrt{2}\, n_2)(p_1 - \sqrt{2}\, p_2) = (n_1 - \sqrt{2}\, n_2)(p_1 + \sqrt{2}\, p_2)$$

and checking that we have $n_1 p_2 = n_2 p_1$. This we solve by our method again, eventually getting (10). Collecting all variables together we have

$$X = adMQT^2, \qquad U = adM\bar{M}T\bar{T}, \qquad V = adQ\bar{Q}T\bar{T}.$$

Next, we practice economy on parameters observing that d absorbs into a (i.e., $d = 1$, without loss of generality), and the three parameters M, Q, T reduce to two only: MT, QT (i.e., $T = 1$, w.l.o.g.).

In effect, we have

$$X = aMQ, \qquad U = aM\bar{M}, \qquad V = aQ\bar{Q}.$$

By separating into "real" and "imaginary" components

$$y = a(m_1 q_1 + 2m_2 q_2), \qquad v = a(m_2 q_1 + m_1 q_2),$$
$$u = a\left(\frac{q_1^2 + m_1^2}{2} - q_2^2 - m_2^2\right), \qquad x = a\left(\frac{q_1^2 - m_1^2}{2} - q_2^2 + m_2^2\right).$$

We get rid of the denominators 2 as before, by a parity agrument: q_1 and m_1 are of like parity so that

$$q_1 = b + c, \qquad m_1 = b - c.$$

Writing at the same time q and m for q_2 and m_2 we get the solution

$$y = a(b^2 - c^2 + 2mq), \qquad v = am(b + c) + aq(b - c),$$
$$u = a(b^2 + c^2 - q^2 - m^2), \qquad x = a(2bc - q^2 + m^2).$$

Certain cubic diophantine equations may be somewhat similarly handled with the identities such as these:

$$x^3 + y^3 + z^3 - 3xyz = (x+y+z)(x+wy+w^2z)(x+w^2y+wz),$$
$$x^3 + y^3 = (x+y)(x+wy)(x+w^2y),$$

and so on, where $w = e^{2\pi i/3}$. We work then in the ring of integers of the imaginary quadratic field $R(w)$. Here it becomes necessary to solve such equivalents of (1) as

$$XYZ = UVW \qquad \text{or} \qquad XYZ = UV.$$

The solutions here (as also in generalizations of the above equations) arise from rectangular arrays by transposition:

$$X = ABC, \qquad U = ADG$$
$$Y = DEF, \qquad V = BEH$$
$$Z = GHT, \qquad W = CFT$$

or

$$X = AB, \qquad U = ACE,$$
$$Y = CD, \qquad V = BDF,$$
$$Z = EF,$$

as can be shown by repeated applications of (1) and (2). For examples of use of Bell's technique in solving cubic equations see Rosenthall [147, 148]. We meet another generalization of (1) to which the multiplicative technique applies: chains of multiplicative equations. A very simple special case occurs in the vibrations of a rectangular membrane; if we want to look for multiple eigenvalues we are led to solving in integers equations like

$$m_1^2 + n_1^2 = m_2^2 + n_2^2 = \cdots = m_s^2 + n_s^2.$$

However, in this case some of the consistency equations are no longer multiplicative but are of the form

$$\sum_{i=1}^{n} x_i y_i = 0.$$

These can be handled as shown in [161], or we can observe that in the vector form the above relation is $x \cdot y = 0$ and solutions are obtained by taking $x = a$, $y = Sa^T$; here a is an integral n-vector, and S an integral $n \times n$ skew-symmetric matrix. For the general theory of multiplicative systems and chains, including such problems as irredundancy of representations, number of arbitrary parameters necessary to get all solutions, and so on, see the references [12, 180] mentioned earlier, and also [161]. Since many intermediate consistency relations have to be solved, the amount of work grows very fast with the size of the problem. On the other hand, the computations appear to lend themselves to a partial mechanization, if not to a complete one.

3. GAUSSIAN BINOMIAL COEFFICIENTS AND OTHER q-ANALOGS

Let us define

$$[a] = \frac{1 - q^a}{1 - q} = 1 + q + q^2 + \cdots + q^{a-1}$$

$$[n]! = [1][2] \cdots [n] = (1 - q)^{-n} \prod_{j=1}^{n} (1 - q^j).$$

Then as $q \to 1$ we have

$$[a] \to a, \qquad [n]! \to n!.$$

With these definitions we obtain the so-called q-generalizations of ordinary concepts. For instance, since the binomial coefficient is given by

$$\binom{n}{k} = \frac{n!}{k!\,(n-k)!},$$

we introduce the q-binomial, or Gaussian binomial, coefficient by

$$\begin{bmatrix} n \\ k \end{bmatrix} = \frac{[n]!}{[k]!\,[n-k]!}$$

getting

$$\begin{bmatrix} n \\ k \end{bmatrix} = \frac{1 - q^n}{1 - q}\,\frac{1 - q^{n-1}}{1 - q^2} \cdots \frac{1 - q^{n-k+1}}{1 - q^k};$$

these hold for $0 < k \le n$, if $k = 0$ we define $\begin{bmatrix} n \\ k \end{bmatrix} = 0$. The quantity q will have a dual role: In some analytical circumstances it will be a variable satisfying $|q| < 1$, but in some algebraic and combinatorial contexts it will be an integer which is prime—more precisely, the characteristic of the ground field. For instance, just as the binomial coefficient $\binom{n}{k}$ counts the number of k-element subsets of an n-element set, so it will turn out that the q-binomial coefficient $\begin{bmatrix} n \\ k \end{bmatrix}$ counts the number of k-dimensional subspaces of an n-dimensional vector space over the finite field $GF(q)$.

Among further analogies between the Gaussian binomials and the ordinary ones we have

$$\prod_{k=1}^{n}(1+q^{k-1}x) = \sum_{k=0}^{n} \begin{bmatrix} n \\ k \end{bmatrix} q^{k(k-1)/2} x^k;$$ (1)

this becomes the ordinary binomial theorem when we let $q \to 1$. In general, the "q-analogs" tend to the "classical results" as $q \to 1$, somewhat as in physics, according to Bohr's correspondence principle, the quantum-physical results are asymptotically equal to the classical ones in the limit of high quantum numbers (or small wavelength). A proof of (1) can be given by employing the device of "telescoping mirroring." This name is with reference to telescoping cancellation and coincidence of vol. 1, pp. 81–94, and the device itself is fairly frequently used in developing identities concerning products, continued fractions, and so on. We start with $f(x)$ which is the L.H.S. of the identity, and replace x by something else, say X, chosen so that there is an almost cyclic repetition of factors between $f(x)$ and $f(X)$; we then make up the missing factors obtaining a functional equation of the type

$$A(x, X)f(x) = B(x, X)f(X)$$

and we use this functional equation to develop a recurrence relation. Here, for instance, we take $f(x)$ to be the L.H.S. of (1) and we let $X = qx$, getting the functional equation

$$(1+x)f(qx) = (1+q^n x)f(x).$$

We use this in the expansion

$$f(x) = \sum_{k=0}^{n} A_k x^k$$

where $A_k = A_k(q)$, and we get the recursion

$$(1-q^k)A_k = q^{k-1}(1-q^{n-k+1})A_{k-1}.$$

From this, and from $A_0 = 1$, we get A_k and we prove the identity (1). With this as a guide the reader may wish to obtain the q-analog

$$[x][x+1] \cdots [x+n-1] = \sum_{j=1}^{n} s_{n-j}(q)[x]^j$$

of Stirling numbers defined by

$$x(x+1) \cdots (x+n-1) = \sum_{j=1}^{n} s_{n-j} x^j.$$

Similar reasoning may be used to prove certain identities of Euler from the theory of partitions:

$$\prod_{j=0}^{\infty}(1+x^{2j+1})=1+\sum_{j=1}^{\infty}\frac{x^{j^2}}{\prod_{k=1}^{j}(1-x^{2k})},\qquad(2)$$

$$\prod_{j=1}^{\infty}(1+x^{2j})=1+\sum_{j=1}^{\infty}\frac{x^{j(j+1)}}{\prod_{k=1}^{j}(1-x^{2k})}.\qquad(3)$$

We introduce the function

$$F(a,x)=\prod_{j=0}^{\infty}(1+ax^{2j+1})=1+\sum_{j=1}^{\infty}c_j(x)a^j\qquad(4)$$

and get, by telescoping mirroring, the functional equation

$$F(a,x)=(1+ax)F(ax^2,x).$$

By comparing the coefficients of powers of x in the above we have the recursion

$$c_j=\frac{x^{2j-1}}{1-x^{2j}}c_{j-1}$$

which allows us to find c_j; hence (4) yields

$$\prod_{j=0}^{\infty}(1+ax^{2j+1})=1+\sum_{j=1}^{\infty}\frac{x^{j^2}}{\prod_{k=1}^{j}(1-x^{2k})}a^j\qquad(5)$$

and in particular, putting first $a=1$ and then $a=x$ we get (2) and (3). The identities (2) and (3), as well as (5), have simple combinatorial interpretations in terms of partitions and their Ferrers graphs (see vol. 1, pp. 222–223). So, for instance, the L.H.S. of (2) enumerates the partitions of n into parts which are odd and unequal, while the R.H.S. enumerates the self-conjugate partitions of n (= the ones whose Ferrers graph has the line with $-45°$ slope as axis of symmetry). Hence the number of partitions of n into odd distinct parts equals the number of self-conjugate partitions. Similar interpretations obtain for equations (3) and (5) [67]. Once we have these interpretations it is possible to reverse the procedure and offer *combinatorial* proofs of *analytic* identities, rather than combinatorial *interpretations* of analytically *proved* identities. For instance, with partitions into odd unequal parts we have the Ferrers graph as in Fig. 1a, which is the usual Ferrers graph modified so as to have a vertical symmetry axis. We now tilt the axis to the inclination of Fig. 1b, and bend

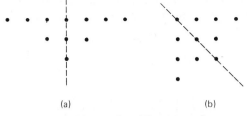

(a)   (b)

Fig. 1. Symmetries of Ferrers graphs.

the two sides, thus transforming the partition into odd unequal parts of
Fig. 1a into the self-conjugate partition of Fig. 1b.

A somewhat more involved application of the telescoping mirroring will
prove analytically the Jacobi identity

$$\prod_{n=1}^{\infty} [(1-x^{2n})(1+x^{2n-1}q^2)(1+x^{2n-1}q^{-2})] = \sum_{k=-\infty}^{\infty} q^{2k}x^{k^2}; \qquad (6)$$

we outline the proof. Starting with

$$f_m(x, q) = \prod_{n=1}^{m} [(1+x^{2n-1}q^2)(1+x^{2n-1}q^{-2})]$$

we have by the symmetry expressed in $f_m(x, q) = f_m(x, q^{-1})$ the relation

$$f_m(x, q) = \sum_{j=0}^{m} a_j(x)(q^{2j} + q^{-2j}) \qquad (7)$$

where the last coefficient $a_m(x)$ is easily found:

$$a_m(x) = x^{1+3+\cdots+(2m-1)} = x^{m^2}. \qquad (8)$$

Next, with some care for factors gained and factors lost, the telescoping
mirroring gives us the functional equation

$$(xq^2 + x^{2m})f_m(x, xq) = (1 + x^{2m+1}q^2)f_m(x, q).$$

From this, comparing coefficients of powers of q we get a recurrence for
$a_j(x)$, and then, employing (8), we obtain $a_j(x)$'s themselves. Now (7)
becomes of the form

$$f_m(x, q) \prod_{j=1}^{m} (1 - x^{2j}) = \text{R.H.S.}$$

and on letting $m \to \infty$ the above becomes (6). Putting in (6) $x = t^{3/2}$ and
$q^2 = -t^{1/2}$ we get the Euler identity (cf. vol. 1, p. 221)

$$\prod_{n=1}^{\infty} (1 - t^n) = \sum_{n=-\infty}^{\infty} (-1)^n t^{(3n^2+n)/2}.$$

Returning now to the Gaussian binomial coefficients we verify the following analogs of binomial-coefficient identities:

$$\begin{bmatrix} n \\ k \end{bmatrix} = \begin{bmatrix} n \\ n-k \end{bmatrix}, \qquad \begin{bmatrix} n+1 \\ k \end{bmatrix} = \begin{bmatrix} n \\ k \end{bmatrix} + \begin{bmatrix} n \\ k-1 \end{bmatrix} q^{n-k+1}$$

and

$$\begin{bmatrix} n \\ 0 \end{bmatrix} - \begin{bmatrix} n \\ 1 \end{bmatrix} + \cdots + (-1)^n \begin{bmatrix} n \\ n \end{bmatrix} = 0$$

if n is odd (the reader may wish to evaluate the L.H.S. when n is even). Somewhat in analogy to the fact that

$$\binom{n}{k} = \frac{n(n-1)\cdots(n-k+1)}{1 \cdot 2 \cdot \cdots \cdot k}$$

is an integer rather than a fraction, we have the fact that the Gaussian binomial coefficient, originally defined as a rational function in q, is really a polynomial in q. The integral nature of the binomial coefficient $\binom{n}{k}$ is easily proved by attaching a meaning and interpreting it as the number of ways of doing something. Something similar may be done with the Gaussian binomial coefficient: We let

$$\begin{bmatrix} n \\ k \end{bmatrix} = \sum_{j=0}^{k(n-k)} c(n, k, j) q^j$$

and we find that the coefficient $c(n, k, j)$ has the following combinatorial interpretation due to Polya [137]. It is the number of paths from $(0, 0)$ to $(k, n-k)$ such that the area under the trajectory of the path and above the x-axis is j; by a path we understand here a sequence of unit displacements, each one being up or to the right. On this it may be verified again that $\begin{bmatrix} n \\ k \end{bmatrix}$ becomes $\binom{n}{k}$ as $q \to 1$ since the *total* number of paths from $(0, 0)$ to $(k, n-k)$, irrespective of the area condition, is plainly the number of ways of choosing the k objects (up displacements) out of the total of n, namely $\binom{n}{k}$. This integralness of the ordinary and of the Gaussian binomial coefficient suggests an occassionally useful method of argument: to prove that a quotient of integers (such as $n(n-1)\cdots(n-k+1)/1 \cdot 2 \cdots k$) or of polynomials over integers (such as the Gaussian coefficient $\begin{bmatrix} n \\ k \end{bmatrix}$ regarded as a function of q) is really integral itself, we attach a combinatorial meaning to the quotient from which it follows that the quotient is the number of ways of doing something, and so it is integral.

Let $V_n(q)$ be an n-dimensional vector space over the finite field $GF(q)$. We observe now the analogy between $\begin{bmatrix} n \\ k \end{bmatrix}$ as the number of k-dimensional subspaces of $V_n(q)$, and $\binom{n}{k}$ as the number of k-sized subsets of an n-set. So, for example,

$$\binom{n}{k} = \text{No. of sequences of } k \text{ distinct elements}$$

out of $1, 2, \ldots, n$/same with $n = k$ \hfill (9)

on the one hand, and on the other

$$\begin{bmatrix} n \\ k \end{bmatrix} = \text{No. of sequences of } k \text{ independent}$$

vectors in $\cdot V_n(q)$/same with $n = k$. \hfill (10)

Here we get the numerator as follows: The first vector can be selected in $q^n - 1$ ways for we can take any nonzero element of $V_n(q)$. The vector we have selected generates q vectors under multiplication by the elements of $GF(q)$ and so for the second vector we have $q^n - q$ possibilities. Similarly, the two vectors we have selected so far, say v_1 and v_2, generate q^2 vectors under independent multiplication by elements of $GF(q)$: $xv_1 + yv_2$, $x, y \in GF(q)$, so the third vector can be picked in $q^n - q^2$ ways, and so on. This is in analogy to the numerator $n(n-1) \cdots (n-k+1)$ of (9), and so we have for the numerator of (10)

$$\prod_{j=0}^{k-1} (q^n - q^j).$$

The denominator of (10) is analogous to that of (9): We use the fact that

$$k! = [n(n-1) \cdots (n-k+1)]_{n=k}$$

or that there are just $k!$ ways of ordering k distinct objects. Thus the denominator of (10) turns out to be

$$\prod_{j=0}^{k-1} (q^k - q^j)$$

and so we recover the original expression for $\begin{bmatrix} n \\ k \end{bmatrix}$.

We finish this section with a brief mention of the related matter of certain so-called q-functional equations. These turn up principally in number-theoretic contexts in connexion with partitions and Lambert series, but also with elliptic theta functions and the q-hypergeometric series (where q-analogs of factorials replace the ordinary ones [84]). First, we have the simple cases of q-difference equations

$$f(x) - f(qx) = \phi(x), \qquad \frac{F(x)}{F(qx)} = \psi(x), \tag{11}$$

where $\phi(0) = 0$, $\psi(0) = 1$ and both are sufficiently regular for small x (for instance, admit Lipschitz constants). We use the additive and multiplicative telescoping cancellation: writing in (11) x, xq, $xq^2, \ldots$ for x and adding or multiplying, we get the solutions

$$f(x) = \sum_{n=0}^{\infty} \phi(xq^n), \qquad F(x) = \prod_{n=0}^{\infty} \psi(xq^n).$$

For instance, if $\phi(x) = x/(1-x)$ we get

$$f(x) = \sum_{n=0}^{\infty} \frac{xq^n}{1 - xq^n}$$

and more generally, if

$$\phi(x) = \sum_{j=1}^{\infty} a_j x^j$$

is a convergent power series then

$$f(x) = \sum_{n=1}^{\infty} \frac{a_n x^n}{1 - q^n}$$

so that corresponding Lambert series is

$$f(q) = \sum_{n=1}^{\infty} \frac{a_n q^n}{1 - q^n}.$$

For the multiplicative case in (11), with $\psi(x) = e^x$ we get

$$F(x) = e^{x/(1-q)},$$

with $\psi(x) = 1 + x$ we have

$$F(x) = \prod_{n=0}^{\infty} (1 + xq^n) = 1 + \sum_{n=1}^{\infty} \frac{q^{n(n-1)/2}}{\prod\limits_{j=1}^{n} (1 - q^j)} x^n$$

and with $\psi(x) = \cos x$ we obtain

$$F(x) = \prod_{n=0}^{\infty} \cos xq^n. \tag{12}$$

This function occurs in the probability theory and in the theory of Fourier series. For instance, let $|q| < 1$ and consider the "random" geometric series

$$S(q) = \pm 1 \pm q \pm q^2 \pm q^3 \pm \cdots,$$

where each sign is $+$ or $-$ with probabilities $1/2$, independent of other signs. $S(q)$ could be related to a symmetric random walk on a line, with geometrically decreasing step-lengths. To obtain the distribution of values of $S(q)$ we use characteristic functions and we recall that for sums of independent random variables the characteristic functions just multiply. Here the nth random variable has the values q^n and $-q^n$ with the probabilities $1/2$, $n = 0, 1, 2, \ldots$; hence the characteristic function of $S(q)$ is (12). Instead of the series $S(q)$ we could consider

$$T(q) = \begin{matrix} 1 \\ 0 \\ -1 \end{matrix} + \begin{matrix} q \\ 0 \\ -q \end{matrix} + \begin{matrix} q^2 \\ 0 \\ -q^2 \end{matrix} + \cdots,$$

where each term has the three possible values with probabilities $1/3$, independent of other terms. Then the characteristic function of $T(q)$ is

$$G(x) = \prod_{n=0}^{\infty} \left[1 - \frac{4}{3} \sin^2 \frac{xq^n}{2} \right]. \tag{13}$$

Recalling the duplication and triplication formulas of vol. 1, p. 91 we find that for $q = 1/2$

$$F(x) = \frac{\sin 2x}{2x}$$

while for $q = 1/3$

$$G(x) = \frac{\sin 3x/2}{3x/2}.$$

The reader may wish to obtain similar formulas for other multiple angles and to interpret them in terms of uniform distribution of the series like $S(1/2)$, $T(1/3)$, and so on.

The following geometrical considerations might help with visualizing infinite products like (12). Let

$$3 \le a(1) \le a(2) \le \cdots;$$

starting with a unit circle inscribe into it the regular $a(1)$-gon, a circle into that, a regular $a(2)$-gon into that, then a circle again, and so on. Now the limit of the radius of the nth circle is

$$P = \prod_{n=1}^{\infty} \cos \frac{\pi}{a(n)}.$$

In particular, taking $a(n) = b^n$ where b is an integer ≥ 3, we obtain a simple interpretation for (12) with $x = \pi$.

To indicate a connection with partitions we consider the q-functional equation of Mahler [106]:

$$\frac{f(x+a) - f(x)}{a} = f(qx), \qquad 0 < q < 1. \tag{13a}$$

In the limiting case of $a \to 0$ this becomes

$$f'(x) = f(qx).$$

Supposing that $f(0) = 1$ and computing the derivatives at 0 we get the McLaurin series

$$f(x) = \sum_{n=0}^{\infty} \frac{q^{n(n-1)/2} x^n}{n!}$$

(which is one of the q-analogs of e^x). For general $a \neq 0$ Mahler shows that if x is large and n is given by

$$q^{n-1} n \leq x < q^n (n+1),$$

then under some mild conditions

$$f(x) = \frac{q^{n(n-1)/2} x^n}{n!} O(1). \tag{14}$$

Now, let $r \geq 2$ be an integer and put

$$\prod_{n=0}^{\infty} (1 - x^{r^n})^{-1} = \prod_{n=0}^{\infty} \sum_{k=0}^{\infty} x^{kr^n} = \sum_{n=0}^{\infty} C_n x^n.$$

Then by equating the coefficients of like powers C_n is found to be the number of representations

$$n = n_0 + n_1 r + n_2 r^2 + \cdots$$

with $n_0, n_1, \ldots$ nonnegative integers. If $f(x)$ is defined by

$$f(x) = C_{nr}, \qquad n \leq x < n+1,$$

then

$$C_{nr} = C_{nr+1} = \cdots = C_{nr+r-1}, \qquad C_{nr} = C_{(n-1)r} + C_{r[n/r]}$$

so that $f(x)$ satisfies the functional equation

$$\frac{f(x-1) - f(x)}{-1} = f\left(\frac{x}{r}\right);$$

this is of Mahler's form (13a) with $a = -1$ and $q = 1/r$. From (14) we derive now the asymptotic estimate

$$\log C_n \sim \frac{\log^2 n}{2 \log r}, \qquad n \text{ large.}$$

4. SOME INFORMAL PRINCIPLES OF COMBINATORICS

A formal, or even informal, definition of the subject of combinatorics needn't concern us. Although it deals with the discrete, the enumerative, and the structural, it nevertheless uses continuous methods, it does more than just count (viz., it orders and it imposes structure), and it is not algebra. It exhibits rather well the importance of a small number of "principles":

(1) principle of $\sum$ and $\prod$,
(2) $p:q$ correspondence,
(3) generating function,
(4) structuration,
(5) inclusion–exclusion,
(6) counting one thing in two different ways,
(7) recursion,
(8) reification,
(9) symmetry removal.

Their applicability extends far beyond combinatorics, to number theory and analysis, and to some extent to most of mathematics. So, we recognize the importance of principles (1), (3), (5) in probability, of (6)—in the form of Cavalieri's principle or Fubini's theorem—in measure theory, of (9) in most of algebraic structure theories (subgroups, quotients, ideals, etc). Our simple illustrative examples will be from combinatorics as well as other fields.

Very imprecisely stated, principle (1) asserts that if A can be done in a ways and B in b ways then A or (exclusive or!) B can be done in $a+b$ ways, while A and B can be done in ab ways. Precise statements are obtained using $|A|$ to denote the size of the set A:

$$\sum: |A \cup B| = |A| + |B| \qquad \text{if } A \cap B = \varnothing,$$
$$\prod: |A \times B| = |A|\,|B|.$$

For $p = q = 1$, principle (2) can be expressed thus: If the size $|X|$ of some X is wanted but is hard to get at, set up a $1:1$ correspondence between X and some suitable Y, and count Y instead. The $1:1$ correspondence extends to the $p:q$ case.

As an illustration of principles (1) and (2) we consider the following problem. Given a convex n-gon P in the plane, we draw all diagonals (including the sides of P) and suppose that no three intersect inside P. What is the number N of distinct triangles which appear? First, we use (1)—$\sum$ to get

$$N = N_0 + N_1 + N_2 + N_3$$

where N_i is the number of triangles with exactly i vertices on the boundary of P (i.e., from among the n vertices of P). Next, we use principle (2) several times and establish $p:q$ correspondences between N_i and certain other sets which are simpler to count. Using Fig. 1a and b, we find for N_2 a 1:4 correspondence between quadrilaterals like $ABCD$ and our triangles (of the type ABX), and for N_1 a 1:5 correspondence between pentagons like $ABCDE$ and our triangles (of the type AXY). In both the cases of N_0 and of N_3 the correspondence is 1:1. To sum up: There is a $p(i):q(i)$ correspondence between combinations of $(6-i)$-tuples out of n vertices, and the numbers of N_i; $p(i) = 1$ while $q(0) = q(3) = 1$, $q(1) = 5$, $q(2) = 4$. Hence

$$N = \binom{n}{3} + 4\binom{n}{4} + 5\binom{n}{5} + \binom{n}{6}$$

supposing that $n \geq 6$ (otherwise we get automatic cut-off: $\binom{k}{b} = 0$ if $k < b$). We mention briefly an alternative way of finding numbers like N above, of highly "brute force" variety but occassionally effective. To compute the number $N(n)$ depending on the parameter n we might show first that $N(n)$ is a polynomial and then compute by explicit enumeration sufficiently many values $N(1)$, $N(2)$, ... to enable us to calculate all the coefficients of the polynomial. This is often effective in calculating the numbers of partitioning of certain regions by n lines or n circles.

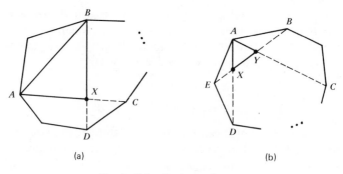

(a)   (b)

Fig. 1. Triangles in a polygon.

A simple and effective use of principle (2) is Euler's method of proving that the number of k-combinations out of n distinct objects, *repetitions allowed*, is $\binom{n+k-1}{k}$. Euler shows that there is a 1:1 correspondence between k-combinations out of n with repetition, and k-combinations out of $n+k-1$ without repetition. The correspondence is thus: a k-combination out of n objects, with repetitions, is a sequence $i_1, i_2, \ldots, i_k$ of k integers, all between 1 and n, and nondecreasing; if we said "increasing" we would have had a k-combination *without* repetitions; augmenting i_j to $i_j + j - 1$ we get the desired 1:1 correspondence. Another, rather more sophisticated, use of principle (2) was provided by Prüfer's proof of Cayley's formula for the number of labeled trees on n vertices (vol. 1, p. 162).

Principle (3), of generating function, is that a sequence $a_0, a_1, a_2, \ldots$ or $a_1, a_2, \ldots$, can sometimes be more easily and usefully handled as a single compound object, the function:

$$f_1(x) = \sum_{n=0}^{\infty} a_n x^n \qquad \text{(ordinary) generating function,}$$

$$f_2(x) = \sum_{n=0}^{\infty} \frac{a_n x^n}{n!} \qquad \text{exponential generating function,}$$

$$f_3(x) = \sum_{n=1}^{\infty} \frac{a_n x^n}{1-x^n} \qquad \text{Lambert series generating function,}$$

$$f_4(x) = \sum_{n=1}^{\infty} a_n n^{-x} \qquad \text{Dirichlet series generating function,}$$

and many other forms. It may be sometimes not obvious a priori which form to use. Usually, f_1 is good in additive problems. For instance, to prove Lagrange's theorem that every positive integer is a sum of four squares of integers:

$$n = n_1^2 + n_2^2 + n_3^2 + n_4^2 \tag{1}$$

we introduce the ordinary generating function f_1 corresponding to the sequence $a_0 = 1$, $a_1 = 2$, $a_2 = a_3 = 0$, $a_4 = 2$, $a_5 = a_6 = a_7 = a_8 = 0$, $a_9 = 2, \ldots$ getting

$$f_1(x) = \sum_{n=-\infty}^{\infty} x^{n^2}.$$

Now we exploit the additive structure and such a simple thing as the index law $x^a x^b = x^{a+b}$:

$$f_1^4(x) = 1 + \sum_{n=1}^{\infty} R_4(n) x^n$$

where $R_4(n)$ is the number of representations of n in the form (1). Here n_i and $-n_i$ are counted separately, and the permutations of n_1, n_2, n_3, n_4 as distinct. It so happens that there is a (perhaps unexpected) relation with the Lambert series:

$$f_1^4(x) = 1 + 8 \sum_{\substack{n=1 \\ 4 \nmid n}}^{\infty} \frac{nx^n}{1-x^n},$$

although the "roots" of it go into the theory of elliptic functions, a proof can be given in reasonably elementary form [67]. We conclude that

$$R_4(n) = 8 \sum_{\substack{d \mid n \\ 4 \nmid d}} d.$$

This not only proves Lagrange's theorem, which is simply $R_4(n) > 0$, but it also gives us the exact number of representations. This numerical feature is rather characteristic: There are many other proofs of Lagrange's theorem but only the above one, with its use of the generating function, gives us the exact information in such a natural way.

The use of the exponential generating function $f_2(x)$ may be illustrated in solving recursions like the following one:

$$d_0, d_1 \text{ given}; \quad d_{n+1} = a \sum_{j=0}^{n} \binom{n}{j} d_j d_{n-j}, \qquad n = 1, 2, \ldots .$$

It is the appearance of the binomial coefficients above that is decisive, for if we put

$$f_2(x) = \sum_{n=0}^{\infty} \frac{d_n x^n}{n!}$$

then

$$f_2^2(x) = \sum_{n=0}^{\infty} \left(\sum_{j=0}^{n} \frac{d_j d_{n-j}}{j! \, (n-j)!} \right) x^n$$

which can be written, multiplying and dividing by $n!$, as

$$f_2^2(x) = \sum_{n=0}^{\infty} \left(\sum_{j=0}^{n} \binom{n}{j} d_j d_{n-j} \right) \frac{x^n}{n!}.$$

So, our recursion will lead us to a very simple differential equation

$$f^2(x) = af'(x) + b,$$

which we solve. Expanding the solution in a power series we get d_n. This was the method used in vol. 1, p. 105, to find the power series for $\tan x$ and $\sec x$.

In order to illustrate the use of Lambert series as generating functions, let us consider the following problem. Given an infinite row of coins marked 1, 2, 3, ..., initially all of them heads up, let us turn over every kth coin starting with the coin marked k. When this is done for $k = 2, 3, 4, 5, ...$, which coins are head up? A simple and direct way of solving this is to observe that the number of times that the nth coin gets turned over does not matter but only its *parity* does. Hence, as once before in the argument for $pe(n) - po(n)$, in vol. 1, p. 222, we do cancelling. Given a divisor d of n, the nth coin will be turned over twice: once on account of d, and once on account of n/d. The only exception is precisely when $d = n/d$ so that n is a perfect square. Therefore the coins remaining heads up are those whose numbers are squares: 1, 4, 9, ...

Though this solution may be simple and in some respects quite satisfactory, it does not easily generalize, nor does it give best insight. Once before, in vol. 1, pp. 250–257, we have solved a problem several times over, each time by a more difficult method, and the most difficult method proved the one we needed for a suitable generalization. Here, too, we may practice the Platonic principle of taking the longest way, and solve the problem again, by a much more difficult method. What we need is a tool that will handle the situation analytically, and a suitable generating function will do it. However, the ordinary or exponential generating functions will not work here; one might say that they do not reflect the structure of the problem, which is multiplicative rather than additive. The right tool is a Lambert series generating function

$$f_3(x) = \sum_{n=1}^{\infty} \frac{a_n x^n}{1 - x^n}.$$

Why this? Expanding the nth term in the geometric series we have

$$f_3(x) = \sum_{n=1}^{\infty} \sum_{k=1}^{\infty} a_n x^{nk}$$

and now the process of collecting like powers begins to reflect the structure of our problem. In fact (serendipity!) our problem has already been solved for us, by means of Lambert series too, though the solution may not be so easily recognizable: the identity of vol. 1, p. 129

$$\sum_{n=1}^{\infty} \frac{(-1)^{b(n)} x^n}{1 - x^n} = \sum_{k=1}^{\infty} x^{k^2}$$

where $b(n)$ is the total number of prime factors of n (but counted with multiplicity: $b(4) = 2$, $b(8) = 3$, $b(12) = 3$, etc.). These "structural" reasons suggest that if we wish to generalize our original simple problem (e.g., to generating similarly cubes, primes, etc.) we ought to give a serious consideration to employing Lambert series.

Examples of Dirichlet series as generating functions are abundant, especially in number theory; several have been given in vol. 1: pp. 110, 124, 128, 231.

It might be parenthetically remarked here that the question of deciding which form of generating function to use is reminiscent somewhat of the problem of which form of integral transform to use on a particular functional equation. In both cases one occassionally gets a hint by examining the form of "multiplication."

Finally, we may consider the relations between different generating functions for one and the same sequence of coefficients. With the ordinary and exponential generating functions

$$f_1(x) = \sum_{n=0}^{\infty} a_n x^n, \qquad f_2(x) = \sum_{n=0}^{\infty} \frac{a_n x^n}{n!}$$

we are led to the idea of the Borel transform [170]:

$$f(x) = \int_0^{\infty} e^{-t} \phi(xt) \, dt. \tag{2}$$

It is verified that if $\phi(u) = u^n$ then $f(x) = n! \, x^n$; hence by the linearity we find that with $\phi = f_2$ we get $f = f_1$. Actually, rather more is true: if the power series for f_1 converges for $|x| < r$ and f_1, as a function of a complex variable, has singularities at $z_1, z_2, \ldots$, then the L.H.S. of (2) will converge in the Borel polygon $B = B(z_1, z_2, \ldots)$. This is, in general, a bigger region than the circular disk $|x| < r$. B is defined as follows: For each z_j let P_j be the half-plane containing the origin O and bounded by the straight line through z_j which is at right angles to Oz_j; then B is the intersection of all the half-planes P_j. Of course, the polygon B may be sometimes improper (i.e., unbounded). The reader may wish to verify that $B = B(z_1, z_2, \ldots)$ can also be described thus: It is the set of all points which are closer to the origin O than to any of the points $2z_1, 2z_2, \ldots$. This suggests some connection with the so-called Voronoi regions: Given a metric space M and a finite or countably infinite set of points $\{p_1, p_2, \ldots\}$ of M, the Voronoi region $V(p_i)$ is given by

$$V(p_i) = \{x : x \in M, |x - p_i| \le |x - p_j|\}.$$

These sets, under the names of Voronoi regions, Dirichlet cells, Wigner–Seitz cells, and so on, turn up in mathematics (packings and coverings, number theory, geometry) as well as in physics, chemistry, and astronomy. Their possible application to social and biological sciences might perhaps be suggested by the fact that they seem to be the mathematical counterpart of the "territory" of the plant, animal, tribe, or nation (recall here the "Lebensraum" of the Nazi geopolitics or the "territorial imperative"). The reader may wish to compare the notion of a Voronoi region with some aspects of graph minimization (vol. 1, pp. 143, 163). It may be possible to generalize the Voronoi regions to the case when the defining system p_1, p_2, ... consists of sets rather than points, or, by a suitable limiting procedure, to the case when it is a continuum (such as a curve or a surface; in these cases the curvature structure appears to play a role).

The structure of the Voronoi regions is particularly simple in the important special case when M is a Euclidean space E (of dimension ≥ 2) and the sequence p_1, p_2, ... of points is fairly uniformly spread: positive constants a, b exist such that no two p_i's are closer than a and no point x in E is further than b from a point p_i. Given two points p_i and p_j, consider the half-space

$$H_{ij} = \{x: x \in E, |x - p_i| \leq |x - p_j|\}.$$

Then the Voronoi region $V(p_i)$ is given by

$$V(p_i) = \bigcap_{j \neq i} H_{ij}$$

and under our assumptions on the sequence p_1, p_2, ..., it is easy to show that only a finite number of values of j need be taken in the above intersection. Hence each $V(p_i)$ is a convex polyhedron.

We indicate an application of plane Voronoi regions by proving the well-known packing theorem: The density of the tightest plane packing by equal nonoverlapping circular disks of radius r is $\pi/\sqrt{12}$ (and is achieved only for the regular hexagonal packing). We take $r = 1$ and we recall that the disks D_1, D_2, ..., of the packing may touch but must not have interior points in common. The packing density is the fraction of the area of the plane covered by the disks, defined as a suitable limit.

Let V_1, V_2, ..., be the Voronoi regions belonging to the point-system consisting of the centers C_1, C_2, ..., of the disks D_1, D_2, ... V_1, V_2, ..., are then nonoverlapping convex polygons covering the plane exactly; for each i V_i contains its disk D_i. Our theorem that the optimal density is $\pi/\sqrt{12}$, is a simple consequence of the fact that the regular hexagonal lattice packing achieves the density $\pi/\sqrt{12}$, once we prove the basic area estimate

$$\text{Area } V_i \geq \sqrt{12};$$

for each disk D_i has the area π. This area estimate is easy to prove in one special case (which we shall use as a pattern for the general proof): when the number s of sides of V_i is ≤ 6. For, let each side be moved, if needed, by a translation till it touches D_i. We get a new polygon W_i, still of ≤ 6 sides, and

$$\text{Area } W_i \leq \text{Area } V_i.$$

Now it is not hard to prove, by a minimum perturbation argument (cf. vol. 1, pp. 148–149) that the area of an s-gon W_i circumscribing D_i is least when W_i is regular. But the area of a regular circumscribing s-gon is a decreasing function of s, and $\sqrt{12}$ is the area of the regular hexagon about D_i. Thus our estimate is proved when $s \leq 6$.

In proving the minimality of the circumscribing regular s-gon we observe a simple duality of extremization, somewhat analogous to the more recondite case of isoperimetric duality of vol. 1, p. 20. Here the area of the polygon corresponds to the integral of the isoperimetric case to be extremized, while the fixed circle into which or about which the polygon is drawn, corresponds to the second integral of the isoperimetric case, which is kept constant. Our duality is this: Suppose that polygons are inscribed to and circumscribed about a fixed circle and let us consider their area. Then the inscribed ones get bigger as the number of sides increases while the circumscribing ones get smaller, *and* for the inscribed ones the area is increased by making two adjacent sides more nearly equal, while for the circumscribing polygons this is reversed.

On the basis of these remarks the general estimate of Area V_i is arranged as follows. First, let A be a vertex of V_i. By the definition of the Voronoi regions A is equidistant from some three centers one of which is C_i, say C_i, C_j, C_k. Since any two of these are ≥ 2 units apart it is not hard to show that the circumradius of the triangle $C_i C_j C_k$ is $\geq 2/\sqrt{3}$, this being its value for the equilateral triangle of side 2.

Hence we get the special case of Blichfeldt's inequality (for which see [145]): each vertex of V_i lies on or outside the circle E_i concentric with D_i and of radius $2/\sqrt{3}$. And, of course, each side of V_i is disjoint from (the interior of) D_i. By adding, at need, new vertices and sides we can transform V_i to a new convex polygon V_i' which

(a)   has all its vertices on E_i,
(b)   contains D_i,
(c)   has area no greater than that of V_i.

The last condition can be achieved, for instance, by arranging V_i' to be a subset of V_i.

Our basic estimate Area $V_i \geq \sqrt{12}$ will therefore be proved if we can show that the least-area polygon P satisfying (a) and (b) has area $\geq \sqrt{12}$. Observe first that the regular hexagon H satisfying (a) and (b) happens to circumscribe D_i. Hence any side of P is no longer than that of H, and in case of equality both touch D_i. Also, by a symmetry reflection we can transpose any two adjacent sides of P keeping (a), (b), the minimality, and the same area. Therefore it can be assumed that all sides of P, tangent to D_i, are consecutive, and so are all others if any. If there are none, $P = H$ and we are finished (for Area $H = \sqrt{12}$). If there are some, there must be at least two. Then let A be the vertex common to some two of them. By moving A on E_i in one of the two directions we can now diminish the area of P, but this contradicts the minimality of P, q.e.d.

G. Voronoi (1866–1908) was a Russian mathematician who worked principally in topics related to number theory; the regions named after him appeared in connection with quadratic forms [177]. Returning now to our subject of generating functions, and transforming (2) into the standard Laplace transform, the reader may wish to invert (2) and express ϕ in terms of f. For us this means expressing the exponential generating function in terms of the ordinary one.

With the ordinary and Dirichlet-series generating functions

$$f_1(x) = \sum_{n=1}^{\infty} a_n x^n, \qquad f_4(s) = \sum_{n=1}^{\infty} a_n n^{-s}$$

we are led to the Dirichlet transform of vol. 1, p. 110; if

$$D_s f(x) = \frac{1}{\Gamma(s)} \int_0^{\infty} z^{s-1} f(xe^{-z}) \, dz$$

then

$$f_4(s) = D_s f_1(x)|_{x=1}.$$

We also note a relation between three types of generating functions: ordinary, Lambert, and Dirichlet: if $a_0 = 0$ then

$$D_s f_3(x)|_{x=1} = \zeta(s) D_s f_1(x)|_{x=1}.$$

We move now on to the combinatorial principle (4), referred to as structuration. This is used sometimes in finding the number $f(n)$ of ways of doing something associated with a finite set X, of n elements. If there is no sufficient structure to enable us to find $f(n)$, we may benefit by complicating the problem (bringing in the process sufficient structure to give us a hold on $f(n)$). We count the number $f(n, k)$ of doing our thing to X, subject to the size condition $|X| = n$, and subject to a further condition referring to the auxiliary parameter k. In many cases we may have

$$f(n) = \sum_k f(n, k)$$

and $f(n, k)$ may be evaluable, for instance, by the principle (7) of recursion. An example of this technique is provided by the problem of counting the number of zigzag permutations on n numbers $1, 2, \ldots, n$, in connection with the power series for $\tan x$ and $\sec x$, vol. 1, p. 104. There, the additional structuring parameter is the place $r+1$ occupied in the permutation $(k_1, \ldots, k_n)$ of $(1, 2, \ldots, n)$ by the largest number n. With this we were able to apply recursion successfully, and so solve the problem.

Examples of the use of principle (5), inclusion–exclusion, occur in several places in vol. 1. Beginning on p. 249, for instance, it was shown how the inclusion–exclusion principle, when applied to multisets ($=$ finite sets indexed by a k-tuple index $(i_1, \ldots, i_k)$ with $1 \le i_1 < i_2 < \cdots < i_k \le N$, rather than by a single index i, $1 \le i \le N$), leads in a rather natural way, by considering index-sharing types, to k-dimensional hypergraphs. There will be some further material on the inclusion–exclusion in the next section, when we come to the generalized Möbius inversion.

The next principle, (6), counting something in two different ways and equating the results, is a common occurrence and, one is almost tempted to add, forms the essence of combinatorics. Examples have been given in vol. 1: applications to integration, p. 175; to series summation, p. 85; to graph minimization (Steiner's problem), p. 140, and so on. We add some simple applications toward obtaining combinatorial identities. The integral

$$\int_0^1 \frac{1-(1-x)^n}{x}\, dx$$

can be evaluated by expanding $(1-x)^n$ in powers of x, or by substituting $1-x = y$ and then dividing, and integrating. Equating the two results we get

$$\sum_{j=1}^n (-1)^{j+1} \binom{n}{j} \frac{1}{j} = \sum_{j=1}^n \frac{1}{j}.$$

A related integral is due to Schlömilch [50]:

$$\int_0^1 \frac{1-(1-x)^n}{x} \log \frac{1}{1-x}\, dx = \sum_{j=1}^n \frac{1}{j^2};$$

expanding the log term, integrating and transforming, we get

$$\frac{\pi^2}{6} - \sum_{j=1}^n \frac{1}{j^2} = \sum_{k=1}^\infty \left[k^2 \binom{n+k}{k} \right]^{-1}.$$

By a similar technique Catalan [23] gets

$$\int_0^1 \frac{t^{2n}}{1+t}\, dt = \int_0^1 t^{2n}(1-t+t^2-\cdots)\, dt$$

$$= \frac{1}{2n+1} - \frac{1}{2n+2} + \frac{1}{2n+3} - \cdots = \log 2 - \sum_{j=n+1}^{2n} \frac{1}{j};$$

hence

$$\int_0^1 \frac{t^{2n}}{1+t} \, dt = C_n - C_{2n} \qquad \text{where} \qquad C_n = \sum_{j=1}^n \frac{1}{j} - \log n.$$

Therefore, by telescoping cancellation we get

$$1 - \gamma = C_1 - \lim_{n \to \infty} C_n = \sum_{n=1}^\infty (C_{2^{n-1}} - C_{2^n}) = \int_0^1 \frac{1}{1+t} \sum_{n=1}^\infty t^{2^n} \, dt,$$

where γ is Euler's constant. Recalling Euler's infinite product of vol. 1, p. 92 we get

$$1 - \gamma = \int_0^1 (1-t) \prod_{n=1}^\infty (1+t^{2^n}) \sum_{n=1}^\infty t^{2^n} \, dt.$$

For a slightly more complicated example we consider the expansion of vol. 1, p. 115:

$$\sum_{n=0}^\infty \frac{(c+n)^n}{n!} w^n = \frac{e^{cz}}{1-z},$$

where $w = ze^{-z}$. If we now substitute ze^{-z} for w, expand both sides in powers and equate the coefficients of like powers of z, we get the identity

$$\sum_{k=0}^n (-1)^k \binom{n}{k} (c+n-k)^{n-k} (n-k)^k = n! \sum_{j=0}^n \frac{c^j}{j!}$$

valid for arbitrary c. By specializing c, differentiating and integrating with respect to it, and so on, we produce a number of further identities. For example, with $c = 0$ we have

$$\sum_{k=0}^n (-1)^k \binom{n}{k} (n-k)^n = n!$$

and with $c = -n$ we get

$$\sum_{k=0}^n \binom{n}{k} k^{n-k} (n-k)^k = \sum_{j=0}^n (-1)^j (n)_j n^{n-j}.$$

Before moving on to principle (7) we recall one of the basic results on transcendental numbers: If x is algebraic and $\neq 0$ then e^x is transcendental (this follows from the generalized Lindemann, or Lindemann–Weierstrass, theorem [158, 98]: if $x_1, x_2, \ldots, x_k$ are distinct algebraic numbers then $e^{x_1}, e^{x_2}, \ldots, e^{x_k}$ are linearly independent over the field of all algebraic numbers). It is easy to deduce from this that the function

$$f(x) = 1 + \sum_{n=1}^\infty \frac{n^n}{n!} x^n, \qquad |x| < \frac{1}{e},$$

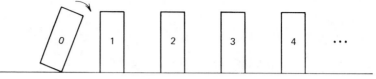

Fig. 1. Simple induction.

has the same property as e^x of having transcendental value for x algebraic and $\neq 0$. For we have now $f(x) = 1/(1-z)$ where $x = ze^{-z}$. Hence, if x and $f(x)$ were both algebraic, the same would have been true for z and e^z. For generalizations of the preceding see W. J. Leveque [99].

The principle (7)—recursion—has a very large number of applications: as induction, as a method of defining functions, as a difference-equation system to be solved for the unknowns, and so on. With respect to induction we may distinguish between (at least) two cases. First, we have the one-variable types schematized by

(A)　$f(0)$ and $f(n) \to f(n+1)$

or

(B)　$f(0)$ and $[f(0), f(1), \ldots, f(n)] \to f(n+1)$.

These can be represented pictorially in Fig. 1 by a row of falling bricks. Then we have also the (usually) harder several-variables case schematized by, say,

(C)　$f(x, 0)$ and $f(0, y)$ and $[f(x, y+1)$ and $f(x+1, y)] \to f(x+1, y+1)$

or higher-dimensional equivalents. A pictorial representation that might help with visualizing (C) is given in Fig. 2. Consider the nonnegative quadrant of the integer lattice in the plane, that is, all points (x, y) where x and y range independently over nonnegative integers. We paint those points of our set which happen to lie on the axes (this corresponds to the

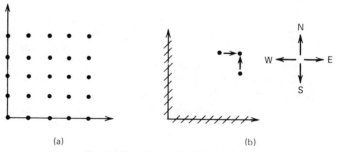

(a)　　　　　　　　　　　　　　　　　(b)

Fig. 2. Two-dimensional induction.

first two parts of (C)). Also, as shown in Fig. 2b, we paint any point whose immediate southern and western neighbors have been painted—this corresponds to the "inductive" part of (C):

$$f(x, y+1) \quad \text{and} \quad f(x+1, y) \rightarrow f(x+1, y+1)$$

just as $f(x, 0)$ and $f(0, y)$ are the "initializing" part of that induction. Then it turns out that all our points get painted (or that the proposition to be proved holds for all nonnegative x and y).

The above suggests a generalization. We take a (finite or infinite) subset X of the integer lattice in the plane, as before, and we distinguish a certain specified nonempty subset X_0 of X. We give also a (finite) number of "transition rules" which generalize our western-and-southern nearest neighbor rule above (Fig. 2b). Now we suppose that the points of X_0 get painted and also all possible points according to our transition rules. We ask: Do all points of X get painted? If not, which ones do? This may be further generalized, dimensionally and otherwise (the reader may wish to compare with the generalized firing squad synchronization problem of vol. 1, p. 170). Also, we may generalize by allowing the paint to be of several colors; this would lead us to generalized dominoes and their use in proofs of unsolvability of special cases of the decision problem [179].

As an example of a problem requiring a form of induction which is of type (C) we consider Ramsey's theorem. Given a set X with $|X| \geq r$, an r-set of X is any one of the $\binom{|X|}{r}$ subsets of X, with exactly r elements. Now:

(RAMSEY'S THEOREM)   Let S be a set with $|S| = n$ and let the set of its r-sets be divided into two disjoint complementary sets A and B. Let $p \geq r$, $q \geq r$, $r \geq 1$; then there is a function $f(p, q, r)$ such that if $n \geq f(p; q, r)$ then either S contains a p-set all of whose r-sets lie in A or a q-set all of whose r-sets lie in B.

There is a simpler, though equivalent, form of Ramsey's theorem which is symmetrical, i.e., has $p = q = r = k$, say; it uses the language of graph theory: There is a function $f(k)$ such that for any graph G with $\geq f(k)$ vertices either G or its complement $\bar{G}$ must contain a complete k-graph. It turns out, however, that in this form there is not quite enough structure to enable us to take an inductive hold (the reader may wish to compare this with principle (4)).

For the function $f(p, q, r)$ we show quite easily that

$$f(p, q, 1) = p + q - 1, \quad f(p, r, r) = p, \quad f(r, q, r) = q;$$

this gives us our "initial" set; then we show, though not quite so easily, that

$$f(p, q, r) \leq f(f(p-1, q, r), f(p, q-1, r), r-1)$$

which gives us a "transition" rule. By the three-dimensional equivalent of the inductive schema (C) $f(p, q, r)$ is now inductively shown to be defined (and finite) for all p, q, r satisfying the conditions of the theorem. F. R. Ramsey (1903–1930) was an English mathematician and logician who died at the age of 26 of smallpox; his brother Michael had been for many years the archbishop of Canterbury.

Besides its inductive form, principle (7)—recursion—is used in combinatorics as a sort of mirroring device by means of which a structure of order $n+1$ reflects that of order n sufficiently well to give us a recursion for counting. We illustrate this on a simple example. By a dendrite let us understand a branching binary structure consisting of a principal branch that splits into two subsidiary or second-order branches; each of these might (though it needn't) split into two third-order branches, and so on. We ask now: What is the number $f(n)$ of distinct n-order dendrites (i.e., with a branch of order n, but no higher)? As shown in Fig. 3a we have $f(1)=1$, $f(2)=1$, $f(3)=2$. To get an inductive hold on $f(n+1)$ we observe that an $(n+1)$-st order dendrite must be of the form illustrated in Fig. 3b where A is an nth order dendrite while B is a kth order dendrite, $k \leq n$. With this observation it is not hard to get the recursion

$$f(n+1) = f(n)\left[\frac{f(n)+1}{2} + \sum_{i=1}^{n-1} f(i)\right]$$

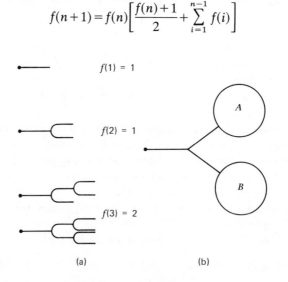

$$f(1) = 1$$

$$f(2) = 1$$

$$f(3) = 2$$

(a)   (b)

Fig. 3. Binary dendrites.

or, eliminating the sum,

$$f(n+2) = f(n+1)\left[\frac{f(n+1)}{f(n)} + \frac{f(n+1)+f(n)}{2}\right]. \tag{3}$$

Now, with $f(1) = f(2) = 1$ it is easy to compute $f(n)$: $f(3) = 2$, $f(4) = 7$, $f(5) = 56$, $f(6) = 2212$, $f(7) = 2,595,782$, and so on. The reader may wish to find an explicit closed-form expression for $f(n)$ or to prove (what is more likely to be the case) that no such simple form exists. To make this statement precise we could borrow some ideas from the Liouville theory of functional complexity (vol. 1, p. 205 et seq.). The reader might wish to begin by showing that $f(n)$ is not a hyper-exponential polynomial. By such a polynomial we understand a function of the form

$$P(n) = P(n, 2^n, 3^n, \ldots, a^n, 2^{3^n}, \ldots, b^{c^n})$$

where $P(x_1, \ldots, x_N)$ is a polynomial with integer coefficients. Our conjecture that there is no closed-form expression for $f(n)$ has to be explained. For, as it turns out, D. W. Matula [109] in fact has proved a more general formula which implies in particular that

$$\sum_{j=1}^{n} f(j) = [2C^{2^{n-2}} - 1]$$

where C is a constant (and $[x]$ is the usual greatest-integer function). Can we still maintain our conjecture? Yes, the above formula is not really of the closed form. It is related to the following "prime number formula" for nth consecutive prime number p_n: There is a real number x such that

$$p_n = [10^{n^2}x] - 10^{2n-1}[10^{(n-1)^2}x]. \tag{4}$$

This may sound exciting till we realize that

$$x = \sum_{n=1}^{\infty} \frac{p_n}{10^{n^2}} = 0.2003000050\ldots \tag{5}$$

so that nothing is really gained. A similar though more recondite result is the well-known formula of Mills: for some real $x > 1$

$$[x^{3^n}] \qquad n = 1, 2, \ldots$$

is always prime. The preceding examples illustrate the need for care in speaking about closed-form formulas and computable numbers. We might be tempted to regard (4) as a closed-form formula and so we seem to get a prime-number representation in closed form. This appears to be justified since the number x is computable to arbitrary accuracy.

If we consider a ternary, quaternary, ..., and generally q-ary splitting in place of the binary one, the formulas for the number $f_q(n)$ of distinct

q-ary dendrites grow rapidly more complicated with q [112]. For instance, already with $q = 3$ we have

$$f_3(1) = f_3(2) = 1, \qquad f_3(3) = 3$$

$$f_3(n+1) = f_3(n)[f_3(n) + f_3(n-1)]\frac{[f_3(n) + f_3(n-1) + f_3(n-2)]}{6}$$

$$+ f_3(n)\frac{f_3(n) + f_3(n-1)}{f_3(n-1) + f_3(n-2)}\left[\frac{f_3(n)}{f_3(n-1)} - \frac{f_3(n-1)}{f_3(n-2)}\right] + \frac{f_3^2(n)}{f_3(n-1)}.$$

This gives us $f_3(4) = 31$, $f_3(5) = 8401$, $f_3(6) = 100,130,704,103$, and so on. However, as pointed out in [109], the recursions are very much simpler if instead of the number $f_q(n)$ of distinct q-ary dendrites of order exactly n we introduce the number $g_q(n)$ of distinct q-ary dendrites of order $\leq n$. We have then $g_q(1) = g_q(2) = 1$ and a simple application of the $n+1$ to n mirroring gives the recursion:

$$g_q(n+1) = \binom{g_q(n) + q}{q}.$$

Now we easily recover our previous quantity $f_q(n)$: $f_q(n) = g_q(n) - g_q(n-1)$. Further material of this type will be found in Chapter 5.

The reification principle, (8), has been discussed in vol. 1, pp. 106, 109, 118, and so on. Informally, it was mentioned that it consists in handling operations and operators as though they were numbers. Several examples were given: Treating the derivative Dy of $y = f(x)$ as though it were a multiplier, the Euler convergence acceleration, Bernoulli polynomials, and so on. One use of reification is to *suggest* answers, that is, to produce candidates which we then test. Some simple examples can be given from linear algebra. Let M be a matrix satisfying the quadratic identity

$$aM^2 + bM + cI = 0, \qquad c \neq 0;$$

then M is invertible. To show this, we "reify" in the above and get

$$\frac{I}{M} = -\frac{a}{c}M - \frac{b}{c}.$$

Now it is simple to show that the well-defined matrix $U = -(a/c)M - (b/c)I$ is the inverse of M: $UM = MU = I$. Similarly, let M be a nilpotent matrix, to show that $I - M$ is invertible we reify and get

$$\frac{1}{I - M} = I + M + M^2 + \cdots,$$

where the series breaks off since $M^n = 0$ for some n by the nilpotency.

Now it is simply verified that

$$U = I + M + M^2 + \cdots + M^{n-1}.$$

is the desired inverse: $(I - M)U = U(I - M) = I.$

5. MÖBIUS INVERSION

(a) The classical Möbius function $\mu(n)$ is defined for positive integers n as follows: $\mu(1) = 1$, if $n > 1$ then let its prime-power factorization be

$$n = \prod_{i=1}^{s} p_i^{a_i}, \qquad p_i \text{ prime, } a_i \geq 1, \tag{1}$$

and we define

$$\mu(n) = 0 \qquad \text{if some } a_i \text{ is } > 1,$$
$$= (-1)^s \qquad \text{otherwise.}$$

It follows that μ is a multiplicative function:

$$\mu(ab) = \mu(a)\mu(b) \qquad \text{if g.c.d. } (a, b) = 1 \tag{2}$$

Next,

$$\sum_{d \mid n} \mu(d) = 1 \qquad n = 1, \tag{3}$$
$$= 0 \qquad n > 1.$$

For $n = 1$ this follows from the definition. For $n > 1$ we assume the form (1) for n and observe that in the summation (3) we need sum only over those divisors d of n which contain no prime powers higher than 1, since otherwise $\mu(d)$ contribution is 0. So, we sum over d's which are products of distinct prime divisors of n:

$$\sum_{d \mid n} \mu(d) = 1 + \sum_{p_i} \mu(p_i) + \sum_{p_i, p_j} \mu(p_i p_j) + \sum_{p_i, p_j, p_k} \mu(p_i p_j p_k) + \cdots. \tag{4}$$

Anticipating somewhat, we remark that some similarity may be detected even at this early stage to the inclusion–exclusion principle. We might have felt happier about such similarity if the terms on the R.H.S. of (4) alternated in sign, and it turns out that they do:

$$\sum_{d \mid n} \mu(d) = 1 - \binom{s}{1} + \binom{s}{2} - \cdots + (-1)^s \binom{s}{s} = (1 - 1)^s = 0. \tag{5}$$

The same reasoning shows that

$$\sum_{d \mid n} |\mu(d)| = 1 + \binom{s}{1} + \binom{s}{2} + \cdots + \binom{s}{s} = (1 + 1)^s = 2^s. \tag{6}$$

By the analogy of (5) and (6) to the binomial identities

$$\sum_{i=0}^{s} (-1)^i \binom{s}{i} = 0, \qquad \sum_{i=0}^{s} \binom{s}{i} = 2^s$$

we note that for n given by (1) $\mu(d)$ appears to play the role of the *signed* binomial coefficient $(-1)^i \binom{s}{i}$. The analogy above will be reinforced when we come to the generalized Möbius inversion. It is now not hard to prove the classical Möbius inversion formula

$$\text{if} \quad g(n) = \sum_{d \mid n} f(d) \quad \text{then} \quad f(n) = \sum_{d \mid n} \mu(d) g\left(\frac{n}{d}\right). \tag{7}$$

The last equation could also be written

$$f(n) = \sum_{d \mid n} \mu\left(\frac{n}{d}\right) g(d)$$

or, preferably, in the transparently symmetric form,

$$f(n) = \sum_{dk=n} \mu(k) g(d).$$

To prove the second equation of (7) we start with the R.H.S., substitute into it for g from the first equation of (7), and simplify by using (3). We get then

$$\sum_{d \mid n} \mu(d) g\left(\frac{n}{d}\right) = \sum_{d \mid n} \mu(d) \sum_{c \mid n/d} f(c) = \sum_{cd \mid n} \mu(d) f(c)$$

$$= \sum_{c \mid n} f(c) \sum_{d \mid n/c} \mu(d) = f(n)$$

since by (3)

$$\sum_{d \mid n/c} \mu(d) = 0 \qquad n/c \neq 1,$$

$$= 1 \qquad n/c = 1.$$

By letting $f(n) = \log F(n)$ and $g(n) = \log G(n)$ in (7) the multiplicative form of Möbius inversion is obtained:

$$\text{if} \quad G(n) = \prod_{d \mid n} F(d) \quad \text{then} \quad F(n) = \prod_{d \mid n} G\left(\frac{n}{d}\right)^{\mu(d)}. \tag{8}$$

As an exercise we evaluate the cyclotomic polynomial $P_m(x)$ defined by

$$P_m(x) = \prod_{\substack{k=1 \\ \text{g.c.d.} (k, m) = 1}}^{m} (x - e^{2\pi i k/m}), \qquad m \text{ is a positive integer.}$$

The product here is over the $\phi(m)$ primitive roots of 1, where $\phi(m)$ is the Euler function:

$$\phi(m) = \sum_{\substack{1 \le d \le m \\ \text{g.c.d. } (d,m)=1}} 1.$$

Since every mth root of 1 is in a unique way a primitive dth root of 1 for some d that divides m, we have by suitable grouping of factors

$$x^m - 1 = \prod_{j=1}^{m} (x - e^{2\pi ij/m}) = \prod_{d \mid m} P_d(x).$$

Hence by the multiplicative form (8)

$$P_m(x) = \prod_{d \mid m} (x^{m/d} - 1)^{\mu(d)}.$$

Expressing the degrees, and equating, we obtain a representation of Euler's function

$$\phi(m) = m \sum_{d \mid m} \frac{\mu(d)}{d}. \tag{9}$$

The reader may wish to compare the coefficients of various powers.

The importance of the Möbius function derives, among other things, from the fact that its Dirichlet series generator is the reciprocal of the Riemann zeta function:

$$\sum_{n=1}^{\infty} \frac{\mu(n)}{n^s} = \frac{1}{\zeta(s)}. \tag{10}$$

This may be simply verified by starting with the Euler product for $\zeta(s)$. We have

$$\frac{1}{\zeta(s)} = \prod_p (1 - p^{-s}) = \prod_p \sum_{k=0}^{\infty} \mu(p^k) p^{-ks}$$

since $\mu(p^a) = 0$ for $a > 1$. Multiplying out the infinitely many series above, then using the multiplicative property (2) and the unique prime-power factorization of integers, we check that the product of all the power series is just the L.H.S. of (10). With (10) we have a very compact way of expressing the Möbius inversion (7) in terms of the Dirichlet generators of $f(n)$ and $g(n)$: Let

$$F(s) = \sum_{n=1}^{\infty} \frac{f(n)}{n^s}, \qquad G(s) = \sum_{n=1}^{\infty} \frac{g(n)}{n^s};$$

then by the comparison of the coefficient of n^{-s} the first equation of (7) goes over into the relation

$$G(s) = F(s)\zeta(s)$$

while the second, inverting, equation of (7) goes over into the reciprocal relation

$$F(s) = G(s) \cdot \frac{1}{\zeta(s)}.$$

We note further that with (10) we have a very simple restatement of the Riemann hypothesis, a restatement that avoids any mention of zeta function or even of complex variables (though not of complex numbers):

$$\sum_{n=1}^{\infty} \frac{\mu(n)}{n^s} \quad \text{converges for Re } s > \frac{1}{2}.$$

A partial Riemann hypothesis is also easily restated: instead of "there exists b, $1/2 < b < 1$, such that every zero of $\zeta(s)$ in the critical strip has real part $\leq b$" we have

$$\sum_{n=1}^{\infty} \frac{\mu(n)}{n^s} \quad \text{converges for Re } s > b.$$

No doubt, these simple and well-known observations have contributed to the recent revival of interest in the Möbius inversion and its restatement and generalization in the very broad setting of partially ordered sets. While no progress appears to have been made in this way toward the Riemann hypothesis, the generalized Möbius inversion is a strong unifying feature in combinatorics as well as a tool of practical usefulness. We shall offer a brief sketch of this theory in Section (c), after giving first some applications of the classical Möbius inversion.

(b) We consider the problem of computing the number $f(n, k)$ of different necklaces where by a necklace we understand a circular string with n beads, each of which may be of one or another of k distinct colors. It remains still to be prescribed what constitutes "different necklaces," and here the principle (9) of the last section—symmetry removal—begins to operate (to ensure that in counting our objects we count as distinct only those that really are so—the statement of distinctness or nondistinctness amounts to the description of symmetries under which the configuration coincides with itself; hence, even without any further or previous knowledge one might guess that the symmetry removal will be concerned with group factorization). Our case is simple—we allow rotation of beads along on the string as the only source of symmetry. So, for instance, the six-bead string BRWBRW has period 3—smallest shift to make it coincide with itself. The *circular* problem of necklaces is reduced to the *linear* problem of strings (straight ones, that is) by noting that

$$\# \text{ necklaces of length } n = \sum_{d \mid n} \frac{1}{d} (\# \text{ strings of period } d) \qquad (11)$$

because: (a) every necklace of period d has d shifts (including the identity) which give us back the same necklace, hence the division by d, and (b) the initial d beads, or any d consecutive ones, determine the whole necklace then. Next,

$$\text{\# strings of length } n = k^n = \sum_{d \mid n} \text{\# strings of period } d$$

because each of the n places is to be filled with a bead of one of the k colors. Hence by the Möbius inversion (7) we get

$$\text{\# strings of period } d = \sum_{x \mid d} \mu(d) k^{d/x},$$

and so by (11)

$$\text{\# necklaces of length } n = \sum_{d \mid n} \sum_{x \mid d} \frac{\mu(d)}{d} k^{d/x}.$$

This can be simplified by using (9) and we have

$$\text{\# necklaces of length } n = \frac{1}{n} \sum_{d \mid n} \phi\left(\frac{n}{d}\right) k^d.$$

The problem of necklace counting is perhaps not quite as frivolous as it might appear. This is so because in genetics and molecular biology we encounter several situations which could be modeled by necklaces—a linear structure (string) with some sort of genetic or biochemical markers (beads). The joining together of the two ends is not an uncommon procedure in modeling; it is the easiest way to avoid nasty end-point troubles by prescribing periodicity. We meet the same procedure in statistical mechanics, setting up a simple two-dimensional Ising model. Here $(m+1)(n+1)$ lattice points of the integer lattice in the plane are taken in a rectangular array

$$R = \{(x, y): 0 \le x \le m, 0 \le y \le n\}$$

Again, to get rid of boundary troubles we identify any two points in R whose x-coordinates differ by m or whose y-coordinates differ by n. We end up with a toroidal array of mn points which exhibits complete homogeneity.

Returning to necklaces as biological models, we observe that the symmetry reduction in necklace counting is somewhat reminiscent of the so-called comma-free codes and the problem of the twenty amino acids. We have here a linear structure—one strand of the DNA helix—on which there is a sequence of discrete markers, each one being of one or another of four kinds. These are the four nucleotides: A–adenine, G–guanine,

U–uracil, C–cytosine. One of the functions of this genetic code is to help with the information necessary to produce, and reproduce, proteins. These are rather big molecules composed of many smaller building blocks, the amino acids, of which there are 20. It is clear that the nucleotide-into-amino acid coding cannot be $1:1$ (because $20>4$) nor $2:1$ (because $20>4^2$). On the other hand, $3:1$ appears to be redundant because $20<4^3$, and we ask what use is made of this redundancy. A plausible hypothesis (Crick, Griffith, and Orgel [38]) is that the coding is by triples of nucleotides to one aminoacid, and we do not want all $4^3 = 64$ triples to function as code words because in our code there are no parsing devices—no commas between the letters, which run on like this: ...AUAGC.... If all possible triples could code then the "decoding" would have to be done from one end and "in series" or consecutively, in three-by-three groups. But there is no reason to suppose that there is a long-range correlation between the nucleotides so that the correct functioning of one triple depends on what happens several hundred nucleotides away. Besides, it stands to reason to have the decoding done "in parallel" so that all triples could start "producing" their amino acids at once. Finally, it is expected that there is some measure of error control: if something goes wrong locally the error should not propagate itself along the nucleotide chain.

For these reasons we introduce the comma-free codes $CF(k, n)$. Such a code consists of strings of k symbols chosen out of $1, 2, \ldots, n$; the defining property is that when any two strings are concatenated,

$$a_1 a_2 \cdots a_k b_1 b_2 \cdots b_k,$$

then no k-string of intermediates

$$a_j a_{j+1} \cdots a_k b_1 b_2 \cdots b_{j-1}, \qquad j = 2, 3, \ldots, k$$

is itself a string of the code. Let $W_k(n)$ be the largest possible number of strings in a code $CF(k, n)$. Using arguments similar to those employed in necklace counting, the reader may wish to prove the bound of Golomb, Gordon, and Welch [62]

$$W_k(n) \leq \frac{1}{k} \sum_{d \mid k} \mu(d) n^{k/d}; \tag{12}$$

the further result that this bound is attained for all n and for all odd k [49, 62] is harder to prove. For $n = 4$ and $k = 3$ the bound is $(4^3 - 4)/3 = 20$ corresponding to the twenty amino acids. It is understandable that this discovery should have brought on a certain genetic-combinatorial excitement; however, on the basis of later work in molecular biology it appears that the nucleotide-amino acid coding is rather more complex than originally imagined.

Our next application of the Möbius inversion is also concerned, though indirectly, with coding theory. We want to find the number $I(n)$ of irreducible monic polynomials of degree n over the finite field $GF(q)$, q a prime ≥ 2. Those polynomials are of interest because they produce certain important error-correcting codes (the so-called Bose–Chaudhuri–Hocquenghem, or BCH codes, [15]). It turns out that

$$\frac{1}{1-qz} = \sum_{n=1}^{\infty} \left(\frac{1}{1-z^n}\right)^{I(n)}. \tag{13}$$

This identity belongs together with the Euler product for the zeta function

$$\sum_{n=1}^{\infty} \frac{1}{n^s} = \prod_{p=2}^{\infty} \left(1-\frac{1}{p^s}\right)^{-1}, \tag{14}$$

the Euler product

$$\frac{1}{1-x} = \prod_{n=0}^{\infty} (1+x^{2^n}), \tag{15}$$

and a number of other less important formulas, such as for instance

$$\frac{1}{1-x} = (1+x+x^2+\cdots+x^9)(1+x^{10}+x^{20}+\cdots+x^{90})\cdots$$

$$= \prod_{k=0}^{\infty} \frac{1-x^{10^{k+1}}}{1-x^{10^k}}. \tag{16}$$

Each such identity is an analytical statement, by means of a generating function, of a law of unique representation. The Euler product (14) is equivalent to the principal theorem of arithmetic: Each positive integer has a unique representation as a product of prime powers. The Euler product (15) shows similarly that a positive integer is a unique sum of distinct powers of 2. The product (16) is an analytical reflection of the simple arithmetical fact that every positive integer can be uniquely written in the ordinary decimal fashion. Equation (13) also expresses a unique representation: a monic polynomial over $GF(q)$ is uniquely factorizable into a product of powers of similar, but irreducible, polynomials. Further, the number of monic polynomials of degree n is the coefficient of z^n of the L.H.S. of (13): q^n. Similarly, the number of those monic polynomials of degree s which are products of the $I(n)$ irreducible ones of degree n, and no others, is the coefficient of z^s in

$$\left(\frac{1}{1-z^n}\right)^{I(n)}.$$

Hence, the R.H.S. of (13) enumerates the products of powers of irreducible monic polynomials; briefly, it enumerates the polynomials by degree, and so, incidentally, is equal to the L.H.S. An interesting fact results if we take the reciprocal of (13)

$$1 - qz = \prod_{n=1}^{\infty} (1 - z^n)^{I(n)};$$

by comparing the coefficients of z^k, $k \geq 2$, we find that for every $k \geq 2$ there are as many products of an even number of irreducible factors, among the monic kth degree polynomials, as there are for an odd number of factors. The reader may wish to compare this with the use of Euler's identity, pp. 220–233 of vol. 1, for partitions. There also we have an analytical identity which reflects an additive representation theorem for positive integers; on taking reciprocals we get an identity which is concerned with the difference of the number of representations of n: those with an even–those with an odd number of summands.

Next, we transform (13) into an identity with Lambert series by taking logarithmic derivatives:

$$\frac{qz}{1 - qz} = \sum_{n=1}^{\infty} \frac{nI(n)z^n}{1 - z^n}.$$

Hence, as with Lambert series, by comparing the coefficients we have

$$q^n = \sum_{d \mid n} dI(d)$$

and applying the Möbius inversion

$$I(n) = \frac{1}{n} \sum_{d \mid n} \mu\left(\frac{n}{d}\right) q^d \tag{17}$$

The reader may wish to compare this with the corresponding identities and numbers for necklace counting and for comma-free codes. Certain well-known propositions of number theory and algebra appear as consequences of (17). For instance, taking n to be a prime p we get a special case of Fermat's theorem: If q and p are primes ≥ 2 then $q^p \equiv q \pmod{p}$. Making some crude estimates in (17) we get

$$I(n) \geq q^n - \sum_{\substack{d \mid n \\ d < n}} q^d \geq q^n - (q + q^2 + \cdots + q^{n-1}) > 0$$

so that $I(n) > 0$ for all $n \geq 1$. In words, there exist irreducible monic polynomials of every given degree n. Since the residue classes modulo such a polynomial form a field with exactly q^n elements, it follows that a finite Galois field $GF(q^n)$ exists for every prime $q \geq 2$ and every $n \geq 1$.

(c) We begin our discussion of the generalized Möbius inversion with a restatement of the inclusion–exclusion principle (p. 254, vol. 1). Let $O = \{o_1, o_2, \ldots\}$ be a finite set of objects and $P = \{p_1, p_2, \ldots\}$ a finite set of properties, which those objects may or may not have. For $T \subseteq P$ we define

$N_\geq(T)$—the number of elements of O that have at least the properties of
    T (possibly more),
$N_=(T)$—the number of elements of O with properties exactly those in T
    and no others,
$N_\leq(T)$—the number of elements of O that have at most the properties of
    T (perhaps not even these but certainly none others).

The inclusion–exclusion results concern the relations between the above numbers and some other related ones, such as

$$\sum_{|T|=t} N_=(T)$$

which is the number of objects in O with exactly t properties (we recall that $|T|$ is the size of the set T—hence the summation in the above is over all sets T of size t). We have

$$N_\geq(T) = \sum_{Y \supseteq T} N_=(Y), \tag{18}$$

$$N_\leq(T) = \sum_{Y \subseteq T} N_=(Y), \tag{19}$$

for an object with at least (or at most) the properties T is the object with exactly the properties Y, for some Y such that $Y \supseteq T$ (or $Y \subseteq T$). We wish to invert the equations (18) and (19), and to express $N_=(T)$, pertaining to exact counting and rather hard to get, in terms of the quantities $N_\geq(T)$ or $N_\leq(T)$ which relate to inclusive counting and are easier to obtain. These inversions are

$$N_=(T) = \sum_{X \supseteq T} (-1)^{|X|-|T|} N_\geq(X), \tag{20}$$

$$N_=(T) = \sum_{X \subseteq T} (-1)^{|T|-|X|} N_\leq(X). \tag{21}$$

In particular, for the especially important number of objects o with no properties we take T to be the empty set $\varnothing$ and get

$$\begin{aligned}
N_=(\varnothing) = N_\geq(\varnothing) &- [N_\geq(p_1) + N_\geq(p_2) + \cdots] \\
&+ [N_\geq(p_1, p_2) + N_\geq(p_1, p_3) + \cdots] \\
&- [N_\geq(p_1, p_2, p_3) + \cdots] + \cdots
\end{aligned} \tag{22}$$

which is (almost) the form of the inclusion–exclusion principle as stated on p. 254 of vol. 1. There are several ways of proving (20) and its dual (21): by induction on $|T|$, by reducing to the special case (22) of $T = \varnothing$ and working with relative properties, and by direct balancing method in which (20) is first rewritten as

$$N_=(T) = \sum_{t \geq |T|} \sum_{|X|=t} (-1)^{t-|T|} N_\geq(X), \tag{23}$$

and then we show that each object of O gets counted the same number of times in either side of (23). The inclusion–exclusion principle is conveniently applicable in the form (23). For instance, to get the number M_n of mismatches (or recontres) we take O to be the set of all $n!$ permutations on the elements $(1, 2, \ldots, n)$. P consists of n properties p_i, where a permutation has the property p_i if and only if it keeps the element i fixed. We take $T = \varnothing$ and we compute the number $M_n = N_=(\varnothing)$ of permutations which move every element. For this purpose we observe that in (23) $N_\geq(X) = (n-t)!$ since it is the number of permutations fixed in at least the $t = |X|$ positions indicated by X. Also, a set with $|X| = t$ elements can be chosen in $\binom{n}{t}$ ways; hence with $T = \varnothing$ in (23) we get

$$M_n = \sum_{t=0}^{n} \sum_{|X|=t} (-1)^t (n-t)! = \sum_{t=0}^{n} (-1)^t \binom{n}{t}(n-t)!$$

$$= n! \left(1 - \frac{1}{1!} + \frac{1}{2!} - \cdots + (-1)^n \frac{1}{n!}\right),$$

which is very nearly $n!/e$ for n large. Incidentally, we obtain a purely combinatorial, rather than analytical, definition of e.

Consider now the following three examples.

A.   Series Summation

Given a function $f(n)$ on positive integers, that is, a sequence $f(1)$, $f(2), \ldots$ and its summation function

$$g(n) = \sum_{m \leq n} f(m),$$

we can invert to get

$$f(n) = g(n) - g(n-1)$$

(the reader may wish to compare this with the practice of telescoping cancellation, vol. 1, p. 81).

B. Inclusion–Exclusion

We have a set $S = \{s_1, \ldots, s_k\}$ and a collection of properties $P = \{p_1, \ldots, p_n\}$. A property p_i is completely known if we know exactly which elements of S enjoy it, hence properties are identifiable with subsets of S. Given a list T of properties, with $T \subseteq P$, we define $N_=(T)$ and $N_\geq(T)$ as before. Now we have the summation formula

$$N_\geq(T) = \sum_{\substack{X \\ T \subseteq X}} N_=(X)$$

which inverts into

$$N_=(T) = \sum_{\substack{X \\ T \subseteq X}} (-1)^{|X|-|T|} N_\geq(X).$$

C. Classical Möbius Inversion

Using the divisibility relation $d \mid n$ on positive integers, we have the following: the summation

$$h(n) = \sum_{d \mid n} f(d)$$

can be inverted to give

$$f(n) = \sum_{d \mid n} \mu\left(\frac{n}{d}\right) h(d).$$

It is this, of course, that gives the name to the whole subject. What is common to all three examples may be emphasized by setting up the table:

	Example A	Example B	Example C
set	pos. integers	subsets of S	pos. integers
relation	usual order $\leq$	subset of: $\subseteq$	divisibility: $\mid$
given function	$f(n)$	$N_=(T)$	$f(n)$
sum function	$g(n) = \sum_{m \leq n} f(m)$	$N_\geq(T) = \sum_{\substack{X \\ T \subseteq X}} N_=(X)$	$h(n) = \sum_{d \mid n} f(d)$

Following the motto of C. G. J. Jacobi (vol. 1, p. 89) in each case we wish to invert the summation. This summation depends on, and is given in terms of, an order relation on the set which is the domain of definition. It is not unreasonable to generalize here and instead of inverting in each case separately for a different relation, to set up *one* inversion for a possibly general order relation. This is exactly the technique of the generalized Möbius inversion: summation and its inversion on posets

(=partially ordered sets). It was started by C. Weisner (1935) and P. Hall (1937) in the group-theoretic setting, and continued by G. C. Rota (1964) [149] in fuller generality.

As usual, we define a poset to be a set on which there is defined a relation $\leq$, of the type of order, satisfying three properties: (a) reflexivity: $x \leq x$ for all x, (b) transitivity: whenever $x \leq y$ and $y \leq z$ then also $x \leq z$, and (c) antisymmetry: if both $x \leq y$ and $y \leq x$ then $x = y$. The last property is quite different from the so-called trichotomy law, which states that for any two elements x, y either $x \leq y$ or $y \leq x$ or $x = y$. This law asserts that any two elements are comparable whereas in our poset two *distinct* elements x, y may exist which are not comparable: neither $x \leq y$ nor $y \leq x$ holds. Hence the element of partialness in our partially ordered set, instead of the total or linear order which we get with trichotomy. On the other hand, a poset P can contain subsystems which are totally ordered by the relation of P; they are called the chains of P. Dually, elements x, y such that neither $x = y$ nor $x \leq y$ nor $y \leq x$ are called incomparable.

A convenient pictorial representation of certain posets P can be given by means of the so-called Hasse diagram, which might be more descriptively called the schema of immediate domination: For each x there is a downward line to any element y such that $y < x$; that is, both $y \leq x$ and $y \neq x$; and further, there is no z strictly between y and x: $y < z < x$ holds for no z. For instance, let P consist of all 2^3 subsets of $\{1, 2, 3\}$ and let $\leq$ mean the usual set-inclusion $\subseteq$. Then the Hasse diagram is given in Fig. 1. We observe that under a simple perspective we have here a picture of the edge scheme of a three-dimensional cube. Similarly, the poset of all 2^n subsets of $\{1, 2, \ldots, n\}$ would yield us an edge scheme of an n-dimensional cube. This is no accident and the correspondence becomes obvious by means of the 2^n selector functions $(\varepsilon_1, \varepsilon_2, \ldots, \varepsilon_n)$ where each ε_i is 0 or 1 and it is 1 if and only if the element i belongs to the subset in question. Another example of a Hasse diagram is for the divisors of 12, ordered by means of the divisibility relation; that is shown in Fig. 2. By an interval $[x, y]$ in a poset P we understand the set

$$\{z: x \leq z \leq y, z \in P\};$$

a poset is locally finite if every interval is finite. The two examples given in Figs. 1 and 2 both show posets with zero, or lowest element; in Fig. 1 it is the empty set and in Fig. 2 the number 1.

As an example of a combinatorially relevant theorem referring to posets we quote Dilworth's theorem, [43, 66]: in a poset P the smallest number of disjoint chains covering all of P equals the largest number of pairwise incomparable elements. This holds for P finite and also for P

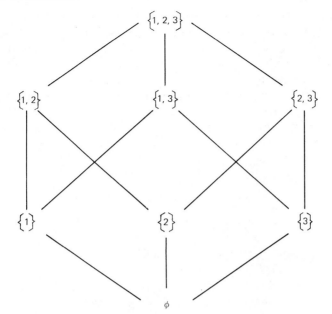

Fig. 1. Hasse diagram for subsets.

infinite if the second number is finite. As an example of an application let us consider $1 + mn$ individuals who are related by ancestry and descent (with the common usage reinforced by the requirement that everybody is his own ancestor as well as his own descendant). Now the relation $x \leq y$ interpreted as "x is a descendant of y" leads to a poset. An application of Dilworth's theorem shows that either there exists a chain of more than n

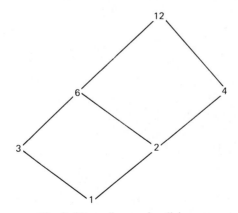

Fig. 2. Hasse diagram for divisors.

individuals each a descendant of the previous one, or a collection of more than m individuals no two of whom are related by descent. However, in some cases similar conclusion follows from the weaker result of Erdös and Szekeres [53]: A sequence of $1 + mn$ distinct real numbers contains either a decreasing subsequence of length $> m$, or else an increasing subsequence of length $> n$. A (relatively) simple proof can be given by exploiting the linear order and by complicating to bring in additional structure. Let $\mathrm{Fal}(i)$ and $\mathrm{Ris}(i)$ be the lengths of the longest decreasing and increasing subsequences starting with the ith one of the $1 + mn$ numbers of the original sequence (which is, say, $s_1, s_2, \ldots, s_{1+mn}$). Consider the map

$$F: i \to (\mathrm{Fal}(i), \mathrm{Ris}(i)), \qquad 1 \le i \le mn + 1.$$

If $i \ne j$ then $F(i) \ne F(j)$. For if $s_i < s_j$ then clearly $\mathrm{Ris}(i) \ge 1 + \mathrm{Ris}(j)$ since an increasing sequence starting with s_j can be legitimately prefaced with s_i. For the same reason we have dually $\mathrm{Fal}(i) \ge 1 + \mathrm{Fal}(j)$ if $s_i > s_j$. We reason now that if the result were false, we would have had $1 \le \mathrm{Fal}(i) \le m$, $1 \le \mathrm{Ris}(i) \le n$ and so F would supply a $1:1$ correspondence between sets of $1 + mn$ elements, and of $\le mn$ elements, which is nonsensical.

As another application of similar type the reader might wish to use Dilworth's theorem on the poset of 2^n subsets of $\{1, 2, \ldots, n\}$ (with set-inclusion order) to show the smallest number of disjoint chains covering this poset is $\max_i \binom{n}{i}$.

Returning to our principal topic we define now the zeta and the Möbius functions for locally finite poset P. First, we consider all real-valued functions defined on intervals of P; these are functions $f(x, y)$ such that $f(x, y) = 0$ if $x \not\le y$. Let the addition $f + g$ and scalar multiplication be defined in the obvious way, and define multiplication fg by convolution: $h = fg$ means

$$h(x, y) = \sum_{x \le z \le y} f(x, z) g(z, y). \qquad (24)$$

On account of local finiteness this is well defined since the summations are always finite. We verify the associativity, commutativity, and distributivity without difficulty. Moreover, there is the identity function, the so-called Kronecker function, given by

$$\delta(x, y) = 1, \qquad x = y;$$
$$= 0, \qquad x \ne y.$$

Under these operations the functions on intervals form an algebra $A(P)$. Having an identity we can speak of inverses and $f \in A(P)$ has an inverse if and only if $f(x, x) \neq 0$ for all $x \in P$. For, let $h = \delta$ in (24) so that

$$f(x, x)g(x, x) = 1, \qquad \text{all } x \in P.$$

Hence $f(x, x) \neq 0$ is a necessary condition. Next, let $f(x, x) \neq 0$ and define the inverse g by induction. First, $g(x, x) = 1/f(x, x)$; then an inductive extension is used for $x < y$. So, in defining $g(x, y)$ we can already suppose $g(z, y)$ to be defined for all z such that $x < z \leq y$; we have then

$$0 = \delta(x, y) = \sum_{x \leq z \leq y} f(x, z)g(z, y)$$

$$= f(x, x)g(x, y) + \sum_{x < z \leq y} f(x, z)g(z, y)$$

and so $g(x, y)$ is *defined* by means of the above:

$$g(x, y) = -\frac{1}{f(x, x)} \sum_{x < z \leq y} f(x, z)g(z, y).$$

In this way we get the right inverse, say g_1. Similar induction with f and g interchanged and $x \leq z < y$ shows that left inverse g_2 also exists. Since both inverses exist we have $g_2 = g_2 1 = g_2(fg_1) = (g_2 f)g_1 = 1 g_1 = g_1$, so that there is just one inverse. We now define the ζ function of $A(P)$:

$$\zeta(x, y) = 1 \qquad \text{if } x \leq y,$$

$$0 \qquad \text{otherwise}$$

Since $\zeta(x, x) \neq 0$ we see that ζ has the inverse and we call it the generalized Möbius function: $\mu(x, y)$ is the inverse of $\zeta(x, y)$. Now we have the principal theorem of the subject: Let $f(x)$ be a function defined on a locally finite poset P with 0-element and let g be its sum function

$$g(x) = \sum_{y \leq x} f(y) \tag{25}$$

(where the summation extends over the necessarily finite set of y's such that $0 \leq y \leq x$), then

$$f(x) = \sum_{y \leq x} g(y)\mu(y, x) \tag{26}$$

The proof of (26) goes through very much as for the classical Möbius inversion: We substitute from (25) for g into (26), interchange summations, and verify. Let

$$S = \sum_{y \leq x} g(y)\mu(y, x)$$

then, by the above,

$$S = \sum_{y \leq x} \sum_{z \leq y} f(z) \mu(y, x)$$

$$= \sum_{z \leq y} f(z) \zeta(z, y) \sum_{y \leq x} \mu(y, x)$$

since nothing is lost or gained by inserting ζ in the above. Therefore

$$S = \sum_{z} f(z) \sum_{z \leq y \leq x} \zeta(z, y) \mu(y, x)$$

$$= \sum_{z} f(z)\, \delta(z, x) = f(x), \qquad \text{q.e.d.}$$

To give some concrete content to the above, the reader is invited to verify that for the poset of subsets of $\{1, 2, \ldots, n\}$

$$\mu(y, x) = (-1)^{|y| - |x|},$$

for instance by using induction on $|x|$, and that for the poset of positive integers under the relation of divisibility

$$\mu(y, x) = \mu\left(1, \frac{x}{y}\right) = \mu\left(\frac{x}{y}\right)$$

the ordinary Möbius function. In either case the generalized Möbius inversion (25)–(26) reduces to what it should: In the first case to the inclusion–exclusion, and in the second case to the classical Möbius inversion. For further examples, consequences and background material see [14].

6. THE LINDELÖF HYPOTHESIS

As was shown in the previous section, the Möbius function, by way of equation (10), affords a simple restatement of the Riemann hypothesis. This remains perhaps the most famous unsolved problem in mathematics. Part of its importance derives from the following connection with the function $\pi(x)$ which is the number of primes $\leq x$: The relation

$$\pi(x) = \int_2^x \frac{dx}{\log x} + O(x^{1/2 + \varepsilon}), \qquad x \text{ large, any } \varepsilon > 0, \qquad (1)$$

is equivalent to the Riemann hypothesis. We recall that even the very much weaker statement, in which the "error term" instead of being best possible as in (1) is merely asserted to be of lower order than the principal, integral, term, is a deep result known as the prime-number theorem

$$\lim_{x \to \infty} \pi(x) \Big/ \frac{x}{\log x} = 1.$$

Just as the Riemann hypothesis can be restated in terms of the Möbius function, so can the prime-number theorem: it is equivalent to the statement that the series

$$\sum_{n=1}^{\infty} \frac{\mu(n)}{n}$$

converges to the value 0 [83].

Since the Riemann hypothesis remains still unproved it is not surprising that it has been weakened in various forms. In particular, we have the Lindelöf hypothesis [171]: For fixed $\sigma \geq 1/2$, t large and any $\varepsilon > 0$

$$\zeta(\sigma + it) = O(t^{\varepsilon}). \tag{2}$$

This is also as yet unproved; for the relation to the Riemann hypothesis and to prime numbers see [171]. It is rather unexpected that the very technical-looking Lindelöf hypothesis (2) can be rephrased in a purely combinatorial form and restated in terms of counting of lattice points (an observation due to Henry Helson). We show this in the present section after assembling first some standard facts on the subject from [83] and [171].

One starts with

$$\zeta^k(s) = \left(\sum_{n=1}^{\infty} n^{-s} \right)^k = \sum_{n=1}^{\infty} d_k(n) n^{-s} \tag{3}$$

where $d_k(n)$ is the number of ways of representing n as a product of k positive integers (cf. vol. 1, pp. 231–232). In particular, $d_2(n)$ is the number of divisors of n, 1 and n included. Define

$$D_k(N) = \sum_{n < N} d_k(n) \tag{4}$$

where the positive number N will be taken, for convenience, to be noninteger; later on, this will avoid troubles with singularities on the path of integration.

In the k-dimensional Euclidean space E^k consider the set P of all positive integer lattice points, that is, all points $x = (x_1, \ldots, x_k)$ where each x_i is a positive integer. Then from its definition (4) $D_k(N)$ has a simple geometrical interpretation: It is the number of positive integer-lattice points of P in the hyperbolically bounded region of E^k

$$H_k(N) = \{(\xi_1, \ldots, \xi_k): 0 < \xi_i, i = 1, \ldots, k, \xi_1 \xi_2 \cdots \xi_k \leq N\}.$$

That is, we have

$$D_k(N) = \sum_{x \in H_k(N)} 1. \tag{5}$$

Next, $D_k(N)$ will be expressed in terms of the zeta function. In view of (3) and (4), what we want first is a way to pick out a specified coefficient in a Dirichlet series, in analogy to the use of residue theorem with power series. Then we want a representation for the sum of the first $[N]$ coefficients of a Dirichlet series. In fact, it is easy to get rather more: an integral representation of a specified partial sum of a Dirichlet series. This is based on the simple discontinuous factor

$$\frac{1}{2\pi i} \int_{c-i\infty}^{c+i\infty} u^w \frac{dw}{w} = 1, \qquad u > 1;$$

$$= 0, \qquad u < 1, \qquad c > 0. \tag{6}$$

If now f is a suitably convergent Dirichlet series

$$f(s) = \sum_{n=1}^{\infty} a_n n^{-s}$$

then from (6) we get the Perron formula [170]

$$\sum_{n < N} a_n n^{-s} = \frac{1}{2\pi i} \int_{c-i\infty}^{c+i\infty} f(s+w)N^w \frac{dw}{w}$$

where $c > 0$ is sufficiently large so that all singularities are to the left of the path of integration. In particular, putting $s = 0$ in the Perron formula and taking $f(w) = \zeta^k(w)$ we get the integral representation

$$D_k(N) = \frac{1}{2\pi i} \int_{c-i\infty}^{c+i\infty} \zeta^k(w)N^w \frac{dw}{w}, \qquad c > 1. \tag{7}$$

Here $c > 1$ because $\zeta(w)$ has its only singularity at $w = 1$. Using the residue theorem we can move the line of integration to the left getting

$$D_k(N) = \frac{1}{2\pi i} \int_{b-i\infty}^{b+i\infty} \zeta^k(w)N^w \frac{dw}{w} + Q_k(N), \tag{8}$$

where $0 < b < 1$ and $Q_k(N)$ is the residue arising from the kth order pole at $w = 1$. It turns out that $Q_k(N) = NP_{k-1}(\log N)$ where P_{k-1} is a polynomial of degree $k - 1$. Equation (8) will be written in the form

$$D_k(N) = \Delta_k(N) + Q_k(N). \tag{9}$$

Now, it turns out that the Lindelöf hypothesis (2) is equivalent to a growth condition on $\Delta_k(N)$

$$\Delta_k(N) = O(N^{1/2+\varepsilon}) \tag{10}$$

for large N and any $\varepsilon > 0$.

For our combinatorial restatement of the Lindelöf hypothesis we remark first that on any vertical path V in the complex plane, given by $\sigma + it$ (σ fixed, $0 < \sigma$, $\sigma \neq 1$, $-\infty < t < \infty$), we can replace

$$\zeta(s) = \frac{1}{1^s} + \frac{1}{2^s} + \frac{1}{3^s} + \cdots \tag{11}$$

by

$$L(s) = \frac{1}{1^s} - \frac{1}{2^s} + \frac{1}{3^s} - \cdots. \tag{12}$$

This is so since

$$L(s) = \zeta(s)(1 - 2^{1-s})$$

and a simple calculation shows that $|1 - 2^{1-s}|$ is bounded on V between two positive constants. Since in the equivalent form (10) of the Lindelöf hypothesis we deal with O-term statements it follows that we can replace in (8)–(9) ζ by L, that is, (11) by (12). The main point of this is that L has no singularities being entire, hence the path of integration can be moved back to the right of point $w = 1$. This might appear to be a circularity but now we can use the Perron formula in reverse, so to say. When this is done $\Delta_k(N)$ is represented in the manner of D_k in (5) as a sum associated with lattice points:

$$\Delta_k(N) = \sum_{x \in H_k(N)} (-1)^{x_1 + x_2 + \cdots + x_k}$$

and we get our combinatorial restatement of the Lindelöf hypothesis: For any $\varepsilon > 0$ and for N large

$$\sum_{x \in H_k(N)} (-1)^{x_1 + x_2 + \cdots + x_k} = O(N^{1/2+\varepsilon}), \qquad k = 2, 3, \ldots .$$

That is, let alternate signatures 1 and -1 be put on positive integer lattice points in E^k, starting with 1 at $(1, 1, \ldots, 1)$ and letting immediate neighbors have opposite signs, then the Lindelöf hypothesis is equivalent to the assertion that for any $k \geq 2$, for any $\varepsilon > 0$ and for N large the sum of signatures in the region bounded by the inequalities

$$0 < x_i, \qquad i = 1, \ldots, k, \qquad x_1 x_2 \cdots x_k \leq N$$

is $O(N^{1/2+\varepsilon})$. For $k = 2$ this assertion is easy to prove. We consider the lattice points (x, y) in the subhyperbolic region $0 < x$, $0 < y$, $xy \leq N$; there are $[N^{1/2}]$ diagonal lattice points (x, x) in the region. With each such diagonal point associate the points of the region which are either exactly above it or exactly to the right of it. The collection of these will be called the track of the diagonal point in question (which we also include in the track). Since the signatures 1 and -1 change from neighbor to neighbor the sum of signatures over any track is 1, 0, or -1. As there are $[N^{1/2}]$ tracks the sum of all signatures is $O(N^{1/2})$, q.e.d. The notion of a track can be generalized to be any sequence of neighbors or even further, any sequence along which the signatures alternate. Now the whole Lindelöf hypothesis is expressible in terms of tracks: the alternatingly signed points of $H_k(N)$ can be represented as a union of $O(N^{1/2+\varepsilon})$ tracks.

The above formulation suggests that we consider sums of alternating signatures over arbitrary convex, star-shaped, or other regions. Also, the concept of signature may be generalized; the generalization is analogous to passing from $\zeta(s)$ and $(1-2^{1-s})\zeta(s)$ to the general L-series.

4

AN APPROACH TO COMPUTING AND COMPUTABILITY

When it comes to these two subjects at least three different levels can be treated: numerical, mathematical, and metamathematical, with respective concerns: numbers, theorems about numbers, and theorems about theorems. Examples of problems typical of these levels are: how to compute $\sqrt{2}$ to the accuracy of at least 10^{-n}? What numbers are iteratively computable by additions alone? Are there well-formulated but unsolvable problems in computing? In this chapter we offer an approach to computing and computability which helps to introduce all three levels together. At first glance the material of this chapter differs so much from that of vol. 1 or from the rest of this volume that the reader may ask what it is doing here. With some reference to the title of this book, we believe that we touch here on an important distinction between "concrete mathematics" with its bent toward individual things, and "abstract mathematics" with its towering aggregates and structure-oriented generalizations. If this sounds too pompous, we can put the matter thus. Our approach (the "concrete") comes by idealizing the act of counting on the fingers of one's hands just as, say, Turing's approach (the "abstract") comes by idealizing the act of calculating with arbitrary symbols on some symbol space. Briefly, the difference of emphasis is that between the hand and the brain; the principal acts are manual in the one case and mental in the other. We start with a short historical sketch which is here partly for "class" and partly to put together some facts, dates, and quotations, each of which separately may be known well enough to the point of being a platitude.

1. DESCENT OF COMPUTERS FROM THE WEAVING MACHINERY

An essential human activity being the making of things—artefacts, let us consider what is involved in their making. At least five factors may be distinguished: (1) raw materials, (2) tools, (3) power necessary to carry out the manufacture, (4) control (= proper sequencing of material operations in space and time), and (5) design. The raw materials are either found naturally or are themselves artefacts; the second possibility leads to longer and longer chains of manufacturing processes. The same is true for tools. Up to relatively recently in history the third factor, power, was provided by man and other animals, with some slight allowance for wind and water. Then the so-called industrial revolution replaced man and animals as sources of power with steam engines. It is the fourth factor, control, that is of interest here.

Let us consider an artefact such as a piece of cloth with a pattern woven into the material rather than dyed on, impressed, or otherwise produced after the cloth is made. Here it is necessary to control the operation of the loom so as to make the threads of certain colors go across over rather than under some other threads. There is the binary— we are almost tempted to say, Boolean—decision, the smallest control act to be made, many thousands of times, for different pairs of threads: whether thread A goes over or under thread B. This used to be done just purely by eye and hand up to about the year 1800, and we can perhaps appreciate, from the repetitiousness of the task, as well as from economic factors, why weaving was the first industry to be automatized. It ought to be added that the right over/under passage for threads is of importance even aside from color and pattern. By changing it one gets different types of cloth with all threads alike, for example, plain weave, twill, and damask.

It was Joseph-Marie Jacquard (1752–1834) of Lyons, France, who invented the Jacquard loom; this was exhibited in Paris in 1801. There is a romantic-sounding story that Jacquard got the idea for his loom while watching his wife comb their daughter's hair. However, this is probably apocryphal and it is more to the point that Jacquard was a weaver and the son of a weaver, that he drew upon a previous loom model of J. de Vaucanson (1709–1782), and maybe even that his wife, during lean times, worked at straw-plaiting. Jacquard's mechanization of the control of pattern weaving was done by providing a pattern tape on which certain spots were either perforated or not. If they were, then metal spindles moved through the holes raising the thread A over the thread B; if the spots were unperforated then A went under B.

There had been previously some mechanisms of the Swiss music-box type, these were fairly elaborate in the case of certain clocks and clockwork mechanisms (such as the mechanical walking lion made by Leonardo da Vinci for Francis the First of France). However, the Jacquard loom was the grand opening of the second industrial revolution: replacement of man as source of control by automata running from a paper tape, even though it took some time since Jacquard's day till the pervasive and insidious encroachment of computers and other automatic control mechanisms of today.

On the computing side, the counting and arithmetic devices are older than the Jacquard loom, being normally dated to the invention of the adding machine in 1642 by the nineteen-year old Pascal (1623–1662). This was followed at some time between 1670 and 1694 by the adding and partially multiplying machine of G. W. Leibniz (1646-1716). However, from some correspondence of Kepler, which became known only in 1957, it would appear that a German mathematician Wilhelm Shickard (1592–1635) of Tübingen, constructed an adding and partially multiplying machine before Pascal [61].

It may be parenthetically added here that the foremost ones among mathematicians were often interested in, and in many cases even actively engaged in the making of, mechanisms and devices, for instance Archimedes, Pascal, Newton, Leibniz, Gauss, Chebyshev, von Neumann. It was only men like G. H. Hardy who have publicly prided themselves on their unsullied purity [68]. However, the dispute between "mental" and "manual" mathematics goes back to antiquity; we have Plato with his strictures against Eudoxus, Archytas, and Menaechmus, blaming them for trying to reduce the Delian problem (= duplication of the cube) to instruments and mechanics, "that thereby the good of geometry was lost, being dragged down into the world of sense from the realm of eternal verities" [4]. Some of these Platonic attitudes have spread by way of Byzantium and Florence to England, giving rise to the so-called Cambridge Platonism. This school may have contributed indirectly to the formation of the view that mathematics should be beautiful and useless, so well exemplified by Hardy's lucubrations and by Cayley's celebrated remark that "Bessel functions are beautiful even if they do have applications."

Under the social and economic pressures more and more need for fast calculating arose from greatly increasing bureaucracy (taxation, state lotteries) and volume of traffic (especially of marine navigation). Calculating machines have been steadily proliferating since the ones of Pascal and Leibniz. However, all these, as well as their more modern descendents— the desk calculating machines of yesterday and today—were highly inert and provided only the analog of hand-loom on which the user wove his own course of calculation. On the other hand, the automatic calculators are much less inert because they contain the whole control sequence for the calculation, stored somewhere inside, just as the woven design is stored on its perforated-paper Jacquard tape. But there is a highly significant difference. It is possible to change the control of the computation during the computation itself, depending on intermediate results, by a reflex process. On the other hand, the unwieldy perforated Jacquard tape remains unchanged throughout the weaving (except for an occassional splice or branching) and so the control is pre-set, there being no "intermediate results" on which it could significantly depend.

The passage from the Jacquard loom to modern automatic calculators occurred partly through the agency of Charles Babbage (1792–1871) who designed the difference engine (1812) and the analytical engine (1833). The last device, though never completed, was the first calculator that was controlled automatically rather than manually. Babbage was an English inventor and mathematician who held for several years Newton's chair at the University of Cambridge (apparently without having held a single lecture [127]). Besides his work on calculating machines he also pioneered an early form of operations research, was active in cryptanalysis, and generally was engaged (as someone wrote about him [127]) in improving the *executive* as distinguished from the *legislative*, powers of mathematical analysis. Babbage's ideas were taken up by Luigi F. Menabrea, an Italian military engineer and mathematician, who wrote the earliest memoir on computing [119], and became in later years one of the first prime ministers of Italy.

Babbage was greatly assisted in his work by his student and friend, Augusta Ada Lovelace (1815–1851) who translated and greatly amplified Menabrea's memoir [103]. Countess Lovelace was the (only legitimate) child of the poet Byron. Lady Byron was rather an accomplished mathematician for her day; her daughter was even more so. In fact, the latter has a claim to being the first computer scientist in history. When she was eighteen, it was recorded of her [127] that "I well remember accompanying her to see Mr. Babbage's wonderful analytical engine. While other visitors gazed at the working of the beautiful instrument with the sort of expression, and I dare say, the sort of feeling, that some savages are said to have shown on first seeing a looking glass or hearing a gun—if indeed, they had as strong an idea of its marvellousness—Miss Byron, young as she was, understood its working, and saw the great beauty of the invention." From Countess Lovelace comes the quotation that "the machine can do only what we know how to order it to perform." This has some considerable bearing on the third industrial revolution: replacement of man as source of design by the automaton.

Here we touch on another aspect of the subject. Machines as tools of production became of interest not only to science but even to literature. There is a line from a novel by Benjamin Disraeli (1804–1881) [46, Bk. 4, Ch. 2], approximately contemporary with Countess Lovelace's dictum, which runs: "And yet the mystery of mysteries is to view machines making machines; a spectacle that fills the mind with curious, and even awful, speculation." This antedates John von Neumann and his self-reproducing automata [24, 85], by more than a century. Together with the then recent appearance of Darwinism, the Disraeli line may well have inspired "The Book of the Machines" in S. Butler's *Erewhon* [25], as well as George Eliot in her *Shadows of the Coming Race*, in [52].

The descent of the whole brood of modern computers from the Jacquard loom, by way of Babbage's engines and Countess Lovelace's programming ideas, is unmistakeable: they are its progeny and the mark of descent is the characteristic ubiquitous perforated paper card. The intervening time has changed some accidents of hardware, from mechanical devices to electron tubes first, and to solid-state elements later, but not the principle of the thing.

2. THEORIES OF COMPUTING

On the theoretical side of computing, we note that it was only in 1936 that A. M. Turing [173] and, independently and almost simultaneously E. L. Post [138], gave an account of what is meant by an effective calculation. Abstracting in a quite transparent way from the symbolic calculating process carried out by a man on some actual symbol space (piece of paper, blackboard, sand, etc.) by means of arbitrary symbols, Turing postulated a linear tape (the symbol space), divided into squares and stretching indefinitely far in both directions. The tape passes through a box that corresponds to man's field of attention and action, so that exactly one square is within the box at any time $t = 0, 1, 2, \ldots$. Each square can carry at any one time one symbol out of a finite alphabet $s_1, \ldots, s_n$ which includes the blank. The box can likewise be at any one time in one or another out of a finite collection of states $q_1, \ldots, q_m$ corresponding to mental states. If at some time t the symbol on the square just then in the box is s_a and the box itself is in the state q_b, then three things happen between t and $t + 1$: s_a is changed to s_A, q_b is changed to q_B, and the box moves ε squares to the right (where $\varepsilon = 1, 0,$ or -1). When the computing job of the machine is finished, if ever, it goes into a state q_F called the final, or quiescent, state and nothing happens thereafter.

This Turing machine, as it came to be known, is entirely specified by: (a) the list of symbols and states, (b) the initial configuration consisting of the specification of those squares that have initially nonblanks, together with their symbols, and the prescription of the initial state and the square initially in the box, and (c) the transition table from the pairs (a, b) to the triples (A, B, ε). The finite totality of all these is the description of the particular Turing machine in question. If the blank symbol is interpreted as 0 and some other symbol as 1, then it may happen that the binary decimal development of some real number x eventually appears on the tape. The number x is then called effectively computable, or Turing computable. An important achievement of Turing was to show that there exist universal Turing machines. These, when presented with a suitably coded description of any particular Turing machine T on their tape, can duplicate the whole computation of T. In this respect any particular Turing machine acts as a "program" while a universal Turing machine is the "computer" into which the program is inserted. By replacing "computing" with "constructing," von Neumann [24] was able to show the existence of universal constructing machines and eventually of self-replicating machines (it may be observed that by a premonition or otherwise, Norbert Wiener who died of a heart condition had worked on nonlinear control and applications of differential equations to the theory of heart action, while von Neumann who died of cancer had worked on the mathematics of self-replication and hence, potentially at least, on the modelling of its pathologies, viz. cancer).

Rather than amplify the foregoing very sketchy outline of Turing's theory, we shall develop a different approach. Turing's method is logic-oriented and the reader with some slight background in logic may perhaps notice that a Turing machine is in some sort a time-dependent, or dynamic, truth-table. As his very terminology shows, Turing emphasizes the role of the mind, or at any rate of the brain with the eye; perhaps for this reason the Turing approach is rather "atomic." Our approach emphasizes the role of the hand and perhaps that is why it is "molecular" rather than "atomic." Indeed, we start by abstracting from the process of counting on one's fingers. Presumably, the primitive arithmetic act of man—if one can rely on certain ancient cave-pictures—is to lower and raise one's fingers and to tick off the hand when all its fingers are "used." Once we have made sure of our supply of fingers and hands, this will lead us to our model of computing. In this connection one may speculate on the hypothetical arithmetic, and even mathematics, of an intelligent dolphin-like or avian species.

We conclude this section with some general remarks. First, like Turing's model, ours will be synchronous: Events happen only at times $t = 1$, $2, \ldots$. An asynchronous computer, for instance such as could be synthesized out of relays and without any clocks, has many attractive features but it is difficult to prevent in it the so-called race condition. Here the whole flow of control of the computation depends on the accidental race between two relays, which will do its job and close first. Second, on the assumption that the number of computationally illiterate, or if preferred, computationally virgin, mathematicians is rather larger than might be supposed, we develop our model from the bottom up and try to bring in certain basic features of actual computing practice. Third, we do not even touch on such alternative means of rigorization of effective computability as Post's systems, calculi of λ-conversion, recursion theory, and Markov's theory of algorithms; all of these are logic-oriented, like Turing's theory, with the possible exception of Markov's approach which is rather more algebraically motivated. All of those distinct theories are of considerable variety of persuasion, approach, and method; the fact that in each case the same set of numbers turns out to be effectively calculable is all the more significant. Fourth, it may strike the reader who is a beginner in the subject that any such definition of "effective computability" is too narrow; for example, only countably many reals are computable and all sorts of convenient closure properties get lost. However, the common experience is that, after a while perhaps, one finds the shoe on the other foot: there is a slight indefiniteness in any (so far) formalization of definite computability, and the concept of effective calculation is really too broad rather than too narrow. Fifth, there is sometimes an odd sort of back-and-forth trade between the current technology and contemporary, or perhaps slightly future, philosophy, with reference to the peak of current technical achievement (clocks, looms, telephone exchanges, Turing universal machines, . . .) serving as the scaffolding for "models of man" (Descartes and clocks, nineteenth century (William James, for example) and telephone exchanges, some of the present-day philosophical hang-ups on computing machines). This dependence is perhaps circular, or rather helical, since a particular philosophy may in turn influence the next developments of technology.

As an example where computer terminology may help in the "modeling of mind" we offer the following hypothesis, which arose from privately collected statistics on the age at which children who *will* become mathematicians learn to speak (those statistics indicate substantial positive correlation between future mathematical ability and relatively late age of learning to speak). In modern computers there are separate parts such as the input equipment which accepts problems, memory units which store information items, central processor unit whose name explains its function, output equipment which brings forth the answers, and others. Those operate in different "languages." For instance, the input equipment may accept problems in the so-called Fortran language which is reasonably similar to the standard mathematical language of elementary algebra and analysis. This is translated by the machine itself into a binary on–off language; incidentally, Fortran is an acronym for FORmula-TRANslation. That binary on–off language is akin to the Morse code, Boolean switching, and the neurophysiology of nerve impulses; it even recalls the basic operation of the Jacquard loom and is more suitable for the central processor use. The answer to the problem obtained by the central processor is then retranslated and printed out. There may be other intermediate levels with their own specific languages but we shall concentrate on the peripheral equipment language, called simply the I/O-language (for input–output, main use is for communication between the environment and the central processor) and the C-language (for central processing).

On the side of neurophysiological–computational analogies, it appears that the brain might have not one but at least two types of memory mechanisms: the fast-acting one (perhaps of the circulating memory type) with the characteristic time scale of the order of minutes, and the slow-acting one (perhaps of the engrammatic type) with the characteristic time scale up to the length of human life. Additionally, there appears to be the possibility of a third mechanism which transfers items of information between the two memories: from the fast-acting small one to the slow but capacious one, and conversely. Some light on such matters is derived from the pathology observations on various memory malfunctions in humans (Korsakov's syndrome, amnesias, agnosias, etc). Similar division of storage occurs in the computing machine where the central processor has its own small but fast-acting core memory while in addition there are other memory systems such as the much bigger but relatively slower-acting magnetic tapes.

It may be hypothesized that the ordinary spoken languages that children learn correspond to the I/O-language of the machine while mathematical reasoning and problem solving may be done in the brain in terms of some C-language. It would appear that other things being equal and good general intelligence being granted we have two possibilities:

(A)   the child who learns to speak late has a better chance to develop his "central processor," perhaps due to linguistic insulation from the environment, or,

(B)   conversely, the child whose "central processor" activity starts earlier is less likely to begin learning the I/O-language right away, on account of more intense inner activity.

Further speculation on the above hypothesis, its possible relation to neoteny and emergence of new mental faculties, or to serial–parallel information processing, and so on, is left to the interested reader.

3. COMPUTING BASED ON COUNTING ON ONE'S FINGERS

Consider a row of locations marked 0, 1, 2, ... For concreteness, let us say that each location is a hole in the ground. Hole 0 is a bottomless pit filled with counters. For concreteness, let us say that these counters are pebbles (calculi). The locations 1, 2, ... are initially empty. There is an attached "program," something like this:

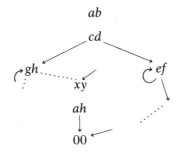

Here all quantities a, b, ... are nonnegative integers and the program is executed as follows. The first instruction is ab: Transfer one counter from location a to location b if possible, that is, if a is not empty just then. The next instruction is cd; if this can be done, that is, if c is not empty at the moment, we do it and then we go for the next instruction along the left arrow, to gh. Otherwise if c happens to be empty so that cd is not executable, we leave c and d alone and go for our next instruction along the right arrow, to ef. Continuing in this way, we eventually reach the final instruction 00—stop. Note that an arrow can lead from an instruction back to itself, as is the case with ef in our example. If both arrows from one instruction lead to another, the latter is just written under the former, as with ab and cd or xy and ah in the example. We now modify our assumption that the locations 1, 2, ... are initially empty. Since any program can be prefaced by a vertical string of instructions

01

01

.

.

.

01

$$0a$$
$$\cdot$$
$$\cdot$$
$$\cdot$$
$$0a$$
$$\cdot$$
$$\cdot$$
$$\cdot$$
$$0z,$$

this modification does not really change anything. We can therefore suppose that finitely many locations are initially nonempty and each one of them contains a finite number of counters. We give now some simple examples in which initial and final contents always refer to locations 1, 2, ... (but never to 0).

(1)   Partial addition.
      Initial contents: $x, y, \ldots$

$$C^{12}$$
$$\downarrow$$
$$00$$

final contents: $0, x + y, \ldots$

Here the effect is that of transfer: contents of location 1 are emptied into location 2.

(2)   Interchange.
      Initial contents: $x, y, 0, \ldots$

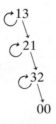

final contents: $y, x, 0, \ldots$

The effect is to interchange the contents of locations 1 and 2; location 3, initially and finally empty, is necessary for the execution as intermediate storage space. We generalize this to the bookkeeping program

$$B\begin{pmatrix} a_1, a_2, \ldots, a_n \\ b_1, b_2, \ldots, b_n \end{pmatrix}$$

which is just

With this the interchange above can be written as

$$B\begin{pmatrix} 1 & 2 & 3 \\ 3 & 1 & 2 \end{pmatrix}.$$

(3) Copy $C(1, 2, 3)$
     Initial contents: $x, y, 0, \ldots$

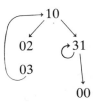

final contents: $x, x + y, 0, \ldots$

This copies the contents of location 1 into location 2, irrespective of what location 2 may contain to begin with. Similarly, we shall use $C(u, v, w)$ for the same program but with locations 1, 2, 3 replaced by u, v, w respectively.

(4) Addition.
     We call the previous, or future, programs subroutines, especially if they are to function together in larger programs by being suitably linked together. Subroutine (3), copy, shows how to execute the

addition $x + y$ if we are not concerned with the preservation of both summands x and y somewhere, as final contents. If we are so concerned, it is possible to synthesize addition by a simple linking of two copy-subroutines thus: Initial contents: x, y, 0, 0, . . .

$$C(1, 3, 4)$$
$$C(2, 3, 4)$$

final contents: x, y, $x + y$, 0, . . .

Here it is understood, *as in all such linking of subroutines*, that the stop instruction, 00, is erased from the first copy, $C(1, 3, 4)$, and all arrows that led to it now go instead of the first instruction of $C(2, 3, 4)$. In the same way we get

(5)   Multiplication $M(1, 2, 3, 4, 5)$.
      Initial contents: x, y, 0 0 0 . . .

$$C(1, 4, 5)$$

final contents x, y, xy, 0, 0, . . .

The reader may wish to check whether this could be achieved with fewer instructions or fewer locations. Also, we note the dissymetry (= effective noncommutativity, roughly) of multiplication: $0 \cdot y = 0$ takes much less time than $x \cdot 0 = 0$; the reader may wish to rectify this.

The preceding examples gave us some performative subroutines: simple programs that do something, such as copy, add, or multiply, the contents of certain locations. We note next a nonperformative subroutine which does nothing to its input numbers, but is nevertheless of considerable usefulness in synthesizing larger programs. In analogy with the well-known Fortran language, it corresponds to a simple IF-statement.

(6)   Magnitude transfer $G(1, 2, 3, 4, 5)$.
      Initial contents: x, y, 0, 0, 0, . . .

$$C(1, 3, 4)$$

$$C(2, 4, 5)$$

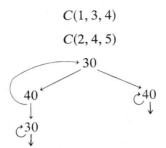

we exit on the left arrow if $x > y$ and on the right arrow if $x \leq y$. The contents of locations 1–5 remain unchanged. For obvious and very important reasons we might call this a control subroutine.

Simple modifications of the above will execute certain forms of subtraction. For instance, we have the "absolute subtraction": Initial contents: x, y, 0, 0, 0, . . .

$$C(1, 3, 4)$$

$$C(2, 4, 5)$$

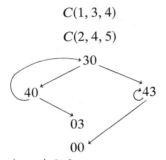

final contents: $x, y, |x - y|, 0, 0, . . .$

Another modification yields the limited subtraction $x \div y = \max (x - y, 0)$:

(7)   Limited subtraction $S(1, 2, 3, 4, 5)$
      Initial contents: x, y, 0, 0, 0, . . .

$$C(1, 3, 4)$$

$$C(2, 4, 5)$$

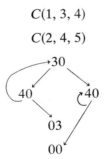

final contents: $x, y, x \div y, 0, 0, . . .$

It may occur to the reader to ask here why the subtractions are relatively complicated as compared to addition. Here it must be remembered that in our system there is no direct way of dealing with negative integers. Recalling the putative role of the hand in our initial abstractions, we might exaggerate somewhat and say that positive and negative integers are similar to the eye but not to the hand. Of course, there is an easy way to handle negative integers as ordered pairs of positive ones. We let $x = (x^+, x^-)$ where x^+ and x^- are ≥ 0 and the usual laws hold

$$(a, b) = (c, d) \qquad \text{if and only if} \qquad a + d = b + c,$$
$$(a, b) + (c, d) = (a + c, b + d),$$
$$(a, b)(c, d) = (ac + bd, bc + ad).$$

This would correspond to splitting each location, a, into two, a^+ and a^-, giving us two rows of locations: $1^+, 2^+, \ldots$ and $1^-, 2^-, \ldots$, with common store 0. Subtraction is now always possible: To subtract the contents of location 1 from 2 we just add the contents of 1^+ to 2^-, and of 1^- to 2^+. Other details for this modification are similarly handled.

As an example of use of the open subroutine $G(1, 2, 3, 4, 5)$ we show how to synthesize from it a simple and not very efficient sorting. Given n nonnegative integers $x_1, x_2, \ldots, x_n$, we recall that $x_{(1)}, x_{(2)}, \ldots, x_{(n)}$ is a monotone nondecreasing rearrangement of the x's.
Initial contents: $x_1, x_2, x_3, 0, 0, 0, \ldots$

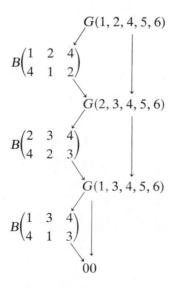

final contents: $x_{(1)}, x_{(2)}, x_{(3)}, 0, 0, 0, \ldots$

4. PROGRAM MODIFICATION

We consider now the following problem. Let n nonnegative integers x_1, $x_2, \ldots, x_n$ be given, as well as the number n itself. Can we write a program similar to the preceding one that would sort the n numbers by nondecreasing magnitude? The main point here is that one single program is to work for any size of n. Obviously our computing scheme must be enlarged: There must be a means of making the program modify itself during the execution. In line with our previous discussion of the Jacquard loom, where the pattern of control is stored on the perforated paper tape, this is as though we now gave the loom some means of punching new holes and covering up the old ones on its tape. We incorporate this self-modifying ability into our programs by allowing not only instructions like

$$ab \qquad a, b \text{ location numbers,}$$

which we shall call simple instructions, but also iterated instructions like

$$\bar{a}b, \qquad a\bar{b}, \qquad \bar{a}\bar{b}$$

In executing these we take $\bar{a}$ to be the location whose number is currently stored in location a. A simple argument shows that there is no need for using multiple iteration.

Returning now to our sorting program for n integers $x_1, \ldots, x_n$ we start with the following block diagram which shows the sequence of steps and helps us to construct the program:

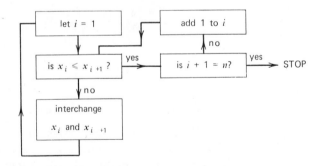

This will sort the numbers $x_1, \ldots, x_n$ by magnitude in a very inefficient way: If any two neighbors are in the wrong order we transpose them and then start from the beginning. It is precisely because we have now the device of iterated instructions, like $\bar{i}$, that we can execute the interchange of x_i and x_{i+1} for general i, as distinguished from, say, interchanging x_1 and x_2 or x_{1000} and x_{1001}. We just reserve some two locations, 2 and 3 in our case, and we store the index i in location 2 and $i+1$ in 3. Then the interchange of x_i and x_{i+1} becomes simply

$$B\begin{pmatrix} \bar{2} & \bar{3} & 4 \\ 4 & 2 & 3 \end{pmatrix}$$

where location 4 is reserved for the execution of this interchange. For technical convenience the numbers $x_1, \ldots, x_n$ will be stored not in locations $1, 2, \ldots, n$ but in another block of n consecutive locations, starting with 8. The first seven locations are reserved for other purposes: 2 and 3 contain the indices of the currently compared pair of x's, 4, 5, 6 are kept as the "arithmetical space," and the essential number n occurs as $n+7$ stored in location 7. The sorting program is now synthesized as follows:

Initial contents: 8, 0, 0, 0, 0, 0, $n+7$, x_1, x_2, ..., x_n, 0, ...

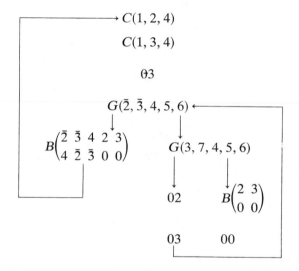

Final contents: 8, 0, 0, 0, 0, 0, $n+7$, $x_{(1)}$, $x_{(2)}$, ..., $x_{(n)}$, 0, ...

This sorting algorithm can be improved in various ways, in particular by obviating the need of returning to x_1 each time a pair of numbers is transposed into right order. It is a rather nontrivial task to minimize the sorting with respect to the number of locations, the complexity of the program, and above all with respect to the average running time assuming, for instance, all $n!$ possible orders as equally likely.

As the above sorting algorithm suggests, it is precisely the feature of iterated instructions which enables us to write down a single algorithm that will work for any n numbers (rather than for 3, 10, or 10^{10}). So, it might be said that the use of iterated instructions extends *bounded* computation to *finite* computation.

5. EXAMPLES OF PROGRAMS

In this section we continue to assemble some simple programs that we shall need as subroutines later on. Then we (attempt to) justify the claim made at the beginning of this chapter, with respect to the numerical level, by developing a program that calculates the decimal development of $\sqrt{2}$, accurate to n decimal places.

(8)   Exponentiation $E(1-8)$.
   Initial contents: x, y, 0, 0, 0, 0, 0, 0, ... (x and y positive integers)

$$C(1, 4, 5)$$

$$C(1, 5, 6)$$

$$C(2, 3, 6)$$

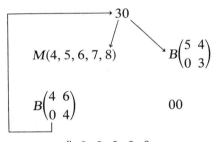

final contents: x, y, x^y, 0, 0, 0, 0, 0, ...

At this point we note something which will occupy us rather considerably in a later section. The program (4), addition, has several loops but no loops-within-loops. The program (5), multiplication, has a doubly embedded loop, and the program (8), exponentiation, shows loop embedding to the depth 3. This of course, is just as expected since the basic arithmetic operation of our computer is the Peano succession: adding 1. Therefore addition is iterated succession, multiplication is iterated addition, and exponentiation is iterated multiplication. In each case iteration means an increase of 1 in the loop-embedding depth. The relation between the magnitude of the number computed, or the growth rate of the function being computed, and the depth of loop-embedding will occupy us soon. Here we recall the Ackermann function $A(n)$ introduced in vol. 1, p. 76.

$A(n)$ was defined by diagonalizing on a sequence $f_1(x, y), f_2(x, y), \ldots$ of functions each of which, as a function of x, increases faster than any bounded iterate of the previous one. Since $A(n) = f_n(n, n)$, it might be thought that none of our programs could possibly compute $A(n)$, given n as the initial contents of location 1, since deeper and deeper loop embedding is needed as we compute successive values of $A(n)$. However, this is true only if we consider simple programs; with iterated programs using iterated locations $\bar{1}, \bar{2}, \ldots$ the matter stands differently as will be shown.

(9)   Division with remainder $D(1, 2, 3, 4, 5, 6, 7, 8)$.

Let x and y be integers, where $x \geq 0$ and $y > 0$, and let $x = qy + r$ with $0 \leq r < y$, so that r is the remainder after dividing x by y and $q = [x/y]$ is the quotient. To compute q and r we have:

Initial contents: $x, y, 0, 0, 0, 0, 0, 0, \ldots$

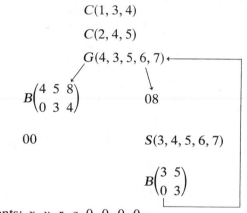

$$C(1, 3, 4)$$

$$C(2, 4, 5)$$

$$G(4, 3, 5, 6, 7)$$

$$B\begin{pmatrix} 4 & 5 & 8 \\ 0 & 3 & 4 \end{pmatrix} \qquad 08$$

$$00 \qquad\qquad S(3, 4, 5, 6, 7)$$

$$B\begin{pmatrix} 3 & 5 \\ 0 & 3 \end{pmatrix}$$

final contents: $x, y, r, q, 0, 0, 0, 0, \ldots$

Here the first two operations merely copy x and y into locations 3 and 4, the loop shown executes the continued subtraction $x \div y, \ldots$, counting the number of subtractions by means of the instruction 08 (this number will become q) and keeping track of whether y exceeds $x \div ky$. As soon as it does, we exit from the loop keeping the last difference as the remainder r.

It is easy to show now that the following programs can be written (and the reader may wish to do so): Euclidean algorithm that computes the g.c.d. of two positive integers x and y, prime-testing program that tests a positive integer n for being prime, a program that computes the nth consecutive prime on being given n, and others of increasing complexity.

(10)   Converting a fraction to decimal form.

Given integers n, x, y where n is positive and $0 < x < y$, we wish to convert the fraction x/y into the standard decimal form keeping n

digits $d_1, \ldots,$ d_n and observing the usual round-off rules with respect to d_n.

Initial contents: $x, y, 10, 5, n, 15, 0, 0, \ldots$

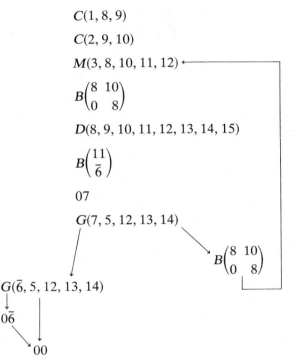

$$C(1, 8, 9)$$

$$C(2, 9, 10)$$

$$M(3, 8, 10, 11, 12) \leftarrow$$

$$B\begin{pmatrix} 8 & 10 \\ 0 & 8 \end{pmatrix}$$

$$D(8, 9, 10, 11, 12, 13, 14, 15)$$

$$B\begin{pmatrix} 11 \\ 6 \end{pmatrix}$$

$$07$$

$$G(7, 5, 12, 13, 14)$$

$$B\begin{pmatrix} 8 & 10 \\ 0 & 8 \end{pmatrix}$$

$$G(\bar{6}, 5, 12, 13, 14)$$

$$0\bar{6}$$

$$00$$

final contents: $x, y, 10, 5, n, 14+n, n, 0, 0, 0, 0, 0, 0, 0, d_1, d_2, \ldots,$ $d_n, 0, \ldots$

Here the large loop executes the continued division with the subroutine $D(8\text{--}15)$, keeping track of the number of divisions and placing each quotient $10r_i/y$ in the right slot as the successive digit; the remainder after each division is multiplied by 10 and serves as the next number to be divided by y. The locations 2 and 3 carry the constants needed for the program; the locations 5 and 7 keep track of the number of passages round the big loop; the location 6, whose contents increases by 1 on each loop-passage, gives us the place in which the next digit is to be stored. The part of the program just above the stop 00 executes the round-off depending on whether or not the last digit is less than 5.

Next, we compute $\sqrt{2}$ to within accuracy of 10^{-n}, using the algorithm of vol. 1, pp. 134–135. Let p_0 and q_0 be positive integers and put

$$p_{k+1} = p_k + 2q_k, \qquad q_{k+1} = p_k + q_k. \tag{1}$$

Then, as was shown, the sequence p_0/q_0, p_1/q_1, ... converges to $\sqrt{2}$ in the alternating manner: Each difference $p_k/q_k - \sqrt{2}$ is less in absolute value than the previous one, and opposite in sign. Hence

$$\left| \frac{p_k}{q_k} - \sqrt{2} \right| < \frac{1}{2} \left| \frac{p_k}{q_k} - \frac{p_{k-1}}{q_{k-1}} \right| = \frac{|p_k q_{k-1} - p_{k-1} q_k|}{2 q_k q_{k-1}}. \tag{2}$$

We exploit this to obtain the stopping rule: Keep computing successive approximations p_k/q_k from (1), each time testing from (2) whether

$$2 q_k q_{k-1} > 10^n |p_k q_{k-1} - p_{k-1} q_k|. \tag{3}$$

As soon as this occurs we stop and keep the last approximation p_k/q_k computed as the answer. Substituting from (1) we find

$$p_k q_{k-1} - p_{k-1} q_k = 2 q_{k-1}^2 - p_{k-1}^2 = p_{k-2}^2 - 2 q_{k-2}^2 = \cdots = \pm(p_0^2 - 2 q_0^2).$$

We shall take $p_0 = q_0 = 1$ and so, by the above, (3) simplifies to

$$2 q_k q_{k-1} > 10^n. \tag{4}$$

To carry this out we have the following:
Initial contents: 10, n, 0, 0, ...

$$E(1, 2, 3, 4, 5, 6, 7, 8)$$

04

05

$$C(4, 7, 8) \leftarrow$$

$$C(5, 7, 8)$$

$$C(7, 6, 8)$$

$$C(5, 7, 8)$$

$$M(5, 7, 8, 9, 10)$$

$$C(8, 9, 10)$$

$$\circlearrowleft 98 \searrow$$

$$G(8, 3, 9, 10, 11)$$

$$\downarrow \qquad \searrow$$

$$B\begin{pmatrix} 4 & 5 & 8 \\ 0 & 0 & 0 \end{pmatrix} \qquad B\begin{pmatrix} 4 & 5 & 6 & 7 \\ 0 & 0 & 4 & 5 \end{pmatrix}$$

00

final contents: 10, n, 0, 0, p_k, q_k, 0, 0, ...

The final program for getting the decimal development is obtained by subtracting q_k from p_k, and then using the subroutine (10) to convert $(p_k - q_k)/q_k$ into the decimal form, with n digits.

6. NUMERICAL, MATHEMATICAL, AND METAMATHEMATICAL LEVELS

It will be shown now that our computing scheme can duplicate the computation of any Turing machine T, including of course the case when T is a universal Turing machine. This will be done in the convincing though not necessarily shortest way: by bodily and literally simulating T in our terms.

In the first place, instead of a single row of locations $1, 2, \ldots$ we may assume any finite number of rows, say k of them, with one common store 0:

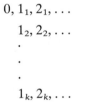

$$0, 1_1, 2_1, \ldots$$
$$1_2, 2_2, \ldots$$
$$\vdots$$
$$1_k, 2_k, \ldots$$

This is arranged by means of the division with remainder, of example (9). Given n, let $n = kq + r$ where $q = 0, 1, 2, \ldots$, and $0 \le r \le k - 1$; we let the nth location of the single row correspond in a 1:1 manner to the $(q+1)$-st location of the $(r+1)$-st row, $(r+1)_{q+1}$. Although we shall not need it, we remark that an infinite row of locations could also be assumed, under any standard 1:1 correspondence between positive integers n and ordered pairs of positive integers (i, j); for instance, with $n \leftrightarrow (i, j)$ where

$$n = j + \frac{(i+j-1)(i+j-2)}{2}.$$

For our demonstration we take $k = 8$. Of the eight rows of locations the first two simulate the tape of the Turing machine, each location corresponding to one square, in this order: $\ldots, 3_2, 2_2, 1_2, 1_1, 2_1, 3_1, \ldots$. Location 1_1 corresponds to the square currently in the box, and instead of moving the box we shall move the simulated tape. However, a stationary tape with a moveable box could also be simulated and the reader may wish to carry out the details.

The next five rows $3_1, 3_2, \ldots$; $4_1, 4_2, \ldots$; $7_1, 7_2, \ldots$ carry the lists of state–symbol and state–symbol–motion combinations in the following fashion. Let the Turing machine T, about to be simulated, have n symbols

$s_1, \ldots, s_n$ and m states $q_1, \ldots, q_m$. We store symbol indices in $3_1, 3_2, \ldots$ and state indices in $4_1, 4_2, \ldots$ in such a way that for every one of the mn pairs (s_i, q_j) there is exactly one value r, $1 \leq r \leq mn$, such that the contents of 3_r and 4_r are i and j, respectively. If the ith symbol and the jth state give rise under the regimen of T to i_1th symbol and j_1th state, we arrange for the contents of the locations 5_r and 6_r to be i_1 and j_1. The location 7_r carries then the motion specification: Its contents is 0, 1 or 2 depending on whether the motion is one square to the left, 0, or one square to the right. These contents of the locations of rows 3–7 shall remain untouched throughout.

The eighth row is the arithmetical space reserved for intermediate computations, except for the first three locations. Of these, 8_1 contains the (index of the) current state of T, while 8_2 and 8_3 indicate how much of the "tape" is occupied at any given time (i.e., they show the largest indices of nonempty locations in rows 1 and 2). Finally, we suppose that the initial contents of the first two rows show the initial condition of the "tape."

We are ready to start our simulation. First, the contents of 1_1 and 8_1 are matched against those of 3_1, 4_1, then 3_2, $4_2, \ldots$ till we find coincidence, say with 3_r, 4_r. The contents of 1_1 is now changed to 5_r and that of 8_1 to 6_r. Next, we execute the correct motion by suitable transfer of the contents of all locations in rows 1 and 2. Since the locations 8_2 and 8_3 tell us how much of the tape is occupied, the transfer operation can be arranged when we know 7_r. Finally, we change the contents of 8_2 and 8_3 according to the motion of the tape. Now the whole operation is repeated again. Among the states of the Turing machine T there is one, say q_F, which is the terminal state. If the contents of 8_1 ever comes to be F, the simulation program stops, after perhaps performing some final clean-up operations. Otherwise, the simulation program never stops, just as T itself would never stop. It is clear from its nature that the simulation program must itself be iterated rather than simple.

It follows that our computing scheme is at least as powerful as that of Turing machines, or as any other scheme equivalent to the Turing machines. The reverse is also true: Whatever can be computed by our scheme is also computable by a Turing machine. The reader who is acquainted with the elements of Turing machine theory, may wish now to complete a direct equivalence proof of our scheme and Turing's, by the reverse simulation: of our computing scheme on a suitable Turing machine.

We come now to the other two "levels" mentioned at the beginning of this chapter. First we consider what has been called "mathematical level," very loosely described as concerned with theorems rather than with numbers. The theorems on which we intend to demonstrate the possible usefulness of our approach will be theorems of number theory. They will be concerned with the arithmetical nature of a real number x (is it algebraic? is it transcendental?) or with the arithmetical relation between two (or more) numbers (are they algebraically dependent? If not, which one is "more" transcendental?) Our guiding line will be something like this: Some information about the arithmetic nature of x can be extracted from the examination of a suitably coded recipe for cooking up a sequence p_0/q_0, p_1/q_1, ... of rational approximations converging to x. Here we recall the modification of our computing scheme, mentioned toward the end of Section 3, which takes care of negative integers with two rows of locations; similar modifications with four rows could take care of Gaussian integers $a + bi$.

We shall generate the sequence $\{p_n/q_n\}$ as follows. There will be an unending cycling program
Initial contents: p_0, q_0, ...

$$P$$

$$\circlearrowleft Q$$

After the preliminary part P, which may be absent, the program gets into an unending loop Q, and the contents of locations 1 and 2 are to be p_n and q_n after the nth cycling round the loop. A sample theorem, whose proof has been outlined in vol. 1, p. 135 is this: If P is absent and Q is of the form

$$Q_1$$
$$Q_2$$
$$\vdots$$
$$Q_s$$

(1)

where each Q_i is an addition subroutine, then $x = \lim_{n \to \infty} p_n/q_n$ is necessarily algebraic. Conversely, for every real algebraic x suitable Q_i's can be found. Briefly, a number is algebraic if and only if it is iteratively computable by additions alone. Here, of course, we include subtractions, on the strength of the previously mentioned modifications for handling negative numbers.

Another theorem, proved in vol. 1, p. 137, was the following. Let the initial contents be p_0, q_0, ... z_0, 0, 0, ... with z_0 in the kth location; let the prefaratory part P be missing, and let Q be of the form (1) but now let each Q_i be an addition or a multiplication subroutine. Suppose that the contents of the locations, after nth cycling, are $p_n, q_n, \ldots, z_n, 0, 0, \ldots$.

Then $x = \lim\limits_{n \to \infty} p_n/q_n$ is algebraic if and only if $k = 2$. Here we understand that the number k of sequences necessary to compute x has been reduced to its minimum. Thus $k \geq 2$ since we must have at least two sequences: one for the numerators p_n and one for the denominators q_n (it being tacitly assumed that x is irrational). A possible interpretation of the "two sequences are equivalent to algebraicity" result is that algebraic numbers are clocklessly computable, and conversely. That is, we do not keep track of the number n of cycles completed, since there is no room for accumulating a third sequence $r_n = n$. The reader might wish to formulate this and then to prove it.

There are many other candidates for quantities which might throw some light on the arithmetic nature of x under our scheme of computing x: numbers associated with the prefaratory part P, with whether P and Q are simple or iterated, the total numbers of locations used in P and in Q, the depth of loop imbedding of Q, and so on. The reader may wish to recall here the contents of vol. 1, pp. 247–249. The central point is that the fairly rich combinatorial structure of our computing scheme affords us the possibility of many "complexity-invariants" to support the statement that the more complex the computation of x, the "more transcendental" the number x.

As for the third level, the metamathematical, this will occupy us in the remainder of the present chapter. Here, in this section, we shall show that there exists a well-defined function $f(c)$, on positive integers to positive integers, which is not computable. As a matter of fact, with some elements of mathematical logic, the existence of such a function can be shown to imply the celebrated incompleteness theorem of Gödel [59], which asserts the existence of propositions both true and unprovable. One might even claim that this is one of the simplest ways to prove Gödel's result, if not the simplest one. It is necessary to define first a computable function. In terms of our computing scheme we call $F(x)$ computable if there exists a program $P = P(F)$, whose computation comes to a halt, and which transforms initial contents $x, 0, 0, \ldots$ into final contents $x, F(x), 0, 0, \ldots$, no matter what x may be.

We examine first some examples of programs that are not halting. Our first example is the one-instruction program

which tries to exhaust the infinite store of counters in 0, and so goes on forever. Our second program also starts with all locations empty,

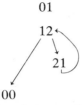

and juggles one counter forever between the locations 1 and 2. The third example is an iterated program
Initial contents 0, 0, . . .

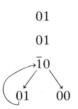

This will keep on hunting forever for the first initially nonempty location; in the process it "visits" every location. Another example, which the reader may wish to encode, is the program that examines successively all rational numbers p/q, p, and q being positive integers, and stops only if it finds p and q such that $p^2 = 2q^2$.

We return now to our well-defined but not computable function $f(c)$. Roughly, this will be defined as the size of the largest computation by a halting program of size c. Here, and in the remainder of this section, we shall assume that the locations are initially empty, unless the contrary is explicitly stated. The size c of a program is simply the number c of instructions in it. We use the letter c because originally the name "command" rather than the more common "instruction" was used. In some ways "command" is better—the computer can operate, so to say, only in the imperative mode rather than the declarative or the interrogative, to borrow some standard grammatical terms. In this connection the reader may wish to recall the statement of Countess Lovelace, quoted in the historical Section 1.

We must now explain what is meant by "the size of the largest computation by a halting program of size c." Let $S(c)$ be the maximum of the final contents after a terminating program $P(c)$ of size c. Recalling the synchronous operation of our computation schemes (= events happen only at times $t = 1, 2, \ldots$) we define similarly $T(c)$ to be the longest running time of such a program $P(c)$. Finally, we define $L(c)$ to be the largest number of locations that are visited by such a program. It ought to be noted that at this point we cannot yet claim that $S(c)$, $T(c)$, $L(c)$ are defined; it might be a priori possible that an infinite sequence of programs $P_1(c)$, $P_2(c)$, ... exists for which the largest computed numbers $S_1(c)$, $S_2(c)$, ... tend to infinity. We show first that the three functions $S(c)$, $T(c)$, $L(c)$ are equi-definable—either all are well-defined or none. This follows simply from the inequalities

$$S(c) \le T(c) \le S(3c), \qquad (2)$$

$$2L(c) \le T(c) \le L(3c). \qquad (3)$$

The first part of (2), $S(c) \le T(c)$, is obvious since just one counter is transferred at a time. To show that $T(c) \le S(3c)$ we consider any program $P(c)$ of size c and replace it by interpolating a new instruction. Specifically, we replace in it any part

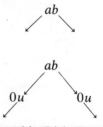

by

where u is any location not used in $P(c)$. The effect is to replace $P(c)$ by a completely equivalent "time-keeper" program; after the computation is finished the running time of $P(c)$ appears as the location contents in location u. The size is now at most three times that of $P(c)$; "at most" refers to the fact that, for example, in

01

01

.

.

.

we replace by

01

0u

01

0u

.

.

.

Since $T(c)$ appears as the final location contents, we conclude that if $S(c)$

is well defined then so is $T(c)$ and (2) holds. The reader may wish to prove (3) in the same way.

To show that all three functions S, T, L are in fact well defined, we have to examine the set of all halting programs of size c. This set is obviously infinite, as can be seen by examining the case $c = 1$:

$$0a \qquad\qquad a = 1, 2, \ldots$$
$$00$$

This example suggests that the infinitude of our halting programs of size c is spurious and in fact, it depends on mere relabeling of locations. Suppose that proper economy is exercised in rewriting programs, so that no unnecessary locations are used and a higher-numbered location never appears where a lower-numbered one would do. Under these conditions, as the reader may wish to show, the number of distinct programs of size c is finite. The same holds, of course, for the halting programs of size c. Hence the three functions $S(c)$, $T(c)$, $L(c)$ are well defined and any one will do as the "size of the computation." However, none of the three is computable. By (2) and (3) it is enough to show this for any one of the three, and we choose for convenience $S(c)$.

Suppose then that there exists a program P that transforms initial location contents $c, 0, 0, \ldots$ into final contents $c, S(c), 0, \ldots$. This P is a fixed-size halting program, say of K instructions, and it is independent of c. We first produce a program Q that starts with all locations empty and ends up with c counters in the first location (and all others empty). It is easy to produce a small Q that will do it, for instance, one with $[\sqrt{c}]$ instructions (for c sufficiently large). Now we preface P with Q, to get a halting program QP. This is of size $K + [\sqrt{c}]$, starts with all locations empty, and produces final contents $S(c)$. However, by the definition of the function S it follows that

$$S(c) \le S(K + [\sqrt{c}]),$$

which is a contradiction since $S(c)$ is increasing with c.

As claimed before, the existence of a well-defined but not computable function $f(c)$ implies Gödel's incompleteness theorem. The reader who is acquainted with the proof of Gödel's theorem may recall that it proceeds by an oblique self-referential procedure: A proposition is constructed which, *under proper interpretation*, asserts its own unprovability. Both the obliqueness and the self-reference are rather clearly exhibited in our argument with $S(c)$. It may be observed here that it is the obliqueness of the self-reference which preserves the correctness of the argument and prevents it from becoming a direct self-referential paradox like Russell's classical paradox (the class of all sets that do not contain themselves both is, and is not, a member of itself) or the ancient paradox of the liar ("The assertion I now make is false").

7. WHAT COMPUTERS CAN AND CANNOT DO

We have just seen that there exist well-defined functions which are not computable, like the function $S(c)$ of the last section. There exist also well-formulated problems which are unsolvable. The classical one in the Turing formulation is: Can one construct a Turing machine T which, on being presented with a suitably coded description of any Turing machine T_1, will decide whether or not the computation of T_1 will stop? Of course, this metacomputation of T must itself stop. This is known as the halting problem and, as it turns out, is unsolvable, no such T exists. The halting problem has been rather thoroughly investigated and is now so central that new unsolvability results, even in unrelated fields, such as group theory, are often obtained by reducing, rephrasing, or interpreting the problem in hand so that it becomes the halting problem.

In our computing scheme the halting problem can be handled once we acquire the means to have programs handle programs. When this is done there is no difficulty in proving the unsolvability of the halting problem in these terms: There exists no halting program P_0 which, on being presented with (a suitably coded description of) any program P, will compute the function $h(P)$ ($= 0$ if P is not halting, $= 1$ if it is). This is proved indirectly. For if such P_0 existed, we could generate successively the finite number of essentially distinct programs of size c, apply P_0 to each in turn, simulate those pronounced to be halting by P_0, and so compute the noncomputable functions $S(c)$, $T(c)$, $L(c)$. There are now two opposite directions in which this metamathematical development can continue:

(a)   smaller and better unsolvability results in which the halting problem is shown to be unsolvable for smaller and smaller subsets of the set of all programs, and

(b)   bigger and better solvability results in which the halting problem is shown to be solvable for bigger and bigger subsets of the set of all programs.

Such two-sided delimitation, from the "outside" and the "inside," is one reasonably precise restatement of a part of the problem of describing what machines can, and what they cannot, do. The possibility of delimitation from the outside is guaranteed by the unsolvability of the general halting problem itself. But we have yet to show that there is a reasonably large class of programs for which the halting problem is in fact solvable. This will be done for the class N of normal programs: Simple programs (i.e., without iterated locations $\bar{a}$) with completely imbedded loop-structure (to be described in more detail later on). It will turn out that these programs correspond rather closely to the primitive recursive functions introduced in vol. 1, p. 77. In particular, each application of the primitive recursion schema corresponds to one deeper level of loop imbedding. Therefore, for the subclass $N(c)$ of N, consisting of the N-programs of size c, the depth of loop imbedding is automatically bounded, by c for instance. This implies a certain bound on the growth, in terms of c, of the biggest number that a program P in $N(c)$ can produce (if it ever stops). The bounding in question parallels exactly the proof that a primitive recursive function cannot increase as fast as the Ackermann function introduced in vol. 1, p. 76. Such a proof is (almost) obviated in our computing scheme. First, each successive use of iteration calls for one more level of loop imbedding. So, for example, the Peano successor function we can generate looplessly, addition (which is its iteration) calls for a simple loop, multiplication (which is iterated addition) needs a loop within a loop, similarly exponentiation requires a triply embedded loop, and so on, as shown in examples (4), (5), (8) in earlier sections. Next, the Ackermann function is a diagonalization of a sequence of functions, each of which grows faster than any bounded iterate of the previous one.

Therefore it is (almost) clear that the biggest number that can be computed by a halting program P in $N(c)$, or equivalently by equation (2) of the last section, the longest time which such a P can run, is bounded by an Ackermann-like function.

But the Ackermann function, or such of its relatives as may be needed, can be computed though, of course, by an iterated rather than a simple program. This will be shown later. Therefore we can compute a bounding function $Z(c)$ which bounds the running time of all halting programs in $N(c)$. We then simulate our program P of $N(c)$ for the length of time $Z(c)$, and if it is still running then it is not halting.

Two jobs must be done before we can tackle the details of the proof. First, we must show how programs can act on programs. Second, we must indicate how one computes Ackermann-like functions. These will be done in the next section, and then the restricted halting problem for N will be shown to be solvable.

8. PROGRAMS ACTING ON PROGRAMS

A complete description of a program P of size c may be obtained as follows. There are c instructions $I_1, \ldots, I_c$, exclusive of the stop I_0. Each instruction I_i consists of two location numbers: $I_i = F(i)\ S(i)$, and $F(0) = S(0) = 0$. In general, $F(i)$ and $S(i)$ can be any nonnegative integers. The numbers $F(i)$ and $S(i)$ can be simple or iterated; to account for this we introduce two further functions:

$$a(i) = 0 \quad \text{if } F(i) \text{ simple,}$$
$$= 1 \quad \text{if } F(i) \text{ iterated;}$$
$$b(i) = 0 \quad \text{if } S(i) \text{ simple,}$$
$$= 1 \quad \text{if } S(i) \text{ iterated;}$$

and in particular, $a(0) = b(0) = 0$. From each instruction I_i $(1 \le i \le c)$ there start two arrows, the right one leads to $I_{N(i)}$ and the left one to $I_{Y(i)}$. Y and N are mnemonics for *yes* and *no*, with reference to the question whether the transfer of a counter involved in I_i can be done. $Y(i) = 0$ if the *yes* arrow from I_i leads to "stop," similarly $N(i) = 0$ for the *no* arrow. We must have certain obvious conditions as consistency relations; for instance, $0 \le Y(i) \le c$ and $0 \le N(i) \le c$, $Y(i) = i$ or $N(i) = i$ may occur but not both, if $a(i) = 1$ then $F(i) \ne 0$ and if $b(i) = 1$ then $S(i) \ne 0$, and so on, but these are unimportant.

We have now our six program functions $F(i)$, $S(i)$, $a(i)$, $b(i)$, $Y(i)$, $N(i)$, the first four defined for $0 \le i \le c$, the last two for $1 \le i \le c$. Together with the number c, these six functions form a complete description of the program P. We also introduce the state functions for P:

$$\phi(t), f_1(t), f_2(t), \ldots .$$

Here $\phi(1) = 1$ and $\phi(t)$ is the number of the instruction being executed at the time t. Similarly, for $t \ge 1$ $f_j(t)$ is the contents of the location j at the time $t - 1/2$. If the program stops at the time T then $\phi(t) = 0$ for $t \ge T$. The reader may wish to verify that

$$\phi(t+1) = Y(\phi(t))\{[1 - a(\phi(t))] \operatorname{sgn} f_{F(\phi(t))}(t) + a(\phi(t)) \operatorname{sgn} f_{f_{F(\phi(t))}(t)}(t)\}$$
$$+ N(\phi(t))\{[1 - a(\phi(t))][1 - \operatorname{sgn} f_{F(\phi(t))}(t)]$$
$$+ a(\phi(t))[1 - \operatorname{sgn} f_{f_{F(\phi(t))}(t)}(t)]\}. \tag{4}$$

We call this the state equation for $\phi(t)$. Similar recursive state equation can be developed for the contents $f_j(t)$. Because of the number of ways in which the jth location may be involved in the instruction $I_{\phi(t)}$ being executed at the time t, the state equation for $f_j(t)$ is lengthy:

$$f_j(t+1) = f_j(t) + [1 - a(\phi(t))] \operatorname{sgn} f_{F(\phi(t))}(t)$$
$$- \operatorname{sgn} f_j(t)[1 - \operatorname{sgn} |F(\phi(t)) - j|]\}$$
$$+ a(\phi(t)) \operatorname{sgn} f_{f_{F(\phi(t))}(t)}(t) \operatorname{sgn} |f_{F(\phi(t))}(t) - j|$$
$$- \{[1 - b(\phi(t))] \operatorname{sgn} |S(\phi(t)) - j| + b(\phi(t)) \operatorname{sgn} |f_{S(\phi(t))}(t) - j|\}$$
$$\{[1 - a(\phi(t))] \operatorname{sgn} f_{F(\phi(t))}(t) + a(\phi(t)) \operatorname{sgn} f_{f_{F(\phi(t))}(t)}(t)\}. \tag{5}$$

The size c of the program and its six program functions allows us a direct encoding of programs P into their Gödel numbers $G(P)$:

$$G(P) = 2^c \prod_{i=1}^{c} [P_{6i-5}^{F(i)} P_{6i-4}^{S(i)} P_{6i-3}^{a(i)} P_{6i-2}^{b(i)} P_{6i-1}^{Y(i)} P_{6i}^{N(i)}] \qquad (6)$$

where $P_1, P_2, P_3, \ldots$ are the *odd* primes $3, 5, 7, \ldots$ in order. To each program P there corresponds a unique Gödel number $G(P)$. For instance, consider the copy subroutine (3) of Section 3:

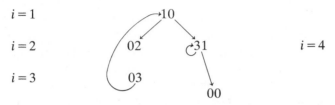

$i = 1$ $i = 2$ $i = 3$ $i = 4$

Here $c = 4$ and the six program functions are given in the tabular form

i	$F(i)$	$S(i)$	$a(i)$	$b(i)$	$Y(i)$	$N(i)$
1	1	0	0	0	2	4
2	0	2	0	0	3	3
3	0	3	0	0	1	1
4	3	1	0	0	4	0

Hence the Gödel number is

$$2^4 \cdot 3 \cdot 13^2 \cdot 17^4 \cdot 23^2 \cdot 37^3 \cdot 41^3 \cdot 47^3 \cdot 61 \cdot 67 \cdot 71^3 \cdot 73 \cdot 89^4.$$

Conversely, given any number g which *is* in fact the Gödel number of a program P, we can "decode" g by means of simple subroutines with prime numbers, and so obtain the six program functions of the program P. In particular, there exists a universal simulating program P_1, with the following property. P_1 accepts as its inputs, in the form of suitable initial contents, Gödel numbers $G(P)$ together with the initial contents of locations that would be necessary for proper execution of P. Working with several rows of locations, just as in the simulation of a Turing machine T, P_1 first "unwinds" the Gödel number $G(P)$ and stores c together with the six program functions in certain six location-rows. Then, employing the state equations (4) and (5), it proceeds to duplicate the whole computation of P, and it stops if P itself is halting. The existence of this universal simulator obviously parallels the existence of universal Turing machines and the reader may wish to write down such a universal simulating program.

With the device of Gödel numbers $G(P)$ we acquire the means of having programs work on programs. For instance, the reader may wish to construct a universal verifying program P_2. This starts with a number g in the location 1 and all others empty, is terminating, and produces final contents g, $h(g)$, 0, 0, ... where $h(g) = 0$ or 1 depending on whether g is or is not the Gödel number of a program. The working of this depends on decoding g according to (6) and then checking whether the (mostly tacit) conventions for the program functions are satisfied. Of course, with the Gödel numbering we have, finally, the means of stating the halting problem precisely: Does there exist a program which accepts Gödel numbers $G(P)$ and decides whether the computation of P stops?

Other programs that work on programs may be presented with Gödel numbers $G(P)$ and required to have as their output various 0–1 functions h. For instance, we might have $h(P) = 0$ or 1 depending on whether P is simple or iterated, or whether P is loopless or has loops. Further examples which are more complicated might be required to accept $G(P)$ and decide whether or not the program P has its loops completely embedded in other loops. If the answer is yes another program might be tacked on to compute the maximal depth of loop embedding.

The reader might wish to examine the possibility of existence, or nonexistence, of various simplifying programs. These accept any Gödel number $G(P)$ and decide, by means of a suitable 0–1 output, whether P can be simplified in some sense. For instance, the reader may wish to prove the nonexistence (e.g., by reducing it to the halting problem) of the simplifying program in the strict sense: To decide, given $G(P)$, whether a simple program exists which computes the same things as P. This is, of course, trivial when P is simple itself, but far from being so if P is iterated. A further question might be: What is the maximum number allowed of iterated location numbers in P so that the simplification *can* be carried out?

We come now to the second part of this section, which deals with Ackermann-like functions. The original function $A(n)$ of Ackermann (somewhat different from the example in vol. 1, p. 76) was given as $A_n = f_n(n, n)$, where

$$f_0(x, y) = x + y \tag{7a}$$

$$\begin{aligned} f_{n+1}(0, y) &= 0 && \text{if } n = 0, \\ &= 1 && \text{if } n = 1, \\ &= y && \text{if } n > 1, \end{aligned} \tag{7b}$$

$$f_{n+1}(x + 1, y) = f_n(f_{n+1}(x, y), y). \tag{7c}$$

We compute successively

$$f_1(x, y) = xy, \qquad f_2(x, y) = y^x, \qquad f_3(x, y) = y^{y^{\cdot^{\cdot^{\cdot^y}}}} \quad (1 + x \text{ occurrences of } y),$$

and

$$f_4(x+1, y) = y^{y^{\cdot^{\cdot^{\cdot^y}}}} \qquad (1+f_3(x, y) \text{ occurrences of } y).$$

Therefore

$$A(1) = 1, \quad A(2) = 4, \quad A(3) = 3^{3^{3^3}}$$

while

$$A(4) = 4^{4^{\cdot^{\cdot^{\cdot^4}}}} \qquad (1 + 4^{4^{4^4}} \quad \text{occurrences of } 4).$$

There is also a simpler function $H(n)$ of the same type, given by Hermes [73]. This too is obtained by the diagonal process: $H(n) = f(n, n)$ where

$$f(0, y) = y + 1 \tag{8a}$$
$$f(x+1, 0) = f(x, 1) \tag{8b}$$
$$f(x+1, y+1) = f(x, f(x+1, y)). \tag{8c}$$

Here we compute successively

$$f(0, y) = y + 1, \qquad f(1, y) = y + 2, \qquad f(2, y) = 2y + 3.$$

Setting $f(3, y) = g(y)$ we have

$$g(y+1) = 2g(y) + 3, \qquad g(0) = 5.$$

Employing the explicit iteration formulas of vol. 1, p. 51 we iterate the function $2x + 3$ y times getting

$$g(y) = f(3, y) = 2^{y+3} - 3.$$

Putting $f(4, y) = h(y)$ we have by (8c)

$$h(y+1) = 2^{3+h(y)} - 3, \qquad h(0) = 13,$$

and so

$$f(4, y) = 2^{2^{\cdot^{\cdot^{\cdot^2}}}} - 3 \qquad (3 + y \text{ occurrences of } 2).$$

Hence

$$H(0) = 1, \quad H(1) = 3, \quad H(2) = 7, \quad H(3) = 61, \quad H(4) = 2^{2^{2^{2^{65536}}}} - 3.$$

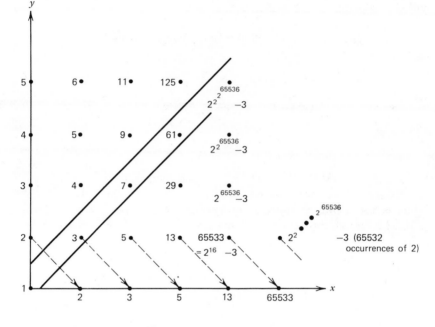

Fig. 1. Construction of $H(n)$.

The graph of Fig. 1 will perhaps make it easier to understand the construction of H. The object here is to write a number—namely $f(x, y)$—at each lattice point (x, y) of the first quadrant; x and y are integers ≥ 0.

We start with the numbers $1, 2, 3, \ldots$ on the y-axis; this initialization corresponds to (8a) and it gives us f on the first column of lattice points. Each next column is started by another initialization: We transfer the number from the previous column along the dotted arrows, corresponding to (8b). Finally, once a new vertical column of numbers is started, it can be continued inductively by (8c). For, suppose that numbers have been written down at the points $(x+1, 0)$, $(x+1, 1), \ldots, (x+1, y)$; then at $(x+1, y+1)$ we write down the number of the column immediately to the left, whose position up on that column is indicated by the last number we have written down in the new column.

For instance, consider the vertical column starting with 3 on the x-axis in Fig. 1. The number above it is 5 because in the left-neighbor column we have, from the bottom up, $2, 3, 4, 5, \ldots$; we do not count the first number, 2, and so the third number is just 5. Similarly, in the sixth column which starts with 65533 on the x-axis, we do the following. The column immediately to the left is

$$13, \quad 65533, \quad 2^{65533}-3, \quad 2^{2^{65533}}-3, \quad 2^{2^{2^{65533}}}-3, \quad \text{etc.}$$

Therefore the second number of the sixth column is the $(1+65533)$-th member of the above sequence:

$$2^{2^{\cdot^{\cdot^{\cdot^{2^{65533}}}}}}-3 \quad \text{(65532 repetitions of 2 in the tower of 2's).}$$

The reader may wish to attempt a degree of comprehension of the next few members of the sixth column, finding out that perhaps our handling ability exceeds our comprehending powers. Nevertheless, in spite of such difficulties, there is a unique number defined at each lattice point. Now, the numbers at the diagonal points give us the successive values of $H(n)$.

Formulas (7c) and (8c) show that A and H are given by *double* recursion: on n and x for A, on x and y for H. The reader may wish to compare this with the scheme of generalized induction described on p. 180. There also we have a set of lattice points, a certain subset of it which was specified as the initial set, and "transition" rules which specified how to add new points to the initial set. However, *there* we had *stationary* transition rules, *here* we have *variable* transition rules.

By sufficiently complicating these variable transition rules it is possible to construct a function $f(x_1, \ldots, x_k)$ of k variables and obtain from it by diagonalization a function $F(n) = f(n, n, \ldots, n)$ given by a k-fold recursion but not by any lower recursion.

On reading the last few paragraphs several questions might occur to the reader. First, it will certainly be granted that in some sense the Ackermann function $A(n)$ is more complicated than its *coordinate functions* $f_0(x, y)$ (addition), $f_1(x, y)$ (multiplication), $f_2(x, y)$ (exponentiation), and so on. Here one might ask: Does the increased speed of growth go necessarily together with the increased complexity? The answer is: obviously no. For, we can define $a(n)$ as follows:

$$a(n) = 1 \quad \text{if } n = A(k) \text{ for some } k,$$
$$= 0 \quad \text{otherwise.}$$

This $a(n)$ takes on the values 0 and 1 only, yet it is possible to compute $A(n)$ from $a(n)$ by a simple double-loop program shown below.

The second question that could be raised is much more serious: Do we really need those Ackermann-like functions? This is far from being a

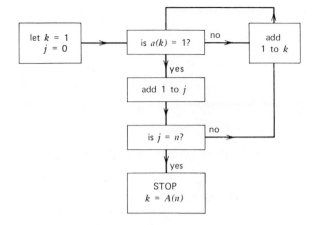

clear question and we shall spend some time on it. Obviously, since those functions serve to distinguish computable functions from primitive recursive functions, we encounter them in the recursion theory or the theory of computability. But, besides such internal consumption, have they some less expected, more remote (and perhaps more "natural") applications? It might be claimed that the answer is yes, and an example of such an application is the proposition that we shall prove in the next section. Namely, for simple programs with completely nested loop structure the halting problem is solvable. For, it will develop that if such a program runs longer than a certain bound, then it must run on forever. To obtain that bound it will be necessary to perform a double recursion: on the size of the program and on the depth of loop embedding. Consequently, the time-bound will be an Ackermann-like function obtained rather naturally. This is a contribution, however impractical, toward developing a criterion that a program must satisfy to be well formed.

The last question we consider arises partly from the preceding. Just how does one compute the Ackermann-like functions? Obviously we must answer this, for the "practical" purpose of being able to prove the solvability of the restricted halting problem. However, there are some wider implications here.

First, it must be observed that our "language"—our method of communicating with our hypothetical computer—is exceedingly primitive. In actual computing practice one could use a more sophisticated tool, for instance, one of the so-called Algol type of programming languages. These are much more powerful than what we have. One reason for it is precisely because they incorporate the recursion features. Hence the defining recursions, what we have called transition rules, can be simply programmed in such languages: We essentially write down the definitions (7) and (8) themselves in a suitable code.

Since we do not have the recursion built into our "language," we must proceed differently. It might even appear that computing Ackermann-like functions is beyond our powers, by the following reasoning. What we want is a program that transforms initial location contents $n, 0, 0, \ldots$ into, say, $A(n), 0, 0, \ldots$. As we have seen, one loop is needed for addition (= computing $f_0(x, y)$), a loop-within-loop for multiplication (= computing $f_1(x, y)$), and so on, and generally, a $(1 + k)$-tuple depth of loop embedding is needed for computing the kth coordinate function $f_k(x, y)$. Therefore it might appear that no program could compute $A(n)$ since it would need a variable depth $n + 1$ of loop nesting.

However, there is an indirect, and quite simple, way of computing Ackermann-like functions which depends on the normal-form theorem of Kleene [94, 73]. This asserts that any recursive function can be constructed out of primitive recursive functions by a single application of the minimalization operation of vol. 1, p. 78. We demonstrate this on the Hermes function $H(n) = f(n, n)$ for which the details are slightly simpler than for $A(n)$.

Suppose first that $f(2, 1)$ were to be computed. We rewrite this as just $2, 1$ and we copy the defining relations (8a–c) in the same bracketless and functionless form as

> **a:** $0, y \rightarrow y + 1,$
>
> **b:** $x + 1, 0 \rightarrow x, 1,$
>
> **c:** $x + 1, y + 1 \rightarrow x, x + 1, y.$

Now the computation of $2, 1$ is done in fourteen steps, starting with the sequence $2, 1$ and transforming it successively by the rules **a, b, c**:

$$2, 1 \xrightarrow{\ c\ } 1, 2, 0 \xrightarrow{\ b\ } 1, 1, 1 \xrightarrow{\ c\ } 1, 0, 1, 0$$

$$\xrightarrow{\ b\ } 1, 0, 0, 1 \xrightarrow{\ a\ } 1, 0, 2 \xrightarrow{\ a\ } 1, 3$$

$$\xrightarrow{\ c\ } 0, 1, 2 \xrightarrow{\ c\ } 0, 0, 1, 1 \xrightarrow{\ c\ } 0, 0, 0, 1, 0 \xrightarrow{\ b\ } 0, 0, 0, 0, 1$$

$$\xrightarrow{\ a\ } 0, 0, 0, 2 \xrightarrow{\ a\ } 0, 0, 3 \xrightarrow{\ a\ } 0, 4 \xrightarrow{\ a\ } 5$$

We transform the sequences by checking the last two numbers in them, and replacing these two numbers by one new number if **a** is used, by two new numbers if **b** is used, and by three new numbers if **c** is used. As soon as only one number is left in the sequence the computation is finished and the final number is just $f(2, 1) = 5$. It is precisely this feature, that the sequence has been reduced to one number, that corresponds to the use of minimalization. It is clear and unambiguous which of the three steps **a, b, c** to use at any stage, in terms of the last two numbers of the previous sequence: **a** if the penultimate number is 0, **b** if the last one is 0, and **c** if neither is 0. The procedure always comes to an end, it is perfectly general, and will compute x, y, that is to say $f(x, y)$, for arbitrary x and y. In particular, it computes $H(n) = f(n, n)$. The reader may wish now to produce, as an exercise, an iterated program that computes $H(n, n)$, given n.

On a more sophisticated level, the reader may wish to recall the general problem of extension to a larger domain a quantity, operation, or a function, defined originally for integers alone (Section 4, p. 59, vol. 1, and other places). In this connection it may be attempted to generalize $f(x, y)$ to *real* values of x and y, and in particular, to extend the Hermes and Ackermann functions $H(n)$ and $A(n)$ to $H(x)$ and $A(x)$ defined for real x. A certain similarity with the problem of extending $n!$ to $\Gamma(x)$ will be observed here: Just as with the factorial, so here also it suffices to define $f(x, y)$ suitably in the square domain $0 \le x, y \le 1$, and then use the defining relations

$$f(0, y) = y + 1, f(x + 1, 0) = f(x, 1), f(x + 1, y + 1) = f(x, f(x + 1, y))$$

to continue to, say, all nonnegative real x and y.

9.   SOLVABILITY OF A RESTRICTED HALTING PROBLEM

Let us consider now the geometrical structures of programs. Some illustrative examples are given in Fig. 1 in which the programs are represented by directed graphs, with instructions as vertices. Figure 1a shows a *lineal* program, a simple forked program is shown in Fig. 1b, and a compound forked program in Fig. 1c. The program of Fig. 1d is a simple loop; here the loop starts and ends on the same instruction r which is the root of the loop. A multiple loop program is shown in Fig. 1e; here all loops are rooted, and the depth of loop nesting is 3, starting with the outer loop that is labeled *rstuvr*. A program with two independent loops, of which one is simple and one is of depth 2 but both are rooted, is shown in Fig. 1f. Figures 1g and h show programs with loops which are simply nested, but not rooted. Finally, in Fig. 1i we have a program with loops which are not simply nested. Certain programs with loops which are simply embedded though not rooted, can be transformed into an equivalent form in which the loops are both simply embedded and rooted. For instance, take the program of Fig. 1g, which is copied again, with labeled

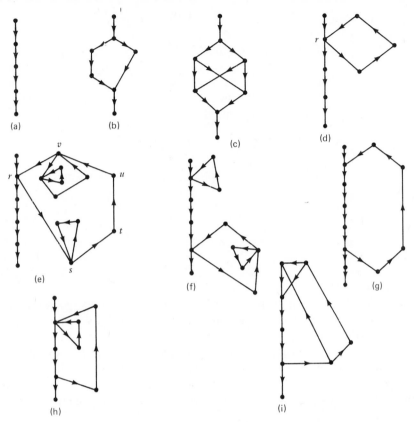

Fig. 1. Examples of programs.

instructions and in simplified form, in Fig. 2a. This is equivalent to the rooted loop of Fig. 2b in which 2_1 is a copy of the instruction 2. The same can be done for the more complicated case of the program of Fig. 1h. First we label the instructions as in Fig. 3a. Then we introduce a neutral instruction u between 1 and 2, as shown in Fig. 3b. Finally, copying as before certain instructions and loops, we transform our program into an equivalent one of Fig. 3c, in which all loops are rooted.

We shall consider henceforward only simple programs, without forking (for simplicity) and with simply embedded loop-structure, in which all loops are rooted. It follows necessarily that any two loops have distinct roots. One final restriction will be made, on the analogy with such standard programming languages as Fortran. The object of loops is to repeat a certain operation by cycling; and we now suppose that the

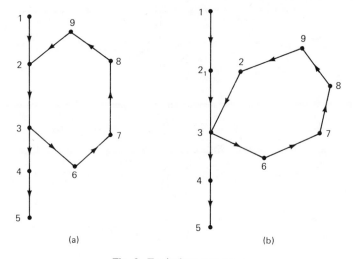

Fig. 2. Equivalent programs.

cycling is *pre-set* so that its conditions do not vary, as we go round the loop. In our terminology this amounts to the following. We suppose that the root of every loop is in the form shown in Fig. 4: an instruction *ab*, with a *necessarily unequal* 0, and that the location *a*, which controls the cycling round the loop, does not occur anywhere else on the loop except at the root; the condition does not apply to the location *b*. This means that, previous to coming to the loop, we accumulate some contents in the

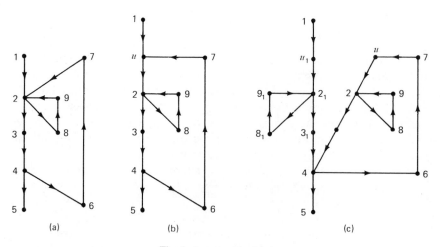

Fig. 3. Loop equivalence.

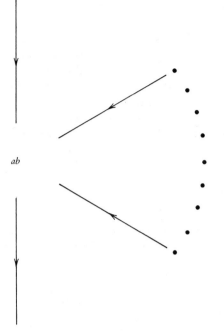

Fig. 4. Pre-set cycling.

location a which serves as the cycling index: we go round the loop as many times as there are counters in location a, when it is empty we exit from the loop.

We shall call any program subject to these restrictions a *normal* program, and N will denote the class of all normal programs. This may seem quite restricted, but to counterbalance this the reader may wish to verify that any primitive recursive function is computable by a normal program.

We show now that the halting problem for a normal program P is solvable. First, the class N is decomposed as follows

$$N = L_0 \cup \overset{\infty}{\underset{i=1}{U}} L_i,$$

where L_0 is the class of lineal programs and L_i is the class of programs with exactly i independent, or outer, loops. Further, we have

$$L_i = \overset{\infty}{\underset{d=1}{U}} L_{id}, \qquad i = 1, 2, \ldots,$$

where L_{id} is the subclass of L_i, consisting of programs with exact maximal depth d of loop nesting.

We suppose now that our program P starts with all locations empty, is of size c and belongs to L_{id}. Let $\psi(c, i, d)$ be the longest time that P can run, provided that it is halting. Suppose that $i = j + 1$ with $j > 0$; let r be the root of the last, $(j+1)$-th, independent loop in P. The running time of P, up to the point of its coming to r, is at most $\psi(c, j, d)$. Therefore the whole part of P, preceding r, can be replaced by a lineal program of size $\leq \psi(c, j, d)$. Together with the last loop, we get a program of size $\leq c + \psi(c, j, d)$, and belonging to L_{1d}. Hence the recursion

$$\psi(c, j+1, d) \leq \psi(c + \psi(c, j, d), 1, d). \tag{1}$$

Next, we obtain a similar recursion on the depth d. Let $d = 1 + b$, $b > 0$, and suppose that P has just one independent loop of depth d; our aim is to bound $\psi(c, 1, b+1)$ in terms of ψ with $b+1$ replaced by b. Let the single outer loop be as shown in Fig. 5. The cycling number, of times

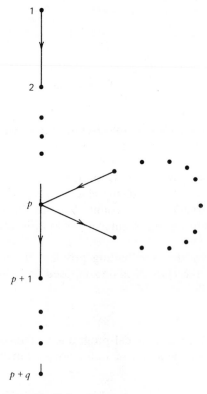

Fig. 5. Loop recursion.

round the loop, is therefore $\leq p-1$. If the loop is opened at its root p it will then form a program of size $\leq c$, with at most $c-1$ independent loops on it, and these have the maximum depth of loop embedding exactly b. Therefore the total running time will be bounded by the time to run the two lineal parts $1-p$ and $p+1-p+q$, plus $p-1$ times the maximal passage time round the loop. This gives us the recursion

$$\psi(c, 1, b+1) \leq c + (c-1)\psi(c, c-1, b). \tag{2}$$

It still remains to start the recursions by bounding $\psi(c, 1, 1)$. For this purpose we imagine the program shown in Fig. 5, of size c, and with just one simple loop. Then the reasoning used in deriving (2) will give us

$$\psi(c, 1, 1) \leq p + q + (p-1)(c-p-q+1) \leq c^2. \tag{3}$$

If we now replace the inequalities in (1), (2), and (3) by equalities, we shall have defined a function $\phi(c, i, d) \geq \psi(c, i, d)$; it is given by its defining system of equations

$$\phi(c, 1, 1) = c^2, \quad \phi(c, 1, d+1) = c + (c-1)\phi(c, c-1, d),$$

$$\phi(c, j+1, d) = \phi(c + \phi(c, j, d), 1, d).$$

This is a computable Ackermann-like function, and it yields us an upper bound on the running time of a halting normal program of size c: $\phi(c, c, c)$. This last function we compute in the same manner in which Hermes function $H(n)$ was computed: by reducing to primitive recursion and minimalization. Once the upper bound $\phi(c, c, c)$ is obtained and computed, we solve the halting problem very simply: the program P being tested is simulated for $1 + \phi(c, c, c)$ units of time; if it is still running then it is not halting.

The reader may wish to examine whether the restriction to normal programs can be relaxed; also, one may consider some variants of the halting problem, for instance, the question whether a program of a certain class will halt with arbitrary initial contents. In the case of normal problems we observe that we can answer the question whether a program halts with any given initial contents, since a lineal program will load empty locations up to any specified initial contents.

Finally, the following question may be asked. As an attempt to classify the fast-increasing Ackermann-like functions by speed of growth, we call two such functions F and G equivalent if

$$F(n) \leq f(G(g(n))), \quad G(n) \leq h(F(k(n))), \quad n = 1, 2, \ldots,$$

where f, g, h, k are primitive recursive. In many bounding problems, such as ours, an Ackermann-like function may be replaced by an equivalent one. The reader may wish to compare in this way the functions $A(n)$, $H(n)$, and $\phi(n, n, n)$.

5

TOPICS IN
APPLICATIONS

In this chapter and in the next one we treat a number of topics which could be loosely classified under applied mathematics, applicable mathematics, and applications of mathematics. No significance need be attached to this tripartite division which we use for inclusiveness (since "applied mathematics" alone might perhaps be considered too narrow).

1. ELLIPTIC INTEGRALS AND ELLIPTIC FUNCTIONS

(a) In this section we consider briefly a number of occurrences of elliptic functions and integrals, with special reference to the reason for their turning up in each problem. To put it roughly, we catalogue, with illustrations, some reasons for ellipticity. In order to explain what is meant let us make a short list of properties each of which suffices to characterize, or recognize, elliptic functions as such:

(A) being a meromorphic doubly periodic function,
(B) satisfying a specific differential equation,
(C) satisfying some other specific functional equation,
(D) having an algebraic addition theorem,
(E) possessing a particular infinite series or product,
(F) inverting an integral of a function which depends rationally on x and on the square root of a cubic or quartic polynomial in x.

Some analogs of these hold for ordinary trigonometric functions which are themselves degenerate limiting cases of elliptic functions. For instance, $y = A \sin kx + B \cos kx$ satisfies the differential equation $y'' = -k^2 y$ and the sine function has the algebraic addition theorem

$$\sin (x + y) = \sin x \sqrt{1 - \sin^2 y} + \sin y \sqrt{1 - \sin^2 x}.$$

It also satisfies a number of functional equations such as

$$\sin^2 2x = 4 \sin^2 x(1 - \sin^2 x) \qquad \text{or} \qquad \sin 3x = 3 \sin x - 4 \sin^3 x.$$

We recognize $\sin x$ either from its McLaurin series

$$x - \frac{x^3}{3!} + \frac{x^5}{5!} - \cdots$$

or from its infinite product

$$x \prod_{n=1}^{\infty} \left(1 - \frac{x^2}{n^2 \pi^2}\right).$$

In analogy to the relation $\sin^2 x + \cos^2 x = 1$ we have two similar relations for the three Jacobian elliptic functions $\operatorname{sn}(x, k)$, $\operatorname{cn}(x, k)$, $\operatorname{dn}(x, k)$:

$$\operatorname{sn}^2(x, k) + \operatorname{cn}^2(x, k) = 1, \; k^2 \operatorname{sn}^2(x, k) + \operatorname{dn}^2(x, k) = 1$$

and as the modulus k approaches 0 we get $\operatorname{sn}(x, 0) = \sin x$, $\operatorname{cn}(x, 0) = \cos x$. In some cases the periodicity of the trigonometric function is an obvious consequence of its representation, for example, with

$$\sec^2 x = \sum_{m=-\infty}^{\infty} (x - n\pi)^{-2}$$

since replacing x by $x + \pi$ leaves the series unchanged. Finally, $y = \sin x$ is the inverse of $x = \arcsin y$ and we have

$$y = \sin x \qquad \text{if} \qquad x = \int_0^y \frac{dy}{\sqrt{1 - y^2}}.$$

Returning to elliptic functions we observe that (A) is the basic definition. It states that the elliptic function is analytic in the whole plane except for isolated poles. Further, there is a fundamental region, namely the period parallelogram, and the function values keep repeating themselves with respect to this parallelogram, after the manner of a wallpaper pattern or a mosaic of tiles. Under (B) we have the differential equation

$$y'^2 = P(y) \tag{1}$$

where P is a cubic or a quartic polynomial. This, of course, is immediately related to (F). For the Weierstrass function $y = p(x)$ P is the cubic $4y^3 - g_2 y - g_3$ (where g_2 and g_3 are constants related to the two periods). For the Jacobian function $y = \operatorname{sn}(x, k)$ P is the quartic $(1 - y^2)(1 - k^2 y^2)$. These two examples serve to distinguish the two simplest types of elliptic functions. For, a simple application of Liouville's theorem shows that with a nonconstant elliptic function the sum of residues in a period parallelogram is 0. Hence we cannot have just one simple pole: There must be either a pole of order at least two as with the Weierstrass p-function, or at least two simple poles as with the Jacobian functions. It may be noticed that any differential equation

$$y'' = P(y), \qquad P \text{ a cubic or quadratic,} \qquad (2)$$

can be solved in terms of elliptic functions; we multiply by y' and integrate, thereby reducing it to form (1). In physical terms this may correspond to passing from the equation of motion to the energy equation. Other differential equations may be reducible to form (1) or (2) by a substitution. For instance, with the differential equation $u'' = A \sin u + B \cos u$ we use a substitution $v = c + u$ and reduce the differential equation to the form $v'' = (C/2) \sin v$. Then we multiply by v' and integrate getting $v'^2 = K - C \cos v$. If now we put $y = \cos v$ we have

$$y'^2 = (1 - y^2)(K - Cy).$$

In connection with differential equations it must be observed here that every elliptic function $f(z)$ satisfies a differential equation of the form $P(f, f') = 0$, where P is a polynomial in two variables (otherwise put, an elliptic function and its derivative are algebraically dependent). This theorem, due to Briot and Bouquet [63], follows from another more general one stating that two elliptic functions $f(z)$ and $g(z)$ with the same periods satisfy a polynomial relation $P(f, g) = 0$. Now, to get Briot and Bouquet's theorem we take $g = f'$.

To prove this, we let Π be a period parallelogram of both f and g and we suppose that $z_1, \ldots, z_m$ are all the poles of f and g in Π. Let the order of the pole at z_j be p_j for f and q_j for g, and put

$$N = \sum_{j=1}^{m} \max(p_j, q_j).$$

Let $P(x, y)$ be a polynomial of degree M without constant term; $P(f(z), g(z))$ is then an elliptic function with the same periods as f and g. Hence its only possible singularities in Π are poles at $z_1, \ldots, z_m$ and the order of the pole at z_j is at most $M \max(p_j, q_j)$.

Let us treat the $M(M+3)/2$ coefficients of P as unknowns; if the principal part of $P(f(z), g(z))$ at z_j is to be 0, certain linear homogeneous equations in these unknowns must hold, their number is $\leq M \max(p_j, q_j)$. Hence, if $P(f, g)$ is to have no singularities at all in Π, a total of at most

$$M \sum_{j=1}^{m} \max(p_j, q_j) = MN$$

linear homogeneous equations must hold. It follows that if $M+3>2N$ then there are fewer equations than unknowns and hence a nontrivial solution exists. Now $P(f, g)$ has no singularities in Π and consequently none in the whole complex plane; by Liouville's theorem it is a constant, q.e.d.

An entirely different kind of differential equation is satisfied by complete elliptic integrals when we differentiate with respect to the modulus [183]. For instance, if

$$y(k) = \int_0^{\pi/2} (1 - k^2 \sin^2 \phi)^{-1/2} \, d\phi$$

then (cf. p. 78) we have

$$\frac{d}{dk}\left[k(1-k^2)\frac{dy}{dk}\right] = ky.$$

For other relations of this type see [183, 64 and 40].

The item (D) on our list of properties of elliptic functions refers to the Weierstrass theorem, which asserts that the only functions with algebraic addition theorems are the elliptic functions and their degenerate cases: exponential and trigonometric functions.

Possessing a specific infinite series or product may be directly related to (A). For instance, the series

$$z^{-2} + \sum_{m=-\infty}^{\infty} \sum_{n=-\infty}^{\infty} [(z + \Omega_{mn})^{-2} - \Omega_{mn}^{-2}], \qquad m^2 + n^2 \neq 0,$$

where $\Omega_{mn} = m\omega_1 + n\omega_2$, is shown to be doubly periodic with periods ω_1 and ω_2 by a simple shift of indices. Since it is obviously a meromorphic function it must represent an elliptic function (in fact, the Weierstrass function p). We note the related infinite product

$$\sigma(u) = u \prod_{-\infty}^{\infty} \prod_{-\infty}^{\infty} \left(1 - \frac{u}{\Omega_{mn}}\right) e^{u/\Omega_{mn} + u^2/2\Omega_{mn}^2}$$

which is plainly analogous to

$$\sin \pi x = \pi x \prod_{\substack{n=-\infty \\ n \neq 0}}^{\infty} \left(1 - \frac{x}{n}\right) e^{x/n} = \pi x \prod_{n=1}^{\infty} \left(1 - \frac{x^2}{n^2}\right);$$

here we have $\sigma'/\sigma = -\int_0^u p(u) \, du$.

With respect to the theta functions the situation is somewhat anomalous: they are sometimes called elliptic functions even though they are not. They have the characteristic defining series: For a complex t with Im $t > 0$

let $q = e^{\pi i t}$, then

$$\theta_4(z, q) = \sum_{n=-\infty}^{\infty} (-1)^n q^{n^2} e^{2niz} = 1 + 2 \sum_{n=1}^{\infty} (-1)^n q^{n^2} \cos 2nz,$$

$$\theta_3(z, q) = \theta_4(z + \pi/2, q) = 1 + 2 \sum_{n=1}^{\infty} q^{n^2} \cos 2nz, \qquad (3)$$

$$\theta_1(z, q) = -i \sum_{n=-\infty}^{\infty} (-1)^n q^{(n+1/2)^2} e^{(2n+1)iz}, \qquad \theta_2(z, q) = \theta_1(z + \pi/2, q).$$

They are quasi-doubly-periodic; for instance

$$\theta_4(z + \pi, q) = \theta_4(z, q), \qquad \theta_4(z + \pi t, q) = -q^{-1} e^{-2iz} \theta_4(z, q).$$

Considered as functions of z and t, all four satisfy the partial differential equation

$$\frac{\partial^2 \theta}{\partial z^2} = \frac{4i}{\pi} \frac{\partial \theta}{\partial t}$$

known as the heat, or diffusion, equation. Even though the theta functions are not elliptic some of their combinations are, and an application of Liouville's theorem shows that every elliptic function f is expressible by theta functions (we construct a combination F of theta functions with the same periods and singularities as f, then by Liouville's theorem f/F is constant). This is of considerable numerical importance since the series of (3) often can be made to converge very fast.

We consider an extreme case in point. Suppose that we wish to calculate

$$S = 1 + 2 \sum_{n=1}^{\infty} (0.999)^{n^2} = \theta_3(0, 0.999).$$

The first few terms of the series are

$$S = 1 + 2(0.999 + 0.99606 + 0.991036$$
$$+ 0.984120 + 0.975298 + 0.964625 + \cdots)$$

showing that even though *eventually* the terms of our series decrease very fast, at the start this is not the case. However, we recall the functional equation of the θ-function from vol. 1, p. 122:

$$\text{if} \qquad \theta(t) = \sum_{n=-\infty}^{\infty} e^{-\pi n^2 t} \qquad \text{then} \qquad \theta(t) = t^{-1/2} \theta\left(\frac{1}{t}\right).$$

Rewriting S we have

$$S = \sum_{n=-\infty}^{\infty} e^{-\pi n^2 A}, \qquad A = \frac{1}{\pi} \log \frac{1}{0.999}.$$

To compute A we use the series

$$\log \frac{1+x}{1-x} = 2\left(x + \frac{x^3}{3} + \frac{x^5}{5} + \cdots\right)$$

where $x = 0.00050025\ldots$, hence

$$A = 0.000318469\ldots.$$

By the functional equation mentioned above

$$S = A^{-1/2}[1 + 2(e^{-\pi/A} + e^{-2^2\pi/A} + \cdots)]$$

with the first few terms

$$S = 56.0359\ldots[1 + 2e^{-3140.02\pi} + 2e^{-12560.08\pi} + \cdots].$$

This gives us several *thousand* digits' accuracy with the first term alone. Taking the transformation by the functional equation into account, the reader may wish to show that the poorest convergence in

$$\sum_{n=-\infty}^{\infty} q^{n^2}$$

occurs when $q = e^{-\pi} = 0.0432136\ldots$; this may be compared with the evaluation of π by means of elliptic functions on p. 168 of vol. 1.

(b)   Some of the fields from which we draw our examples are geometry, probability, mechanics, radio engineering. We now note that elliptic integrals and functions occur in problems drawn from those fields for a variety of "reasons." For instance, in geometry they may turn up simply and naturally because certain geometrical concepts lead to integrals with integrands containing square roots, as in (F): so, we have

arc length $\int \sqrt{1 + y'^2}\, dx,$

surface area $\int\int \sqrt{1 + z_x^2 + z_y^2}\, dx\, dy,$

solid angle $\int\int [(x - x_0)^2 + (y - y_0)^2 + (z - z_0)^2]^{-3/2}$

$$\times (x\, dy\, dz + y\, dx\, dz + z\, dx\, dy).$$

Further occurrences in geometry are in the determination of geodesics on surfaces of revolution [165]. If the surface is given by $x = r \cos \phi,$

$y = r \sin \phi$, $z = f(r)$ then the equation of the geodesics depends on the integral

$$\int \frac{(1 + f'^2(r))^{1/2}}{r(r^2 - a^2)^{1/2}} \, dr.$$

For instance, let the torus T be generated by rotating a circle C of radius a about a straight line whose distance from the center of C is $b(\geq a)$. Then, using the theorem of Clairaut (see p. 99) we find that a geodesic of T is given by

$$\phi = \phi(r) = K \int \frac{dr}{r\sqrt{r^2 - c^2} \sqrt{a^2 - (r - b)^2}}.$$

In mechanics elliptic functions and integrals often occur as a direct result of laws of motion and so as under (B): on account of differential equations. For instance, the *exact* pendulum equation is

$$x'' = -k^2 \sin x, \tag{4}$$

x being the angle the pendulum makes with the direction vertically down; this is a simple and direct consequence of Newton's law $F = ma$. As was mentioned before, we multiply (4) by x' and integrate, getting eventually t as an elliptic integral in terms of x, and inverting, x as an elliptic function of time t. The first-order approximation $\sin x \cong x$ takes us from (4) to the simple harmonic motion equation $x'' = -k^2 x$ but the third-order approximation $\sin x \cong x - x^3/6$ leads from (4) to

$$x'' = -k^2 \left(x - \frac{x^3}{6} \right).$$

This too can be solved for x in terms of elliptic functions, much as (4) itself.

Another occurrence of elliptic functions in mechanics is in connection with spinning rigid bodies. Here also differential equations play an essential part. Let A, B, C, be the moments of inertia of the rigid body about its principal axes, and let ω_1, ω_2, ω_3 be the components of angular velocity about the principal axes. Then the Euler equations give us

$$\omega_1' = \frac{B - C}{A} \omega_3 \omega_2, \qquad \omega_2' = \frac{C - A}{B} \omega_1 \omega_3, \qquad \omega_3' = \frac{A - B}{C} \omega_1 \omega_2. \tag{5}$$

Hence it is natural that the solutions [64] should involve the Jacobian elliptic functions sn, cn, dn since these are plainly related to (5): the derivative of each is proportional to the product of the other two

$$\text{sn}' \, x = \text{cn} \, x \, \text{dn} \, x, \qquad \text{cn}' \, x = -\text{sn} \, x \, \text{dn} \, x, \qquad \text{dn}' \, x = -k^2 \, \text{sn} \, x \, \text{cn} \, x.$$

In much the same way as with the pendulum, elliptic functions arise in the problem of beams deflected under compression. Just as with the pendulum, if the deflection is small and the problem is linearized, we get the simple harmonic equation. Otherwise, with the full and exact formulation we get differential equations which lead to elliptic functions.

(c) As was mentioned before, elliptic integrals and functions occur often in geometry because of the integrands in the expressions for arc length and surface area. Historically, the earliest occurrence, which gave the subject its name, was in the rectification of the arc of an ellipse $x = a \cos \phi$, $y = b \sin \phi$, $0 \leq \phi \leq \phi_0$; the arc length is

$$L = a \int_0^{\phi_0} (1 - k^2 \sin^2 \phi)^{1/2} \, d\phi, \qquad k = \left(1 - \frac{b^2}{a^2}\right)^{1/2}.$$

Also, elliptic integrals occur in the rectification of the sine curve $y = A \cos x + B \sin x$, the lemniscate $r^2 = a^2 \cos 2\theta$, generalized cycloid $x = t - a \sin t$, $y = 1 - a \cos t (a \neq 1)$, the cubic $y = ax^3 + bx^2 + cx + d$, the hyperbola, the catenary, and so on [48]. In three dimensions elliptic integrals occur for areas cut out of spheres by cylinders, for lateral areas of certain cones, for the steradian content of conical beams, and so on.

For the area S of the ellipsoid

$$\frac{x^2}{a^2} + \frac{y^2}{b^2} + \frac{z^2}{c^2} = 1, \qquad c < b < a,$$

we find by the standard formula

$$S = 2 \iint_E \left[1 - \frac{a^2 - c^2}{a^4} x^2 - \frac{b^2 - c^2}{b^4} y^2\right]^{1/2} \Big/ \left[1 - \frac{x^2}{a^2} - \frac{y^2}{b^2}\right]^{1/2} dx \, dy, \qquad (6)$$

where E is the elliptical region

$$E = \left\{(x, y): \frac{x^2}{a^2} + \frac{y^2}{b^2} \leq 1\right\}.$$

To reduce (6) to elliptic integrals we use the method of Catalan [29, 63], which expresses certain double (or multiple) integrals by means of ordinary single integrals. The essence of this method is a good choice of the element of integration.

To illustrate the importance of such a choice, we consider the simple problem of finding the volume V common to two circular cylinders of radius a whose axes have a point in common and make the angle α, $0 < \alpha \leq \pi/2$. With an injudicious choice of the volume element the problem of calculating V could be somewhat messy. However, let us take as the volume element that part of our solid which lies between the planes at the distance z and $z + dz$ from the plane of the axes of the cylinders. Since our solid is cut by the z-plane in a rhombus with the heights $2(a^2 - z^2)^{1/2}$ we find for the volume of our element $4 \operatorname{cosec} \alpha (a^2 - z^2) dz$. Integrating this from $z = -a$ to $z = a$ we get $V = 16 \operatorname{cosec} \alpha \, a^3 / 3$. This carries over to the intersection of certain elliptic cylinders. For instance, in the above example let us replace our cylinders by the elliptic ones with semi-axes a, b_1, and a, b_2; here $0 < b_1 \le a$, $0 < b_2 \le a$ and the cylinders just fit between the planes $z = -a$ and $z = a$. Proceeding as before we find $V = 16 \operatorname{cosec} \alpha \, a b_1 b_2 / 3$. However, if the elliptic cylinders just fit between the planes but obliquely rather than at right angles, then V is expressible by an elliptic integral.

Returning to the Catalan method we let

$$I = \iint_D f(x, y) \, g(x, y) \, dx \, dy \tag{7}$$

and we suppose that

$$\min_{.D} f(x, y) = m, \qquad \max_{.D} f(x, y) = M.$$

Let

$$F(v) = \iint_{R(v)} g(x, y) \, dx \, dy,$$

where $R(v)$ is the region

$$R(v) = \{(x, y): m \le f(x, y) \le v\}.$$

This amounts to preparing a suitable element of integration in (7), associated with the region between the level curves of the function f. We have now

$$I = \int_m^M v \, dF(v) = MF(M) - \int_m^M F(v) \, dv.$$

If the upper bound M should happen to be infinite, then this is replaced by

$$I = \lim_{M \to \infty} \left[MF(M) - \int_m^M F(v) \, dv \right].$$

We apply the preceding to (6) taking $g \equiv 1$. Writing the integrand in (6) as

$$f(x, y) = \left[1 + \frac{c^2 x^2/a^4 + c^2 y^2/b^4}{1 - x^2/a^2 - y^2/b^2} \right]^{1/2}$$

we find the values $m = 1$, $M = \infty$. The contour curve $f(x, y) = v$ is the ellipse $E(v)$:

$$\frac{x^2}{A^2} + \frac{y^2}{B^2} = 1,$$

where

$$A^2 = \frac{a^2(v^2 - 1)}{v^2 - 1 + c^2/a^2}, \qquad B^2 = \frac{b^2(v^2 - 1)}{v^2 - 1 + c^2/b^2}.$$

With $v = 1$ $E(1)$ reduces to a single point, the origin, and so the region $R(v)$ is the ellipse $E(v)$ together with its interior. Hence $F(v)$ is its area

$$F(v) = \pi AB = \frac{\pi ab(v^2 - 1)}{[(v^2 - 1 + c^2/a^2)(v^2 - 1 + c^2/b^2)]^{1/2}}.$$

Let $\phi(v)$ be the function under the square root in the denominator above; applying (8) we get

$$S = 2\pi ab \lim_{M \to \infty} \left[\frac{M(M^2 - 1)}{\sqrt{\phi(M)}} - \int_1^M \frac{v^2 - 1}{\sqrt{\phi(v)}} \, dv \right].$$

The square bracket contains an indeterminate form of the type $\infty - \infty$. However, it is verifed that

$$\frac{v^2 - \phi(0)v^{-2}}{\sqrt{\phi(v)}} = \left[\frac{\phi(v)}{v} \right]',$$

$$\int_1^M \frac{v^2}{\sqrt{\phi(v)}} \, dv = \phi(0) \int_1^M \frac{dv}{v^2 \sqrt{\phi(v)}} + \frac{\sqrt{\phi(v)}}{v} \bigg|_1^M.$$

Hence, taking some simple limits, we get

$$S = 2\pi ab \left[\frac{c^2}{ab} + \int_1^\infty \frac{dv}{\sqrt{\phi(v)}} - \phi(0) \int_1^\infty \frac{dv}{v^2 \sqrt{\phi(v)}} \right] \tag{9}$$

and so S has been reduced to elliptic integrals.

From (9) we deduce the elementary-function expressions for the surface area of an ellipsoid of revolution

$$S = 2\pi a^2 + \frac{2\pi a}{\sqrt{a^2-1}} \text{ arg cosh } a \qquad \text{(oblate, semi-axes } a, a, 1$$
$$\text{and } a > 1),$$

$$S = 2\pi + \frac{2\pi a^2}{\sqrt{a^2-1}} \text{ arc sec } a \qquad \text{(prolate, semi-axes } a, 1, 1$$
$$\text{with } a > 1).$$

Also, we can get from (9) approximations to S when a, b, c are nearly equal. For instance, with $a = 1 + A\varepsilon$, $b = 1 + B\varepsilon$, $c = 1 + C\varepsilon$, ε small, we have

$$S = 4\pi\{1 + \tfrac{2}{3}(A+B+C)\varepsilon + \tfrac{1}{15}[(A+B+C)^2$$
$$+ 2(AB+AC+BC)]\varepsilon^2 + O(\varepsilon^3)\}$$

For further approximations, extension to ellipsoids in n dimensions, and so on, see [97]. Before moving to the next topic, we remark on the connection between elliptic integrals and Legendre polynomials. This will be illustrated by deriving another expression for the area S of the ellipsoid with semi-axes a, b, c. Let

$$\alpha^2 = 1 - \frac{b^2}{a^2}, \qquad \beta^2 = 1 - \frac{c^2}{a^2},$$

then (9) can be written as

$$S = 2\pi ab\left[\frac{c^2}{ab} + \int_1^\infty \left(1 - \frac{\alpha^2\beta^2}{v^2}\right)\frac{dv}{\sqrt{(v^2-\alpha^2)(v^2-\beta^2)}}\right].$$

We introduce in the integral the variable $u = \sqrt{\alpha\beta}/v$ and we observe the generating function of the Legendre polynomials

$$\frac{1}{\sqrt{1-2xh+h^2}} = \sum_{n=0}^\infty P_n(x)h^n.$$

With these, the above formula for S gives us

$$S = 2\pi ab\left[\frac{c^2}{ab} + \sum_{n=0}^\infty P_n\left(\frac{\alpha^2+\beta^2}{2\alpha\beta}\right)(\alpha\beta)^n\left(\frac{1}{2n+1} - \frac{\alpha\beta}{2n+3}\right)\right].$$

(d) In mechanics the elliptic functions and integrals arise occasionally when we treat a physical problem in its exact formulation rather than linearizing it. One such example, that of a pendulum, was already mentioned. Another example is the deflection of a homogeneous bar under compression. The basic physical relation here states that if $y = f(x)$ is the deflected curve C of the beam, then the deflection y is proportional to the curvature of C:

$$\frac{1}{\rho} = -k^2 y.$$

In the linearizing treatment we assume that the deflection is small and C is a shallow curve so that $y'^2 \ll 1$ and instead of the exact form

$$\frac{y''}{(1 + y'^2)^{2/3}} = -k^2 y \tag{10}$$

we get merely the simple harmonic equation $y'' = -k^2 y$. But for large deflections we cannot neglect y'^2 and we use (10) as it stands, getting elliptic functions. Since the independent variable x is absent we introduce $y' = p(y)$ as the new variable getting

$$pp'(1 + p^2)^{-3/2} = -k^2 y.$$

Integrating this we have

$$(1 + p^2)^{-1} = C_1 - k^2 y^2.$$

Solving for p and integrating again we have

$$\int \frac{(C_1 - k^2 y^2)\, dy}{\sqrt{1 - (C_1 - k^2 y^2)^2}} = C_2 + x$$

This expresses x as an elliptic integral in y; inverting the relation we get y in terms of the Jacobian elliptic functions of x. For the shapes of various beam curves C, the so-called elasticas, see [64].

Another mechanical application of elliptic functions occurs with the shape of a uniform heavy skipping rope which rotates so fast that gravity effects are neglibible against the centrifugal force. Let w be the weight per unit length, b the largest deflection, ω the angular velocity, s the arc length, and $t = t(s)$ the tension at s. Then, with the geometry of Fig. 1, the equation of motion gives us, after separating into vertical and horizontal components,

$$\frac{d}{ds}\left(t\frac{dx}{ds}\right) = 0, \qquad \frac{d}{ds}\left(t\frac{dy}{ds}\right) + \omega^2 w y = 0. \tag{11}$$

The reader may recognize again (as in (10)) the expressions characterizing curvature. Just as in the previous example of deflected bar, we could

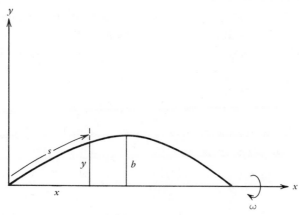

Fig. 1. Skipping rope.

linearize here. This would amount to putting $x \cong s$ getting us a very shallow skipping rope and eventually a simple harmonic equation from (11): $dx/ds \cong 1$, $dy/ds \cong dy/dx$, $t = \text{const.}$ and so on, leads to $y'' = -k^2 y$. With the exact treatment we first integrate the first equation of (11) and find that $T = t \, dx/ds$ is constant (i.e., the horizontal component of tension stays fixed). This is used to eliminate t from the second equation of (11). Integrating the result we get

$$\frac{ds}{dx} = C - \frac{\omega^2 w}{2T} y^2.$$

The constant C can be determined since the slope must be horizontal when y attains its maximum b (as in Fig. 1) and then $ds/dx = 1$:

$$\frac{ds}{dx} = 1 + \frac{\omega^2 w}{2T}(b^2 - y^2). \tag{12}$$

Next, we observe the identity

$$y'^2 = \left(\frac{ds}{dx} - 1\right)\left(\frac{ds}{dx} + 1\right);$$

both factors of the R.H.S. are quadratic in y by (12). Hence we get the characteristic differential equation for the Jacobian elliptic functions

$$y'^2 = \frac{\omega^2 w}{T}(b^2 - y^2)\left[1 + \frac{\omega^2 w}{4T}(b^2 - y^2)\right]$$

and eventually $y = b \, \text{sn} \, Kx$.

(e) We turn now to some occurrences of elliptic functions and integrals in probability. The first one is of a simple formal character and is met occasionally in plane problems when some angle α is normally distributed with the density

$$f(\alpha) = \frac{1}{\sqrt{2\pi\sigma^2}} e^{\alpha^2/2\sigma^2}, \qquad -\infty < \alpha < \infty.$$

We ask: What does this density become when we do not distinguish between angles differing by an integer multiple of 2π? Briefly, we reduce the Gaussian mod 2π. If the new density is $g(\alpha)$ then

$$g(\alpha) = \sum_{n=-\infty}^{\infty} f(\alpha + 2n\pi)$$

so that

$$g(\alpha) = \frac{1}{\sqrt{2\pi\sigma^2}} \sum_{n=-\infty}^{\infty} e^{-(\alpha+2n\pi)^2/2\sigma^2}.$$

This can be expressed in terms of the theta function $\theta_4(z, q)$ of (3) and we get

$$g(\alpha) = \frac{1}{\sqrt{2\pi\sigma^2}} e^{-\alpha^2/2\sigma^2} \theta_4\left(-\frac{\pi}{2} + \frac{i\pi\alpha}{\sigma^2}, -\frac{2\pi^2}{\sigma^2}\right). \tag{13}$$

For the next example of probabilistic ellipticity we note briefly some elements of the theory of recurrent events in a series of trials [55]. Let

$$p_j = \text{Prob}(E \text{ occurs for the first time at the } j\text{th trial}),$$

$$u_j = \text{Prob}(E \text{ has occurred at the time of the } j\text{th trial}),$$

where $j = 1, 2, \ldots$. Define also formally $u_0 = 1$. Then

$$u_j = u_0 p_j + u_1 p_{j-1} + \cdots + u_{j-1} p_1 \tag{14}$$

since E has occurred by the time of the jth trial if and only if it did occur for the first time at the ith trial, $1 \le i \le j$. In that case its occurrence cannot be changed by the subsequent $j-i$ trials.

With the generating functions

$$P(z) = \sum_{j=1}^{\infty} p_j z^j, \qquad U(z) = \sum_{j=0}^{\infty} u_j z^j$$

the recursion (14) becomes

$$P(z) = 1 - \frac{1}{U(z)}.$$

Now, $P(1) = \text{Prob}(E \text{ ever occurs})$ and so E is certain to occur some time if and only if $U(1) = \infty$. Otherwise the probability of its occurrence is only $1 - 1/U(1)$.

Consider now the symmetric random walk on the n-dimensional integer lattice. That is, a point starts from the origin of the n-dimensional Euclidean space and moves to one of its $2n$ nearest neighbors with equal probabilities $1/(2n)$. Let E be the event of returning to the origin. Then it can be shown that

$$U(z) = \frac{1}{(2\pi)^n} \int_{-\pi}^{\pi} \cdots \int_{-\pi}^{\pi} \left(1 - \frac{z}{n} \sum_{j=1}^{n} \cos \phi_j \right)^{-1} d\phi_1 \cdots d\phi_n.$$

It is now not difficult to show that $U(1) = \infty$ only if $n = 1$ or $n = 2$. That is, return to the origin is certain only for linear or planar random walks; we remark that therefore return *infinitely often* is also certain then (hence the name *recurrent event*). To calculate the probability of return to the origin in three dimensions it is necessary to evaluate the Watson integral

$$I_1 = \frac{1}{(2\pi)^3} \int_{-\pi}^{\pi} \int_{-\pi}^{\pi} \int_{-\pi}^{\pi} \frac{da\, db\, dc}{1 - (\cos a + \cos b + \cos c)/3}.$$

This has been done by G. N. Watson [181, 10]

$$I_1 = \frac{4}{3\pi^2} (18 + 12\sqrt{2} - 10\sqrt{3} - 7\sqrt{6}) K^2 \left(\frac{2 - \sqrt{3}}{\sqrt{3} - \sqrt{2}}\right),$$

where $K(k)$ is the complete elliptic integral of the second kind

$$K(k) = \int_0^{\pi/2} \frac{d\phi}{\sqrt{1 - k^2 \sin^2 \phi}}.$$

Accordingly, the probability of ever returning to the origin is now only $0.340537 \cdots$ rather than 1.

The reader may ask: Just why do the elliptic integrals occur here? In the previous applications, or those to come, there is a clear reason: in one place elliptic integrals and functions enter because of an integral with a square root of a quartic, elsewhere on account of a typical differential equation, or because of a typical infinite series or product. But why here? The question is far from being clear. Perhaps some light could be thrown on it by the considerations toward the end of Section 2 of Chapter 2, involving the Hadamard products of power series. Here we just recall that $K(k)$ is the Hadamard square of a very simple function:

$$K(k) = f(k) \circ f(k), \quad \text{where} \quad f(k) = \sqrt{\frac{\pi}{2}} a(1 - k^2)^{-1/2}.$$

For the third application of elliptic functions to probability we need a brief summarization of a continuous-time process known as the birth process. Suppose that at the time $t = 0$ there are i individuals each of which may give birth to replicas of itself, and let $p_n(t)$ be the probability of having exactly n of them at the time t (so that $p_n(0)$ is 0 if $n \neq i$, and 1 if $n = i$). We are interested in the probability $p_n(t + \Delta t)$ of having the total of n individuals at the time $t + \Delta t$. Since they do not die, either none were born in the interval $[t, t + \Delta t]$ and there were n at time t, or there were only $n - 1$ to begin with at time t and one was born during the time $[t, t + \Delta t]$. There are also other possibilities, namely $n - j$ were present at the time t and j were born later with $j > 1$, but we dispose of all these by assigning to them the probability $o(\Delta t)$.

To describe the process we must still specify the probability of a birth in time $[t, t + \Delta t]$ when n individuals were present at time t; we take this to be $\lambda_n \Delta t + o(\Delta t)$. That is, the probability is proportional to the time interval, but the constant of proportionality depends on n. Now we can connect $p_n(t + \Delta t)$ with $p_n(t)$ by applying some simple laws on combination of events, and we get

$$p_n(t + \Delta t) = \lambda_{n-1} \Delta t \, p_{n-1}(t) + (1 - \lambda_n \Delta t) p_n(t) + o(\Delta t), \qquad n > 0,$$

$$p_0(t + \Delta t) = (1 - \lambda_0 \Delta t) p_0(t).$$

By dividing by Δt and letting $\Delta t \to 0$ we get the differential equations

$$p_0' = -\lambda_0 p_0, \qquad p_n' = \lambda_{n-1} p_{n-1} - \lambda_n p_n, \qquad n = 1, 2, \ldots.$$

These equations can be consecutively solved [11], and we get

$$p_n(t) = (-1)^n \prod_{i=0}^{n-1} \lambda_i \sum_{j=0}^{n} \left[e^{-\lambda_j t} \Big/ \prod_{\substack{k=1 \\ k \neq j}}^{n} (\lambda_j - \lambda_k) \right].$$

With $\lambda_n = \lambda n^2$ we speak of the quadratic birth process; p_0 falls away now and we have

$$p_n(t) = \sum_{j=1}^{n} \frac{(-1)^j (2j)! \, (n-1)!}{(n+j)! \, (n-j)!} e^{-\lambda j^2 t}, \qquad n = 1, 2, \ldots.$$

For certain technical reasons the "defect"

$$p_\infty(t) = 1 - \sum_{n=1}^{\infty} p_n(t)$$

is of interest here; following P. W. M. John [87], the reader may wish to show that

$$p_\infty(t) = 1 + 2 \sum_{n=1}^{\infty} (-1)^{n-1} e^{-\lambda n^2 t} = \theta_4(0, e^{-\lambda t}).$$

Finally, theta functions occur in probability in connection with the theorems of Kolmogorov, Smirnov, and Renyi. In all these cases as well as in the quadratic birth process above, and in Polya's proof of the functional equation of θ_4 (vol. 1, pp. 123–124) theta functions occur because of limit properties of binomial quotients. So, for instance, we have

$$\lim \frac{p}{2^{2m}} \binom{2m}{m+kp} = \frac{1}{\sqrt{\pi t}} e^{-k^2 t}$$

if m and p tend to ∞ so that $m/p^2 \to t$.

(f) Many occurrences of elliptic integrals depend on the mapping theorem of Schwarz-Christoffel: The upper half-plane is mapped conformally onto the interior of the polygon with vertices $w_1, w_2, \ldots, w_n$ and external angles $\pi a_1, \pi a_2, \ldots, \pi a_n$ by the function $w = f(z)$ given as

$$w' = A \prod_{j=1}^{n-1} (z - z_j)^{-a_j}$$

Here z_j corresponds to w_j, $j = 1, \ldots, n-1$ and w_n corresponds to the point at infinity. If now the polygon is a rectangle, we have $a_1 = a_2 = a_3 = a_4 = 1/2$; hence the mapping function is given by the typical differential equation leading to elliptic functions. However, in some simple cases the ellipticity of a function under such conditions may be deduced directly from the definition, bypassing any conformal mappings. For instance, the electrostatic potential corresponding to a rectangular metal enclosure R, with several point charges inside, is the real part of an elliptic function f. For, we can apply the method of images and reflect the charges in the sides of R (reflecting also the reflections, etc.). Now, the function f is clearly meromorphic being analytic except at the charges and their images; further, f is doubly periodic, by the construction. Hence f is elliptic. This analogy, based on the principle of evaluating one thing in two ways (potential, by the Schwarz-Christoffel, and as above) can be used systematically, to obtain series expansions of elliptic functions [64].

Finally, as our last application of elliptic functions, we consider a problem in communication engineering: how to conduct several independent telephone conversations over the same cable, and how to maximize the number of such conversations. The idea here is that one conversation takes up a certain minimal frequency range, say $0-f_0$ cycles, for a reasonably faithful transmission over the cable. The second channel, that is, the second conversation-band of frequencies, is prepared by taking the band $0-f_0$ and *adding* $f_0 + \Delta$ cycles to it, making it in effect the $(f_0 + \Delta) - (2f_0 + \Delta)$ band of frequencies. We send both conversations along the cable, without undue mutual interference. If now we are in possession of a suitable filter, we can arrange to isolate the conversations (i.e., their frequency bands) one from the other at the opposite end of the cable, and then we *reconvert* the $(f_0+\Delta)-(2f_0+\Delta)$ band into a $0-f_0$ band by subtracting.

It is obvious that filtering is crucial in the above arrangement. By a filter we understand here a device which offers different resistance to different frequencies, and so acts in effect to stop or let through certain selected frequency bands. In our problem the central question is how to construct good band-pass filters: These are to be such that very little impedance is offered to frequencies in a certain range f_1-f_2 and possibly high impedance is offered to all other frequencies. If $f_1 = 0$, we call this a low-pass filter. Since a band-pass filter can be constructed from a low-pass prototype by a suitable frequency transformation, we shall consider the low-pass case only. We recall that angular frequency ω and frequency f are connected by the relation $\omega = 2\pi f$.

The design idea for a low-pass filter is as follows. With reference to Fig. 2a we are given the four primary design parameters: ω_0, ω_A, d_0, d_1. These

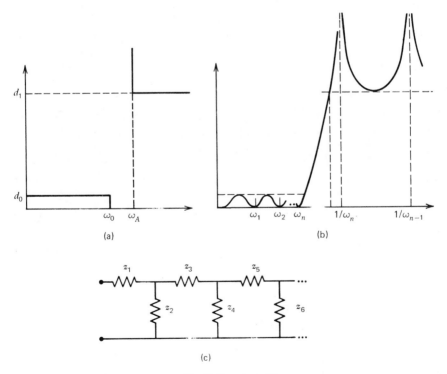

Fig. 2. Low-pass filter.

determine for us the extent $0-\omega_0$ of the pass-band, the extent $\omega_A -\infty$ of the reject-band, the maximum tolerable attenuation d_0 in the pass-band, and the minimum tolerable attenuation d_1 in the reject-band. With the method we are describing the attenuation function is of the form

$$F(\omega) = \omega \prod_{j=1}^{n} \frac{\omega^2 - \omega_j^2}{1 - \omega_j^2 \omega^2}, \tag{15}$$

where the units have been so chosen that $\omega_0 \omega_A = 1$. We observe the important reciprocation property

$$F\left(\frac{1}{\omega}\right) = \frac{1}{F(\omega)};$$

this holds for the product in (15) since it holds for each factor separately. A consequence of the reciprocation property is that, whatever the meaning of "optimal design," once we have designed optimally in the pass-band, the design will be automatically optimal in the reject-band as well. Before going into the matter of optimal design we consider the important question of realizing an impedance function such as that given by (15). It would take us too far away from our object if we were to discuss network synthesis, but we could make the following observations. First, instead of F given by (15) it is more convenient to consider its square, plotted in Fig. 2b, where $n = 2$. The curve has zeros at $\omega = 0$, ω_1, $\omega_2, \ldots$ and a ripple of maxima between the zeros, in the pass-band. In the reject-band the situation is reciprocal and poles at ∞, $1/\omega_1$, $1/\omega_2, \ldots$ replace the zeros, with minima in between. Attenuation function of this type is realized by the ladder four-pole network of Fig. 2c. This is a purely reactive network, that is, one of ideal inductances and capacitances. Therefore each parallel resonance is equivalent to a short and so it corresponds to a pole in the reject-band. Similarly, series resonances correspond to zeros in the pass-band. It is easy to verify that the impedance of the network of Fig. 2c is given by the continued fraction

$$Z = z_1 + \cfrac{1}{\cfrac{1}{z_2} + \cfrac{1}{z_3 + \cfrac{1}{\cfrac{1}{z_4} + \cdots}}}$$

and under proper conditions this is valid even for an infinitely long ladder network. For instance, we have [132]:

$$\frac{e^{1/y}-1}{e^{1/y}+1} = \cfrac{1}{\cfrac{1}{2y}+\cfrac{1}{6y+\cfrac{1}{\cfrac{1}{10y}+\cfrac{1}{14y+\cdots}}}}$$

so that the above impedance is realized by an infinitely long network with series resistances $z_1 = 0$, $z_3 = 6y$, $z_5 = 14y$, $z_7 = 22y, \ldots$ and parallel resistances $z_2 = 1/(2y)$, $z_4 = 1/(10y)$, $z_6 = 1/(18y), \ldots$.

The optimality criterion is now simply stated: with the quantities ω_1, $\omega_2, \ldots$, of (15) at our disposal, subject to $0 < \omega_1 < \cdots < \omega_n < \omega_0$, we choose them so as to minimize the maximum of the absolute value of $F(\omega)$, $0 \le \omega \le \omega_0$.

This brings us to the subject of the minimax approximation, known also as the Chebyshev or L^∞ approximation. In the general L^P case we look for the best approximation to a given function $f(x)$ by a member $R(x)$ of a certain set $\mathcal{R}$ of functions. Here

$$\left(\int_I |f(x) - R(x)|^p \, dx\right)^{1/p}$$

is taken as the measure of approximation, the so-called L^P-norm, and I is the interval on which we work. In the limiting case of $p \to \infty$ the above quantity becomes just

$$\max_{x \in I} |f(x) - R(x)| \qquad (\text{or ess. sup } |f(x) - R(x)|).$$

Hence the names L^∞ or minimax: minimizing the maximum deviation

$$\min_{R \in \mathcal{R}} \max_{x \in I} |f(x) - R(x)|.$$

We state now the fundamental minimax approximation result due to Chebyshev [2]. Given an interval I of the real axis, finite or infinite, given continuous functions $f(x)$ and $s(x)$ on I, and given nonnegative integers n and m, we wish to determine

$$R(x) = \frac{s(x) \sum\limits_{i=0}^{n-\nu} b_i x^i}{\sum\limits_{j=0}^{m-\mu} a_j x^j} = \frac{s(x)B(x)}{A(x)}, \tag{16}$$

where $0 \le \mu \le m$, $0 \le \nu \le n$, and B/A is in its simplest form, so as to approximate to $f(x)$ best in the minimax sense:

$$\text{to determine } H(R) = \min_{R} \max_{x \in I} |f(x) - R(x)|.$$

Chebyshev has solved this problem by proving that the extremal $R(x)$ is uniquely characterized by the following property: The number N of consecutive points of I where the difference $f(x) - R(x)$ assumes the extreme value $\pm H(R)$, with alternating signs, is not less than $m + n + 2 - \min(\mu, \nu)$ if $R \not\equiv 0$, and not less than $n + 2$ if $R \equiv 0$.

The following plausibility argument may help with the above. We have at our disposal $m + n + 2$ degrees of freedom; that is, the coefficients of A and B (the $-\min(\mu, \nu)$ term takes care of the case when true degrees are less than the apparent ones). How shall we use them best? That is, how can we achieve best minimax approximation? First, we arrange for as many zeros in the deviation $E(x) = f(x) - R(x)$ as possible. These zeros are simple so that the graph of $E(x)$ goes positive and negative, attaining maxima and minima in between. Since the approximation is in the minimax sense, we now exercise our control and use up the $m + n + 2$ degrees of freedom, so as to achieve *equal* ripple in $E(x)$: equal (in magnitude) alternating positive and negative extrema. Why *equal* ripple? Because if one extremal value dominated, then we could diminish it by adjusting the coefficients in $R(x)$; but in this process some *other* extremal values would *increase*.

As a simple example of use of Chebyshev's theory we determine the polynomial $T(x)$ of degree n with the leading coefficient 1, so as to minimize

$$\max_{-1 \le x \le 1} |T(x)|.$$

This will be done for the sake of exercise but also to provide an important clue to our problem of filter optimization. We can rephrase our problem in terms of minimax approximation: to find the best minimax approximation to x^n on $[-1, 1]$ by a polynomial *of degree $n-1$*. The answer is: the best minimax approximation is when $T(x)$ is the nth Chebyshev polynomial

$$T_n(x) = 2^{1-n} \cos n \text{ arc cos } x. \tag{17}$$

For, the Chebyshev theorem shows that the minimax is attained at $(n-1) + 2$ points. It is simply verified to be the case for (17): The minimax $\pm 2^{1-n}$ is attained with alternating signs at $x = \cos k\pi/n$, $k = 0, 1, \ldots, n$.

We notice that aside from the normalizing factor 2^{1-n} the Chebyshev polynomial of (17) is just cos n arc cos x, and as such it acts as the n-tuplication function for cos x. Since cos has the right kind of Chebyshev alternation behavior with respect to equal but alternating maxima and minima, we might attempt to solve our optimal filter design problem in the same way. Namely, we should apply (yet again) the conjugation principle so as to find a function $f(x)$ for which $F(\omega)$ of (15) acts as M-tuplication:

$$f(Mx) = F(f(x)) \quad \text{or} \quad F = fMf^{-1}.$$

Moreover, $f(x)$ should be a periodic function of similar behavior (with respect to extrema) to that of cos x. If we succeed in finding such a function then the equal ripple condition of Chebyshev is easily taken care of, the quantities ω_j of (15) are determined, and hence the filter network parameters are found. After this build-up the reader should have no difficulty guessing that an elliptic will do the job for us. In fact, recalling the addition theorem of sn x:

$$\text{sn}(u+v) = \frac{\text{sn } u \text{ cn } v \text{ dn } v + \text{sn } v \text{ cn } u \text{ dn } u}{1 - k^2 \text{ sn}^2 u \text{ sn}^2 v}$$

and the facts that sn x is an odd function while the others are even, we find that

$$\text{sn}(u+v)\text{sn}(u-v) = \frac{\text{sn}^2 u - \text{sn}^2 v}{1 - k^2 \text{sn}^2 u \text{ sn}^2 v},$$

which already begins to acquire formal resemblance to (15). It is now a matter of verification to check that the sn function will do for us, both with respect to Chebyshev alternation and to satisfying the right functional equation: We let $\omega_0 = k$ be the modulus, then the values ω_j are

$$\omega_j = \sqrt{k}\left(\frac{2j}{2n+1} K(k)\right),$$

$K(k)$ being the complete elliptic integral; for details the reader may be referred to [30] and [131].

P. L. Chebyshev (1821–1894) was a Russian mathematician active in analysis, number theory, differential geometry, as well as in many specialized fields, for example, the theory of analog mechanisms and linkages. Among his other achievements is a very early proof that if $\pi(x)$ is the number of primes $\leq x$ and if the limit

$$\lim_{x \to \infty} \pi(x) \bigg/ \frac{x}{\log x}$$

exists, then this limit must be 1, as well as a determination of b (near 0.06) such that

$$1-b<\pi(x)\Big/\frac{x}{\log x}<1+b, \qquad x \text{ large.}$$

2. TRANSPORT EQUATIONS

(a) One of the inventions of the most recent world war was radar, a device for "looking" at metal objects by means of a beam of electromagnetic radiation. This beam is projected from an antenna, scattered by the object, and the scattered echo is picked up by the antenna again. For better image either the antenna must be enlarged or the wavelength diminished. Eventually the wavelength reached the value 1.2 cm and since the radiation of that wavelength propagates in the atmosphere poorly, due to the presence of water vapor, the 1.2 cm radar was a failure as a detection instrument. However, it was an excellent tool for meteorological research.

Among the things it helped to discover and explain was the so-called bright-band. This is a fairly sharply limited narrow range of height in a vertical precipitation column, returning a much higher electromagnetic echo than the regions above or below. It turns out, roughly, that the precipitation starts from high up in the cloud as snow, it melts at a certain height, and below this melting height it continues down as rain. In the snow region the precipitation is present as feathery ice-crystals, large in size relative to a water drop of equal mass, but with the low dielectric constant of ice. In the rain region there are small water drops, but here we have the high dielectric constant of water. On the other hand, in the intermediate bright-band region where the melting occurs, we get both the large cross section of the feathery ice-crystals *and* the high dielectric constant due to a thin layer of water covering the melting ice. Since the radar echo is proportional, among other things, to the dielectric constant and to the sixth power of the radius of an equivalent sphere, we obtain thus a simple and natural mechanism for the greatly increased radiation returned from the melting region.

Obviously one gets in this way some meteorologically valuable data, for instance the height where the melting occurs. To estimate the total amount of water present in a cloud we must know the density function $f(m)$ of the distribution of the raindrop masses m. For, as mentioned above, the radar echo is proportional to the second power of the mass m (= sixth power of the radius) so that the total amount of water depends on the form of the function $f(m)$. If it so happened that the echo were proportional to the cube of the drop radius, that is, to the first power of the mass, then it would be (theoretically) possible to calibrate one's meteorological radar instrument so as to obtain directly the total volume of water in the part of the cloud looked at, without any knowledge of the form of the function $f(m)$. But as it happens, we need here $f(m)$; the same is true in visibility and fog studies in which we are interested in the sum of the squares of the radii, that is, of the 2/3 powers of the masses.

(b) In an attempt to study the evolution of the function $f(m)$ the following simple model was set up. As it turned out later, this model was also applicable to a variety of other situations. We start with a large number of mass particles randomly distributed in a volume of space, and subject to some random mixing in which pairs of particles meet and coalesce. Let $f(x, t)\, dx$ be the average number per unit volume of particles in the mass range x to $x + dx$ at the time t. The assumptions on the randomness of the coalescence process, its dependence on the masses coalescing, and so on, are summed up by stating that

$$f(x, t)f(y, t)\phi(x, y)\, dx\, dy\, dt \tag{1}$$

is the average number per unit volume of coalescences between particles in the mass ranges x to $x + dx$ and y to $y + dy$ during the time interval t to $t + dt$. The function $\phi(x, y)$ is nonnegative and symmetric: $0 \leq \phi(x, y) = \phi(y, x)$; it takes care of such factors as the mass dependence of mobility, of capture cross section, and so on. Roughly, it describes the probability of a coalescence between a mass x and a mass y. Since the mass system is conservative so that no particles enter or leave it, and there is no breakdown of larger masses into smaller ones, the conservation of mass gives us

$$\frac{\partial f(x, t)}{\partial t} = F - D, \tag{2}$$

the characteristic form of scalar (here: mass) transport equations. Here F is the rate of formation of masses x out of pairs y and $x - y$, while D is the rate of disappearance of masses x due to their coalescence with some y so as to form a mass $x + y$. Expressing F and D in terms of f and ϕ leads us to the simplest transport equation

$$\frac{\partial f}{\partial t} = \frac{1}{2}\int_0^x f(y, t)f(x - y, t)\phi(y, x - y)\, dy$$

$$-f(x, t)\int_0^\infty f(y, t)\phi(x, y)\, dy. \tag{3}$$

The symmetry-reducing factor 1/2 ensures that a coalescence of y with $x - y$ counts once only, not twice. For the special cases when $\phi(x, y) = c$, $\phi(x, y) = c(x + y)$ or $\phi(x, y) = cxy$, with a constant c, solving (3) provides an exercise in the elements of differential equations and, in particular, in the use of Laplace transform. The possible occurrence of the latter could have been predicted in advance from the additive form of the coalescence process, and certainly from the consequent occurrence of the convolution term in the R.H.S. of (3).

For instance, with $\phi(x, y) = c$ we let

$$\psi(p, t) = \int_0^\infty e^{-px} f(x, t)\, dx$$

and then (3) becomes

$$\frac{\partial \psi(p, t)}{\partial t} = \frac{c}{2}\, \psi^2(p, t) - c\psi(p, t)\psi(0, t). \tag{4}$$

This is a simple nonlinear differential equation of the Bernoulli type; we solve it by first putting $p = 0$ in (4) and solving for $\psi(0, t)$, then substituting this solution into (4) and solving for $\psi(p, t)$. The initial conditions give us

$$\psi(p, 0) = \int_0^\infty e^{-px} f(x, 0)\, dx$$

where $f(x, 0)$ is the known initial distribution of particle masses.

It might be asked whether (3) can have steady-state solutions, that is, solutions $f(x, t)$ independent of t. It is easy to show, multiplying (3) by x and integrating, that

$$\frac{\partial}{\partial t} \int_0^\infty xf(x, t)\, dx = 0$$

which proves that the total mass is indeed conserved. Similar procedure will prove that for the existence of a steady-state solution we must have $\phi(x, y) = 0$ for some values of x. Physical reasoning is here quite easily translatable into mathematical terms: if $\phi(x, y) > 0$ whenever x and y are > 0, then there is a net transport of mass out of the mass interval $[0, x]$. Also, we note that if

$$N(t) = \int_0^\infty f(x, t)\, dx$$

is the total number of mass particles per unit volume, then integrating (3) we get

$$\frac{dN}{dt} = - \int_0^\infty \int_0^\infty \phi(x, y) f(x, t) f(y, t)\, dx\, dy \tag{5}$$

so that N necessarily decreases unless $\phi = 0$ somewhere.

With respect to the existence of solutions of (3) it can be proved [113, 114] that unique and well-behaved solution $f(x, t)$ exists for all $t \geq 0$, under some minimal assumptions on ϕ. These involve boundedness: we observe that the unique existence for all $t \geq 0$ breaks down if the kernel ϕ increases too fast. For instance, take $\phi(x, y) = c(x + y)$ where c is a positive constant. Then (5) becomes

$$N' = -2cMN \tag{6}$$

where M is the constant total mass:

$$M = \int_0^\infty x f(x, t) \, dx.$$

Hence

$$N(t) = N(0) \, e^{-2Mct},$$

so that the total number of particles decreases exponentially without reaching the value 0. On the other hand, if we take $\phi(x, y) = cxy$ then instead of (6) we have by (5)

$$N' = -cM^2,$$

so that

$$N(t) = N(0) - cM^2 t.$$

It follows that the total number of particles decreases to 0 in finite time $t_0 = N(0)/cM^2$ thereafter becoming negative. The significance of this is, of course, that with the coalescence kernel $\phi(x, y)$ growing as fast as cxy, all the mass particles in the system coalesce together in finite time t_0. Hence the randomness assumptions implicit in the formulation of the transport equation (3) no longer apply as t approaches t_0 and so the whole model loses its validity.

The unique well-behaved solution $f(x, t)$ valid for all $t \geq 0$, referred to before, is actually real-analytic in t for all $t \geq 0$. The method of proof employs the continuation principle of vol. 1, p. 236. In this context the reader may wish to study a simple model of mathematical time-irreversibility by proving that in general no continuation is possible for all times $t \leq 0$. At a further remove this may even suggest something rather strange: A study of the behavior of $f(x, t)$ with respect to complex values of the time variable t. We may consider half-planes (or other regions) free of singularities.

(c) We shall consider next some simple generalizations of the transport equation (3). First, it is physically reasonable as well as mathematically expedient, to assume that masses x not only coalesce in pairs into bigger ones, but also break down, singly, into smaller ones. The breakdown mechanism may be mathematically modeled by supposing that a non-negative function $\psi(x, y)$ describes the probability that a mass x breaks down and a mass in the range $y - y + dy$ is a breakdown fragment. We have therefore

$$\int_0^x y\psi(x, y) \, dy \le x, \qquad \psi(x, y) = 0 \qquad \text{if} \qquad y > x.$$

The mass conservation condition (2) becomes now

$$\frac{\partial f(x, t)}{\partial t} = F_c - D_c + F_b - D_b. \tag{7}$$

Here the subscripts c and b are mnemonics for "coalescence" and "breakdown" just as F and D stand for "formation" and "disappearance." F_c and D_c are the old F and D terms from (2), while F_b is the rate of formation of particles of mass x from the breakdown of larger ones, and D_b is the rate of disappearance of x's due to their breakdown into smaller masses. Expressing F_b and D_b by ψ we get the augmented transport equation

$$\frac{\partial f}{\partial t} = \frac{1}{2} \int_0^x f(y, t) f(x - y, t)\phi(y, x - y) \, dy - f(x, t) \int_0^\infty f(y, t)\phi(x, y) \, dy$$
$$+ \int_x^\infty f(y, t)\psi(y, x) \, dy - \frac{f(x, t)}{x} \int_0^x y\psi(x, y) \, dy. \tag{8}$$

We observe that this can be written as

$$\frac{\partial f}{\partial t} = A_2(f) + A_1(f) - B_2(f) - B_1(f), \tag{9}$$

where the A's and the B's are positive operators (that is, $A_i(f) \ge 0$ and $B_i(f) \ge 0$ if $f \ge 0$). Further, we emphasize that the coalescence operators A_2 and B_2 are quadratic while the breakdown operators A_1 and B_1 are linear. The reader may wish to verify that not only is the whole operator $A_2 - B_2 + A_1 - B_1$ of (9) conservative but so are even its quadratic and linear parts separately. The length of the equation (8) as well as the quadraticity of $A_2(f) - B_2(f)$ suggest an improved notation which turns out to be valuable in more detailed work. Namely, we recall how the norm induces the inner product:

$$x \cdot y = \tfrac{1}{2}(|x + y|^2 - |x|^2 - |y|^2,$$

and we define a sort of inner product operator (f, g) by

$$(f, g) = \tfrac{1}{2}[A(f + g) - A(f) - A(g)],$$

where

$$A(f) = A_2(f) - B_2(f).$$

Writing also $L(f)$ for $A_1(f) - B_1(f)$, to accentuate the linearity, we have now (8) in the simple and useful form

$$\frac{\partial f}{\partial t} = (f, f) + L(f).$$

The foregoing suggests a more general form of the scalar transport equation:

$$\frac{\partial f}{\partial t} = A(f) - B(f), \tag{10}$$

where A and B are again positive operators each of which has an expansion into an operatorial power series

$$A(f) = \sum_{n=1}^{\infty} A_n(f), \qquad B(f) = \sum_{n=1}^{\infty} B_n(f). \tag{11}$$

Here $A_n(f)$, as well as $B_n(f)$, is derived from some n-linear operator $A_n(f_1, \ldots, f_n)$ by equating the argument functions: $A_n(f) = A_n(f, \ldots, f)$. The summands $A_n(f)$ and $B_n(f)$ are what physicists would term n-tuple or n-fold interactions. The reader might conjecture that each operator $A_n(f) - B_n(f)$ could be expected to be conservative. There appears to be an important qualitative difference between the solutions $f(x, t)$ of transport equations with no higher than pair interaction, that is quadratic, terms, like (3) or (8), and those in which cubic or higher terms are present [116]. Namely, as was mentioned before, $f(x, t)$ for (3) (and also for (8)) is real-analytic in t for $t \geq 0$; on the other hand, when cubic or higher terms are present in (10) then it appears that $f(x, t)$ is only of the class C^{∞} in t.

In the context of the generalization (10)–(11) of the transport equations, the reader may wish to attempt an extension of the Weierstrass theorem on the approximation of continuous functions by polynomials. To describe the intended extension we shall first rephrase the Weierstrass theorem itself. Let R be the real line and F the space of all continuous functions on R to R. An n-linear function on $R \times \cdots \times R$ (n times) to R is just $cx_1 \cdots x_n$; equating here the arguments $x_1, \ldots, x_n$ we get the nth power cx^n. Let us call any such function (for $n = 1, 2, \ldots$) a diagonal function. Now the Weierstrass theorem is expressible as follows: Any function $f \in F$ is arbitrarily well approximable, in the topology of uniform convergence on compact sets, by a linear combination of diagonal functions. The operators $A_n(f)$ of (11) are formed in the same way as our diagonal functions: by forming first a multilinear expression and then equating the arguments. Our hypothetical extension of the Weierstrass theorem deals with the question of approximating, in some suitable topology, nonlinear operators by truncated sums of the form (11): $\sum_{n=1}^{N} A_n(f)$. That is, we wish to approximate to nonlinear operators by linear combinations of operatorial powers $A_n(f)$ in the same way in which Weierstrass theorem guarantees arbitrary approximation to continuous functions by polynomials.

We show next how the single integrodifferential transport equation (3) can be used to handle an infinite system of differential equations. This system is due to von Smoluchowski [162, 32], and describes the coagulation process in colloids. The "principle" here is that just as an infinite system of linear equations is sometimes equivalent to a single integral equation, so an infinite system of differential equations is sometimes equivalent to a single integrodifferential equation.

In the coagulation theory we start with identical particles, say of unit mass; at later times these will combine to form double, triple, ... and in general n-tuple masses. Let $x_n(t)$ be the number of n-tuples at the time t. Then it is found [32] that under certain simplifying assumptions

$$x'_n(t) = \frac{1}{2} \sum_{k=1}^{n-1} x_k(t)x_{n-k}(t) - x_n(t) \sum_{k=1}^{\infty} x_k(t), \qquad (12)$$

where $n = 1, 2, \ldots$, and with the initial conditions

$$x_k(0) = 0 \quad \text{if} \quad k > 1, \quad x_1(0) = N \text{ is given.}$$

Chandrasekhar [32] solves (12) inductively by first summing over n, in order to be able to determine the total

$$\sum_{n=1}^{\infty} x_n(t).$$

Once this is known, the sum is substituted into (12) and the equations of (12) are solved one by one. However, we can also solve (12) in a simple manner by letting

$$f(x, t) = \sum_{n=1}^{\infty} x_n(t)\delta(x - n),$$

where δ is the Dirac delta function. It will be found that f satisfies the simplest transport equation (3) with $c = 1$. Hence f, and then $x_n(t)$, can be found by solving this special case of (3) with the help of Laplace transforms, as shown before.

(d)   We note now, very briefly, a somewhat unexpected consequence of the specific form of the convolution term in (3), that is, of the first integral on the R.H.S. Our concern will be with the so-called positivity sets of the solution $f(x, t)$ of (3). We suppose throughout this subsection that ϕ satisfies a suitable convenient condition for our purpose, for instance, that it is sufficiently smooth and bounded between two positive constants. We observe, and we can prove, that the quantity $f(x, t)$ being a density is nonnegative. Also, it can be shown that if $f(x_1, t_1) > 0$ then $f(x_1, t) > 0$ for all $t \geq t_1$. This is so because, to put it roughly, f cannot fall off faster than exponentially in time. We let X be the nonnegative x-axis and we decompose X into the initially positive and initially zero parts:

$$P = \{x: x \geq 0, f(x, 0) > 0\}, \qquad Z = \{x: x \geq 0, f(x, 0) = 0\}.$$

It is now guaranteed that for $x \in P$ $f(x, t) > 0$ for all $t \geq 0$; we inquire about the behavior of $f(x, t)$ for $x \in Z$. Let

$$Z_n = \left\{ x: x \geq 0, \frac{\partial^k f(x, 0)}{\partial t^k} = 0 \quad \text{for} \quad k = 0, 1, \ldots, n-1; \frac{\partial^n f(x, 0)}{\partial t^n} > 0 \right\}$$

for $n = 1, 2, \ldots$, and call Z_n the nth positivity set. The totality of these gives us some information about the initial behavior of $f(x, t)$ for small t. As remarked in [47], this type of information may be of some importance in the numerical solution of transport equations.

Given subsets A, B of X we define their vector sum by

$$A + B = \{x: x = a + b, a \in A, b \in B\} \tag{13}$$

and we use the (not quite consistent) notation

$$A^1 = A, \qquad A^2 = A + A, \ldots, \qquad A^{n+1} = A^n + A.$$

The reader might guess at this point what is coming: The Laplace convolution term in (3) is so strongly related to the vector sum of sets that for our simple equation (3) all positivity sets can be determined [115, 47]. It turns out that

$$Z_n = Z \cap \left(P^{n+1} - \bigcup_{i=1}^{n} P^i \right).$$

The reader may wish to find other functional equations, perhaps with the convolution terms other than of the Laplace type, for which the analogs of positivity sets can be defined and determined. Presumably, the set operations which would then replace the set sum of (13) would be adapted to the type of the convolution present.

(e) The simple mass coalescence phenomenon leading to (3) may be described as 2/1 transport: *Pairs* of masses combine to form *single* larger ones. It is not hard to generalize this to $n/1$ transport. Here n is an integer ≥ 2 and we decree that not pairs but n-tuples of masses coalesce to form single larger masses. We generalize $\phi(x, y)$ to a suitable function $\phi(x_1, \ldots, x_n)$ and we replace (1) by

$$N(x_1, \ldots, x_n, t) \, dx_1 \cdots dx_n \, dt = \phi(x_1, \ldots, x_n) \, dt \prod_{i=1}^{n} [f(x_i, t) \, dx_i].$$

(14)

Then the conservation of mass will give us, as before with (2) and (3).

$$\frac{\partial f(x, t)}{\partial t} = \frac{1}{n!} \int \cdots \int_{\substack{0 \leq x_i \\ \sum x_i = x}} N(x_1, \ldots, x_n, t) \, dx_2 \cdots dx_n$$

(15)

$$- \frac{1}{(n-1)!} \int_0^\infty \cdots \int_0^\infty N(x, x_2, \ldots, x_n, t) \, dx_2 \cdots dx_n.$$

Here the symmetry-reducing factorials prevent multiple counting of a single event. Substituting from (14) into (15) will give us the multiple $n/1$ transport equation for a conservative mass system without mass breakdown. This can be further generalized to the case of $(n_1 + n_2 + \cdots)/1$ transport. Here we have a sequence of integers $n_1, n_2, \ldots$, finite or infinite, with $2 \leq n_1 < n_2 < \cdots$; for each $n = n_i$ an n_i-tuple coalescence mechanism can be given by means of a suitable counterpart of (14) (with ϕ_{n_i} replacing ϕ). Now we suppose that all of these multiple coalescence mechanisms are going on simultaneously. This, incidentally, provides a definite example of a generalization of the type (10)–(11). Also, it allows us to study the competitive effectiveness of several $n/1$ transport mechanisms of different multiplicities n, which are going on simultaneously. Let us consider a simple case of such simultaneous multiple transport in which all the coalescence kernels are constant:

$$\phi_{n_i}(x_1, \ldots, x_{n_i}) = c_{n_i}, \qquad i = 1, 2, \ldots.$$

Let the transport function F be defined as the exponential generating function

$$F(z) = \sum_{n=2}^{\infty} \frac{c_n}{n!} z^n$$

where $c_n = c_{n_k}$ if n is in the sequence $n_1, n_2, \ldots$ and happens to be n_k, and $c_n = 0$ otherwise. The counterpart of the transport equation (3), satisfied now by $f(x, t)$, is rather lengthy. However, the Laplace transform

$$g(p, t) = \int_0^\infty e^{-px} f(x, t)\, dx$$

is easily shown to satisfy

$$\frac{\partial g(p, t)}{\partial t} = F[g(p, t)] - g(p, t) F'[g(0, t)].$$

In particular, for the 0-th moment $N(t) = g(0, t)$ we get the differential equation

$$N'(t) = F(N(t)) - N(t) F'(N(t)).$$

Somewhat similar equations may be written down for the higher moments

$$N_k(t) = \int_0^\infty x^k f(x, t)\, dx = (-1)^k \frac{\partial^k g(p, t)}{\partial p^k}\bigg|_{p=0}. \tag{16}$$

The above may serve to illustrate the well-known advantage of using integral transforms to solve functional equations: Even if we are unable to get the unknown function itself, we may still be able to determine some of its moments. This is so because these moments are just special values of the derivatives of the transform of that unknown function, as is shown in our example by (16).

Another type of generalization of the 2/1 transport is the 2/2 transport. Here two masses, say x and y, undergo a sequence of breakups and recombinations, to end up as two masses again, say u and v. Supposing a linear dependence of u and v on x and y, observing the mass conservation, and using symmetry and indistinguishability of particles, we find that

$$u = a(x + y), \qquad v = (1 - a)(x + y), \qquad 0 \le a \le \tfrac{1}{2},$$

and the transport equation is

$$\frac{\partial f(x, t)}{\partial t} = \frac{1}{2a} \int_0^{x/a} f(y, t) f\!\left(\frac{x}{a} - y, t\right) \phi\!\left(y, \frac{x}{a} - y\right) dy$$

$$+ \frac{1}{2(1 - a)} \int_0^{x/(1-a)} f(y, t) f\!\left(\frac{x}{1 - a} - y, t\right) \phi\!\left(y, \frac{x}{1 - a} - y\right) dy \tag{17}$$

$$- f(x, t) \int_0^\infty f(y, t) \phi(x, y)\, dy.$$

It is an elementary exercise in multiple integration to verify that here both the total mass

$$\int_0^\infty x f(x, t)\, dx$$

and the total number

$$\int_0^\infty f(x, t)\, dx$$

of particles, stay constant in time, just as expected. In the limit as $a \to 0$ we get the ordinary 2/1 transport and (17) becomes (3), provided that the limiting value of the first integral is taken as 0. In the other limiting case, when $a = 1/2$, we get the averaging transport. Here the masses x and y give rise, on meeting, to two offspring of the average parental mass $(x+y)/2$. The transport equation (17) assumes now a simpler form

$$\frac{\partial f(x, t)}{\partial t} = 2\int_0^{2x} f(y, t)f(2x - y, t)\phi(y, 2x - y)\, dy$$
$$- f(x, t)\int_0^\infty f(y, t)\phi(x, y)\, dy.$$

As a possible application of this type of transport (for a change, to social rather than physical phenomena) we consider the "socialist game." Here we have a distribution of a scalar quantity, this time not mass but "wealth": $f(x, t)\, dx$ is the number of individuals (per unit area, say) whose "worth" at time t is between x and $x + dx$ monetary units. It is decreed that two individuals must average out their wealth on meeting. That is, if they were worth x and y before meeting, they will each be worth $(x+y)/2$ after. Since $\phi(x, y)$ is, roughly, the probability that they meet, and so average, we assume as before that $\phi(x, y)$ is nonnegative and symmetric. Beyond that, it could be of a variety of forms; for instance, we could have $\phi(x, y) = g(|x - y|)$ where g is decreasing. Now the solution $f(x, t)$ of (18) is the density of the wealth distribution and so (18) allows us to study, analytically or numerically, the approach to uniformity under decreed equalization. For instance, we could investigate the dependence of the standard deviation $\sigma(t)$ of $f(x, t)$ on ϕ.

Basing ourselves on the previous extension of the simple 2/1 transport first to the $n/1$ and then to the $(n_1 + n_2 + \cdots)/1$ case, we could now develop the rather unwieldy $(n_1 + n_2 + \cdots)/(m_1 + m_2 + \cdots)$ generalization. This might illustrate the generalizing possibilities inherent in the right notation, if nothing else.

(f)   All the different transport types considered so far refer to one kind of mass particles. We shall now introduce a model of transport with k different types, or species, of mass particles and, accordingly, with k different density functions $f_i(x, t)$, $i = 1, \ldots, k$. The transport will be of the 2/1 type although higher types could be also considered (at the cost of considerable notational complications). We suppose then that there is 2/1 transport that is, however, interspecific. That is, a particle of mass x and type i can coalesce with a particle of mass y and type j, to give a particle of mass $x + y$ and type s. To indicate the allowed types of coalescence we introduce the symbols

$$\varepsilon_{ij}^s = 1 \quad \text{or} \quad 0 \tag{19}$$

depending on whether the coalescence of types i and j into s can, or cannot, occur. We may suppose that for each i and j there is at most one index s such that $\varepsilon_{ij}^s = 1$. However, it could be argued on probabilistic grounds that the ith and the jth kind particles could coalesce into several different kinds of particles, with different probabilities. This amounts to replacing (19) by

$$\varepsilon_{ij}^s \geq 0 \quad \text{and} \quad \sum_s \varepsilon_{ij}^s \leq 1 \text{ for fixed } i \text{ and } j. \tag{20}$$

We have also the coalescence kernels $\phi_{ij}(x, y)$ for which we could require the symmetries

$$\phi_{ij}(x, y) = \phi_{ji}(x, y) = \phi_{ij}(y, x). \tag{21}$$

The system of transport equations is now obtained as before, by equating the rate of change to the difference which is the rate of formation minus the rate of disappearance, and we get

$$\frac{\partial f_s(x, t)}{\partial t} = \sum_{i,j} \varepsilon_{ij}^s d_{ij} \int_0^x f_i(y, t) f_j(x - y, t) \phi_{ij}(y, x - y) \, dy \tag{22}$$

$$- \sum_{i,j} \varepsilon_{is}^j f_s(x, t) \int_0^\infty f_i(y, t) \phi_{is}(x, y) \, dy, \qquad s = 1, 2, \ldots, k.$$

Here d_{ij} is the necessary symmetry-removing factor defined to be 1 if $i \neq j$ and $1/2$ if $i = j$.

There are many possible applications of the above multiple transport, and some of its modifications, to physical, biological, and sociological contexts. To show some possibilities, and to be able to indicate later some necessary modifications in our transport mechanism, we first consider a very special case of (22) in which $\phi_{ij}(x, y)$ are all constant, say, c_{ij}. Introducing the Laplace transforms

$$g_s(p, t) = \int_0^\infty e^{-px} f_s(x, t) \, dx, \qquad s = 1, 2, \ldots, k, \tag{23}$$

we get from (22)

$$\frac{\partial g_s(p, t)}{\partial t} = \sum_{i,j} \varepsilon_{ij}^s d_{ij} c_{ij} g_i(p, t) g_j(p, t)$$

$$- \sum_{i,j} \varepsilon_{is}^j c_{is} g_i(0, t) g_s(p, t), \qquad s = 1, 2, \ldots, k. \tag{24}$$

From (23) it follows that $g_s(0, t)$ is just the normalized total number of the individuals of the sth kind at the time t:

$$g_s(0, t) = \int_0^\infty f_s(x, t) \, dx = N_s(t), \text{ say.}$$

So, putting $p = 0$ in (24) we get

$$N_s' = \sum_{i,j} \varepsilon_{ij}^s d_{ij} c_{ij} N_i N_j - \sum_{i,j} \varepsilon_{is}^j c_{is} N_i N_s, \qquad s = 1, \ldots, k. \tag{25}$$

This system is a special case of a class often met: Differential equations describing quadratic competitive rate process. These describe a variety of situations, from binary-collision chemical kinetics for k compounds [125] to the k-species generalization of the Lotka–Volterra equations for prey–predator interaction [126]. We emphasize here the strict relation between quadraticity of the R.H.S. of (25) in the variables N_i and the binary nature of the interactions, whether between reacting molecules or hungry animals. If, say, triples of molecules were needed or if a wolf could not eat a sheep without some metabolic involvement or commensality of the fox, then the R.H.S. of the suitable variant of (25) would have to be a cubic. In fact, we would then require the interspecific 3/1 transport, referred to at the beginning of this subsection.

(g)   Our interest in this final subsection is in the LVT (Lotka–Volterra theory) and, especially, in possible adaptations of the multiple transport model to generalize the LVT equations. We begin with a very brief exposition of the standard LVT. In the simplest case one has $k = 2$: there are just two species, prey and predator, with numbers $N_1(t)$ and $N_2(t)$ which we want to study as functions of time. N_1 and N_2 satisfy the simple LVT system

$$N_1' = a_1 N_1 - b N_1 N_2,$$
$$N_2' = -a_2 N_2 + c N_1 N_2, \tag{26}$$

where a_1, a_2, b, c are positive constants. Lotka [102] and Volterra [176] derive, or rather justify, this as follows. The constants a_1 and a_2 represent the natural rates of growth (a_1, for prey) or decay ($-a_2$, for predator) for each species when isolated from the other. When both are together the prey numbers decrease and those of the predator increase, and the corresponding rates are proportional, with the constants of proportionality b and c, to the possible number of interspecies encounters, which is nearly N_1N_2. The system (26) generalizes for k species to

$$N_i' = c_iN_i + v_i^{-1} \sum_{j=1}^{k} b_{ij}N_iN_j, \qquad i = 1, \ldots, k. \tag{27}$$

Here the c_i have the same meaning as a_i and a_2 in (26)—the natural rate constants, or the Malthusian parameters, for the k species when in complete isolation. The constants b_{ij} are the rates of increase of the ith species on its encounters with the jth one. So, $b_{ij} > 0$ if the ith species preys on the jth one, $b_{ij} < 0$ shows the reverse, and $b_{ij} = 0$ if they do not interact. The quantities v_i are Volterra's equivalence numbers: In the pair encounters of the ith and jth species v_j/v_i is the ratio of the jth one's gain to the ith one's loss (or conversely). The quantities b_{ij} are antisymmetric

$$b_{ij} + b_{ji} = 0 \tag{28}$$

and it is customary to suppose that $b_{ii} = 0$ for all i.

The system (27) has a first integral (an analogous statement for a mechanically derived system of differential equations would be something like the constancy of energy or momentum). Let $N_j(t)$ have the "steady-state" value n_j and let us assume that no n_j is 0. Then by definition we have

$$c_iv_i + \sum_{j=1}^{k} b_{ij}n_j = 0, \qquad i = 1, \ldots, k. \tag{29}$$

Introducing the "normalized logarithms"

$$u_j(t) = \log\left[\frac{N_j(t)}{n_j}\right],$$

we can write (27) as

$$v_iu_i' = \sum_{j=1}^{k} n_jb_{ij}(e^{u_j} - 1).$$

Multiplying by $n_i(e^{u_i} - 1)$ and summing over i gives us

$$\left(\sum_{i} v_in_i(e^{u_i} - u_i)\right)' = \sum_{i,j} b_{ij}n_in_j(e^{u_i} - 1)(e^{u_j} - 1)$$

and by the antisymmetry (28) the R.H.S. above is 0. Hence we get our first integral

$$\sum_{i=1}^{k} v_in_i(e^{u_i} - u_i) = c, \tag{30}$$

where c is easily shown to be a positive constant.

For $k = 2$ the first integral (30) enables us to solve the LV equations graphically. With a simple linear change of variables we can write (26) as

$$x' = Ax(1-y), \qquad y' = -By(1-x) \tag{31}$$

and it can be shown (cf. pp. 205–212 of vol. 1) that $x(t)$ and $y(t)$ are not elementary functions. However, we exploit the first integral as follows. From (31) we get

$$\frac{A(1-y)\,dy}{y} + \frac{B(1-x)\,dx}{x} = 0$$

and integrating this gives us

$$x^B y^A e^{-Ay} e^{-Bx} = K, \tag{32}$$

where K is a constant of integration. Equation (32) is usefully rewritten as

$$u^B v^A = K, \qquad \text{where } u = xe^{-x}, \qquad v = ye^{-y}. \tag{33}$$

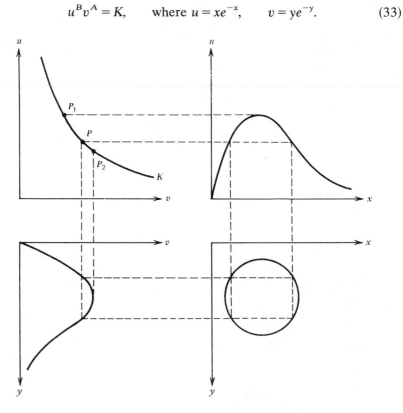

Fig. 1. Graphical determination of trajectories.

It is quite easy to plot the dependence of u and v, of u and x, and of v and y (the reader may recall here the graph and the series of vol. 1, p. 115). Now it is possible to put together those three graphs, as shown in Fig. 1, to get the dependence of x and y. The two maxima, of $u = u(x)$ and $v = v(y)$, projected on the uv-curve show that only the arc $P_1 P_2$ of the curve is of interest. Any point P on that arc gives us, by means of the vertical and horizontal lines shown in Fig. 1, four points on the xy-locus, and so there is no difficulty in plotting this locus. In particular, it is a closed curve so that $x(t)$ and $y(t)$ are periodic. This periodicity, or fluctuation, in the numbers of prey and predator was noticed by Volterra [176] who also gave the graphical construction shown above. By constructing the xy-locus for different values of k we get a family of trajectories of the system (31).

We have provided several other examples of locus plotting by the Volterra method, with special reference to coordinate extrema, inflection points, and so on. Such plottings seem to be a good exercise in developing geometrical and graphical intuition. The loci plotted are given in Figs. 2–4.

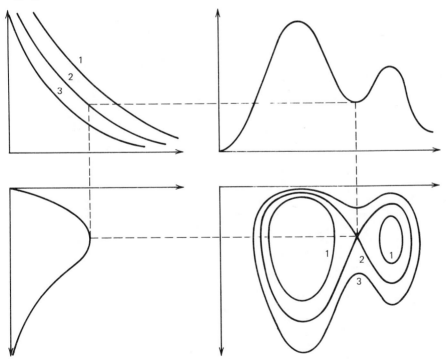

Fig. 2. Curve plotting.

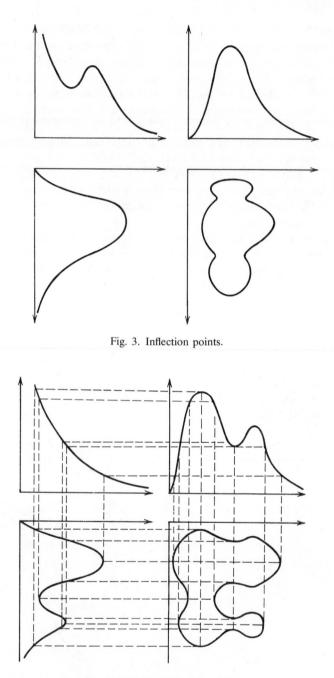

Fig. 3. Inflection points.

Fig. 4. Multiple inflection points.

The LVT attempts to deal with k interacting animal populations by means of the quadratic competitive rate process described by (27). For a number of reasons the equations (27) are not altogether satisfactory. This was recognized by Volterra himself who attempted to bring in "hereditary" influences and to express the competing rates in terms of equations which are not "local." For a single population $y(t)$ we have the Malthusian unbounded growth governed by the equation $y' = ay$. Or, we can postulate growth with a saturation level, governed by the logistic or Verhulst equation [175]

$$y' = ay\left(1 - \frac{y}{b}\right).$$

Volterra generalized this further [182] to

$$y' = \left[a - \lambda y + K \sin ct - \int_0^t y(\tau)f(t-\tau)d\tau\right]y + E,$$

where a is the Malthusian differential growth rate, λ is related to the saturation level of Verhulst, the term $K \sin ct$ represents a forced oscillation, the integral term represents hereditary effects including "self-poisoning by accumulated waste products," and E represents immigration. Also, Volterra had integrodifferential generalizations of the system (26).

Since our moment equations (25) for the 2/1 transport system (24) look formally like the k species LVT equations (27), we might suggest the following. A better approximation to a model of k competing species might perhaps be obtained by introducing k functions like our transport functions $f_i(x, t)$. Here x would be something like "biological mass" or "demographic energy" (a term due to Volterra) and the species interactions might be described by integrodifferential equations involving those f_i's. The LVT equations (27) might perhaps be the moment equations obtained by integrating over the parameter x of distribution so that

$$N_i(t) = \int_0^\infty f_i(x, t)\, dx$$

is the total number at time t of individuals of the ith species. Of course, several simultaneous parameters of distribution could also be considered. In terms of our transport model this would correspond to a simultaneous transport of several scalar quantities at once.

3. BRANCHING

(a) The basic situation we consider here is this: Starting from a point called the root we have a segment which is the principal trunk (or the 0-th order branch); this either terminates at its other end or it splits there into several first-order branches, each of which again either terminates or splits into second-order branches, and so on. For a complete branching specification of the resulting tree we need its depth, which is the highest order of branch present, and a description of the type of branching. To ensure a certain variety for the latter we define the so-called (s, k)-branching. Here an integer $k \geq 2$ is fixed, s is an integer satisfying $1 \leq s \leq k$, and it is supposed that each branch (principal trunk included) either terminates or it splits into at least s and at most k next order branches. The values $s = 1$ and $s = k$ are allowed, though $s = 1$ strains the terminology somewhat.

In applications or models the branching itself may be either "abstract" or "concrete." In the abstract case one possibility is something like taxonomy, that is, scientific classification. For instance, in the example of biology the main trunk is all the living organisms, the two first-order branches are the animal kingdom and the plant kingdom and so on, down to individual species (or even smaller units). There is no single system of categories or taxa (sing. *taxon*, from Greek *taxis*—arrangement) of universal currency; the most compendious one has depth 19: kingdom, subkingdom, phylum, subphylum, superclass, class, subclass, infraclass, cohort, superorder, order, suborder, superfamily, family, subfamily, tribe, genus, species, subspecies. Another possibility with the abstract branching is the so-called decision tree when a primary decision may entail several possibilities each leading to a secondary decision, and so on, as for example in a game such as chess. Something similar occurs in mathematical linguistics with some types of sentence-structure analysis.

In the "concrete" cases there is an actual branching, either in space or in time. In the first case we get something like an actual tree, a dendrite of a nerve cell (from Greek *dendron*—tree), or perhaps a chain of a chemical compound. In the second case we have something like a genealogy or similar family-history, or a picture of a particle shower.

From the point of view of graph theory a branching structure is a singly rooted tree, the root being the distinguished vertex from which the 0-th order branch starts. We recall that the valency of a vertex is the number of edges incident to it; a tip is a vertex of valency 1; the distance between two vertices is the smallest number of edges, that is, branches, joining them. Thus an (s, k)-branching of depth d is a singly rooted tree in which every vertex has the valency 1 or else, between $1 + s$ and $1 + k$, and where the maximum distance from the root to a vertex (necessarily a tip) is $d + 1$. It is also possible to regard branchings as directed trees in which branches have arrows, from lower to higher order; for an (s, k)-branching one would say then that the in-degree, or the fan-in number, of a vertex which is not a tip is 1 and its out-degree, or fan-out number, is between s

and k. However, nothing is gained by this complication in our case.

Supposing the branching structure to be imbedded in the Euclidean three-dimensional space, we distinguish several types of properties and quantities of interest. We have first the topological, or combinatorial, type. This refers to the invariance of the property under homeomorphisms and it can be illustrated in the usual fashion by supposing that the branching structure is made of suitably joined rubber threads. Next, we have the free isometric type referring to the invariance of properties under isometries (isometric = distance-preserving). These are the transformations that keep the lengths of branches fixed, and can be visualized with the branching structure made of perfectly elastic and movable but inextensible wires joined at the vertices. The bound isometric type is the same except that all tips stay fixed. Finally, in the geometrical type we refer to invariance under rigid motions: The wire of the previous example is now completely rigid as well as inextensible.

An example of a combinatorial question is this: What is the number $V(s, k, d)$ of distinct (s, k)-branchings of depth $\leq d$? Questions of the free isometric type are easily formulated, especially for structures branching in time. In fact, since time "runs straight," any continuous question referring to time-branching must be free isometric, for example, average length or duration of a branch, the expected time for the demise of the last surviving nth generation member (or equivalently, for the branching of the last or latest nth order branch).

For an example of a bound isometric type of problems we consider the dendrites of a neuron or nerve cell. In our terms a dendrite is just a $(2, 2)$-branching structure of small depth. In physiological terms it is a binary-branched fiber that conducts nerve impulses from the tips T toward the root R and the cell body, as shown in Fig. 1. A very sketchy and superficial description of dendrite activity follows. The nerve impulses start as trains of uniform frequency, say f_0, at the tips T_i and travel, say with uniform speed v_0, toward the root R (actually, lower-order branches are bigger in dendrites and somewhat faster conducting). Nerve transmission is such that after each single pulse goes through a fiber, this particular nerve fiber gets depleted of certain chemical substances temporarily, and so it cannot transmit anything during the following short time interval Δ. This is called the absolute refractory period and lasts on the order of several miliseconds. Hence arises the upper bound $1/\Delta$ on any frequency f_0. We can also express it thus with the reference to Fig. 1: at each non-tip vertex V_i there is a gate that stays shut for the time Δ after passing a pulse through, before it opens again. However, during that time another pulse may arrive at V_i, from the other branch, and this will just get lost. Suppose now that pulses originate at all tips $T_1 - T_5$ with the

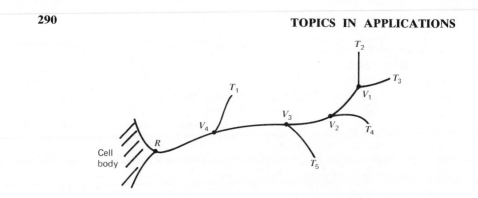

Fig. 1. Schematic dendrite.

frequency f_0, according to some reasonable probability law. Let it be assumed that $5\Delta f_0 < 1$. The pulses are conducted from the tips T_i to R and it is clear that depending on phase relations and arrival times involving f_0, v_0, Δ and the branch length, the average frequency of the pulse arriving at R can be anywhere between f_0 and $5f_0$. Clearly the problems involving the frequency statistics at R are of the bound isometric type. The reader should have no difficulty in formulating problems of the last, geometrical, type; a particular example will be given in Section (d).

We observe that the main mathematical tool in branching, both for discrete and continuous problems is the natural induction on depth d: by erasing the 0-th order branch and the first branching vertex we convert a single branching structure of depth $\leq d+1$ into several ones of depth $\leq d$. This induction from $\leq d+1$ to $\leq d$ is in appearance as well as in substance much more symmetric and, as we shall show, preferable to the induction from depth $d+1$ exactly by the same process to several branchings of depth $\leq d$, at least one of which is of depth d exactly. Thus, it is easier to count objects of depth $\leq d$ than those of depth d. However, there is really no loss of generality here since

$$\# \text{ of depth } d = (\# \text{ of depth } \leq d) - (\# \text{ of depth } \leq d-1). \quad (1)$$

Of course, the reader might wish to do some counts in two ways, both directly with depth d, and indirectly with depth $\leq d$ and the use of equation (1), and then to equate the results.

There is a certain interplay between the discrete and continuous features of branching. We illustrate this, taking as our first example the biological taxonomy again. Some newer biochemical experiments allow one to measure the degree of "compatibility" between certain chemicals occurring in the animals of two different species, say S_1 and S_2. This yields a positive distance-like number and the smaller it is, the closer S_1 and S_2 (should) lie on the phylogenetic tree in taxonomy. Further, if we assume some type of evolutionary mechanism we might attempt to estimate the time lapsed since the differentiation of S_1 and S_2.

A similar example occurs in linguistics when we suppose that languages arise by separation from parental stocks which themselves have split off yet older common stocks, and so on. This leads us at once to a branching or tree structure, as a picture of the process. We might introduce numerical distance-like indices for pairs of languages, analogous to the biochemical degrees of compatibility. Such indices could be obtained in different ways based on, say, comparing the word lists in the two languages (similarity, common origin, mutual comprehensibility, etc). From the set of such indices we could try to derive first the combinatorial nature of the family tree for the languages in question (this amounts to determining the sequence of successive splittings). Then, we could also attempt to estimate the isometric features such as times lapsed since the various splits. So, for instance, we might conjecture that in the tree for Indo-European languages Spanish and Portuguese should be very close, French quite close to both, German at a moderate distance to these three, Lithuanian or Hindi at a fair distance from the preceding four; Albanian, Etruscan, or Hittite at quite a large distance from all others etc. In an attempt to determine the split-off times, one could use here the known or estimated times of the dispersal of Aryan tribes from their presumed home in Southeast Europe.

These bio-taxonomical and linguistic examples suggest that it may be of interest to consider mathematically branching which is gradual rather than sudden. For it is very likely that both subspecies on their way to separate into species and dialects in the process of becoming languages, may still interact significantly for various lengths of time.

A third example, of some vague similarity to the bio-taxonomy and linguistics, is the so-called Dewey decimal bibliographic classification system of book subjects by their topics. This will be familiar, in small part at least, to anyone using a public library of any size.

(b) We consider now the combinatorial problem mentioned before: to find the number $V(s, k, d)$ of topologically different (s, k)-branchings of depth $\leq d$. The numbers $V(2, k, d)$ may be of interest with decision or search trees, where it is required that each non-tip vertex should present an alternative between at least two outcomes. The numbers $V(1, k, d)$ may be of interest in genealogies of various kinds. The numbers $V(k, k, d)$ refer to complete branchings: Each branch either dies off or it branches out into its full quota of k subsidiaries, and so on. With reference to Fig. 2a and b we count explicitly and find that

$$V(s, k, 0) = 1, \qquad V(s, k, 1) = k - s + 2. \qquad (2)$$

Next, we prepare for the use of depth induction and we consider an (s, k)-branching of depth $\leq d$ (with $d \geq 1$) shown in Fig. 2c. There are the primary limbs $P_1, \ldots, P_k$, each of which is an (s, k)-branching of depth

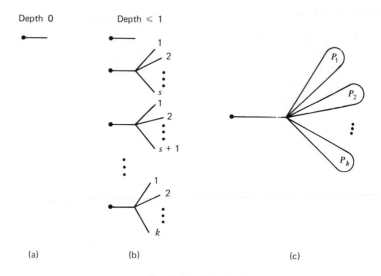

(a)   (b)   (c)

Fig. 2. Depth induction.

$\leq d - 1$ or, possibly, an empty set. By hypothesis, the number j of nonempty selections for P_i's is at least s. For each such selection we have $V(s, k, d-1)$ possibilities and clearly the repetitions are allowed since two primary limbs may be identical. Selections differing in the order of P_i's are not counted as distinct. Hence we are interested here in combinations with repetitions of j objects out of $V(s, k, d-1)$ objects; this number is

$$\binom{V(s, k, d-1) + j - 1}{j}$$

Since the number j varies from s to k the total number of selections of primary limbs is

$$\sum_{j=s}^{k} \binom{V(s, k, d-1) + j - 1}{j}. \qquad (3)$$

Before continuing we prove a combinatorial identity which will be useful:

$$\sum_{j=0}^{q} \binom{N+j-1}{j} = \binom{N+q}{q}. \tag{4}$$

This is easily provable in several ways, for instance by an induction on q. A longer but not uninstructive alternative proof is by applying the principle of counting one and the same thing in two different ways. So, the L.H.S. of (4) may be interpreted as the sum of ways of choosing j objects out of N, where $0 \le j \le q$ and choices refer to combinations with repetition. That is, we choose $\le q$ objects out of N. The R.H.S. of (4) may be interpreted as the number of combinations with repetition of exactly q objects out of the total of $N+1$ objects. If one of the $N+1$ objects is considered to be the empty set, we see that both counts are equal; hence (4).

With (4) we rewrite (3) as

$$\binom{V(s, k, d-1)+k}{k} - \binom{V(s, k, d-1)+s-1}{s-1}. \tag{5}$$

Does this number of primary-limb selections represent *all* possibly (s, k)-branchings of depth $\le d$? No, exactly one possibility is missed, namely the 0-th depth structure of Fig. 2a corresponding to the allowed possibility that the principal branch itself dies out. Thus we add 1 to (5) and get the recursion

$$V(s, k, d) = 1 + \binom{V(s, k, d-1+k)}{k} - \binom{V(s, k, d-1)+s-1}{s-1} \tag{6}$$

which together with (2) enables us to compute the numbers $V(s, k, d)$. As a small check, which may be quite comforting at the time of derivation, we verify that the quantities $V(s, k, 1)$ and $V(s, k, 0)$ of (2) satisfy (6).

The reader may wish to compare the simple recursive formula (6), based on the "$\le d+1 \to \le d$" sort of induction, with the much clumsier treatment of pp. 182–184 based on the "$=d+1 \to \le d + \cdots$" induction. Other counts of distinct branchings, referring to the numbers of tips, branches, vertices, and so on, could be similarly attempted.

Formula (6) can be generalized somewhat. Suppose that the quantities s and k describing the branching are allowed to vary with the depth. We ask now for the number $V(s_1, \ldots, s_d; k_1, \ldots, k_d)$ of topologically or combinatorially distinct branchings of depth $\le d$ in which the main trunk either dies off or branches out into at least s_1 and at most k_1 1-*st* order branches, and so on, and, in general, each $(j-1)$-th order branch either dies off or splits into at least s_j and at most k_j branches of order j. It is supposed, of course, that the numbers s_j and k_j satisfy the obvious conditions $1 \le s_j \le k_j$, $j = 1, \ldots, d$. Now (6) generalizes to

$$V(s_1, s_2, \ldots, s_d; k_1, k_2, \ldots, k_d) = 1 + \left(\frac{V(s_2, \ldots, s_d; k_2, \ldots, k_d) + k_1}{k_1} \right)$$
$$- \left(\frac{V(s_2, \ldots, s_d; k_2, \ldots, k_d) + s_1 - 1}{s_1 - 1} \right)$$

which enables us to compute V recursively.

The depth induction is also used in another discrete problem, that of lexicographical branching. Here we are interested in the vertices-representing individuals—rather than in branches which represent descent. The branching is now complete: a main branch with the ancestral vertex at which branching occurs into exactly k first-order branches leading to the k vertices of the first generation. From each of these there are k branches to the k vertices of the second generation, and so on. Each k vertices of common descent are ordered by some precedence, order, or seniority, which is hereditary. Thus there are k^n vertices of the nth generation marked $1, 2, \ldots, k^n$ where 1 is "the eldest son of the eldest son of $\cdots$" and k^n is "the youngest son of the youngest son of $\cdots$" Let $1 \le i, j \le k^n$, we put $d(i, j) = 0$ if $i = j$ and otherwise $d(i, j)$ is half the distance between the vertices i and j. The reader may wish to verify that $d(i, j)$ is a metric. More precisely, it is the kinship or affinity metric: $d(i, j) = 1$ if i and j are distinct "sons of the same father," $d(i, j) = 2$ if they are "first cousins," and so on. Our interest is in getting the generating function

$$f_n(x, y) = \sum_{i=1}^{k^n} \sum_{j=1}^{k^n} d(i, j) x^i y^j, \tag{7}$$

which may help with finding certain averages relating to degrees of kinship, lexicographic distances, and so on. We start the induction by calculating

$$f_1(x, y) = \sum_{i=1}^{k} \sum_{j=1}^{k} d(i, j) x^i y^j = \sum_{i=1}^{k} x^i \sum_{j=1}^{k} y^j - \sum_{s=1}^{k} (xy)^s.$$

With the abbreviation $F(m, x)$ for

$$\sum_{i=0}^{m} x^i = \frac{1 - x^{m+1}}{1 - x}$$

we have then

$$f_1(x, y) = [F(k, x) - 1][F(k, y) - 1] - [F(k, xy) - 1]. \tag{8}$$

The depth induction will give us $f_{n+1}(x, y)$ in terms of $f_n(x, y)$; to apply it we use the same drawing Fig. 2c as before, supposing now that the primary limbs $P_1, \ldots, P_k$ are k structures of depth n, arranged in the order of seniority with P_1 eldest. The k^n tips of P_i have the numbers

$$k^n(i-1)+s, \quad s=1, 2, \ldots, k^n, \quad i=1, 2, \ldots, k$$

and for $d(i, j)$ two cases are distinguished:
(a) i and j lie in the same primary limb,
(b) i and j lie in different primary limbs.
For (b) $d(i, j) = n+1$ whereas for (a) there is no need to recompute $d(i, j)$—it comes out of the induction. With this the reader may wish to prove the recursion

$$f_{n+1}(x, y) = f_n(x, y)[1 + x^{k^n} y^{k^n} + \cdots + x^{(k-1)k^n} y^{(k-1)k^n}]$$
$$+ (n+1) \sum_{i_1=1}^{k} \sum_{i_2=1}^{k} \sum_{s_1=1}^{k^n} \sum_{s_2=1}^{k^n} x^{k^n(i_1-1)+s_1} y^{k^n(i_2-1)+s_2},$$

which leads to

$$f_{n+1}(x, y) = f_n(x, y) F(k-1, x^{k^n} y^{k^n})$$
$$+ (n+1)[F(k^n, x) - 1][F(k^n, y) - 1][F(k-1, x^{k^n}) F(k-1, y^{k^n})]$$

$$\hspace{8cm} (9)$$

$$- F(k-1, x^{k^n} y^{k^n})].$$

This together with (8) enables us to compute $f_n(x, y)$. The appearance of the powers k^n and x^{k^n}, y^{k^n} suggests the possibility of a connection with Galois fields and polynomials over these, for the case of prime k.

(c)   We pass now on to applications of the depth induction to some isometric properties of branching, referring to statistics of branch length. For convenience we suppose a branching in time: At $t=0$ the single parental individual is born and at some later time it splits into k individuals of the first generation, each of these splits later into k individuals of the second generation, and so on. The integer k is fixed and the lifetimes (=intervals between birth and fission) are independent identically distributed random variables with the density $f(x)$. Let $N = k^n$ and let $T_1, \ldots, T_N$ be the splitting times for the N individuals of the nth generation. Put $X_n = \min(T_1, \ldots, T_N)$, $Y_n = \max(T_1, \ldots, T_N)$ and let $g_n(t)$ and $h_n(t)$ be respective density functions for the random variables X_n and Y_n.

Suppose that the original ancestor splits at the time t_0. Then by the definition of $g_1(t)$

$$\int_0^t g_1(u) \, du = \int \cdots \int_{t_0 + \min(t_1, \ldots, t_k) \le t} f(t_0) f(t_1) \cdots f(t_k) \, dt_0 \cdots dt_k. \tag{10}$$

The $(k+1)$-tuple integral in (10) may be written as

$$\int_0^t f(t-v) \left[\int \cdots \int_{\min(t_1, \ldots, t_k) \le v} f(t_1) \cdots f(t_k) \, dt_1 \cdots dt_k \right] dv,$$

which is

$$\int_0^t f(t-v) \left[1 - \int_v^\infty \cdots \int_v^\infty f(t_1) \cdots f(t_k) \, dt_1 \cdots dt_k \right] dv.$$

If we now introduce the functions

$$\int_t^\infty g_1(t) \, dt = G(t), \quad \int_t^\infty f(u) \, du = F(t)$$

then it follows that (10) becomes

$$G_1(t) = F(t) - \int_0^t F'(t-v) F^k(v) \, dv, \tag{11}$$

which gives us the distribution of X_1. To determine similarly

$$\int_t^\infty g_n(u) \, du = G_n(t)$$

we apply again the depth induction. Let $I_1, \ldots, I_k$ be the k individuals of the first generation and let τ_j be the earliest appearance of an $(n-1)$ times removed descendant of I_j. Suppose as before that the ancestor has fissioned at the time t_0 and put $t_j = \tau_j - t_0$. Then by the depth induction $t_1, \ldots, t_k$ are independent and identically distributed with the density $g_{n-1}(t)$. Hence

$$\int_0^t g_n(u) \, du = \int \cdots \int_{t_0 + \min(t_1, \ldots, t_k) \le t} f(t_0) g_{n-1}(t_1) \cdots g_{n-1}(t_k) \, dt_0 \cdots dt_k.$$

We process this in the same way as before getting eventually

$$G_n(t) = F(t) - \int_0^t F'(t-v) G_{n-1}^k(v) \, dv. \tag{12}$$

Provided that $g_0(t)$ is defined as $f(t)$, we find that (12) includes (11) as a special case.

The functions $h_n(t)$ are determined similarly. We let

$$\int_0^t h_j(u)\, du = H_j(t), \quad \int_0^t f(u)\, du = \phi(t), \quad h_0(u) = f(u)$$

and we find

$$\int_0^t h_n(u)\, du = \int \cdots \int_{t_0 + \max\,(t_1,\,\ldots,\,t_k)\,\leq\, t} f(t_0) h_{n-1}(t_1) \cdots h_{n-1}(t_k)\, dt_0 \cdots dt_k.$$

This leads us to

$$H_n(t) = \int_0^t f(t-v) H_{n-1}^k(v)\, dv, \quad n = 1, 2, \ldots. \tag{13}$$

The reader may wish to determine similarly the distributions of other order statistics of the times $T_1, \ldots, T_N$. As a help in this and an aid to getting (12) and (13) the reader may wish to consider Fig. 3a–c. In each case the point P has coordinates (v, v, v) and the three drawings show the subsets of the first octant corresponding to the three order-statistics

$$C_1 = \{(x_1, x_2, x_3) : m_3(x_1, x_2, x_3) \leq v\} \quad \text{in} \quad a,$$
$$C_2 = \{(x_1, x_2, x_3) : m_2(x_1, x_2, x_3) \leq v\} \quad \text{in} \quad b,$$
$$C_3 = \{(x_1, x_2, x_3) : m_1(x_1, x_2, x_3) \leq v\} \quad \text{in} \quad c.$$

It will be noted that: (a) the domains of integration we have met, or shall meet, are precisely regions of these types, and (b) each region can be represented as an intersection of octant-shaped regions. We illustrate this for the case of the domain C_2 of Fig. 3b. Let T denote the first octant and $T(q)$ its translate with the origin moved to the point Q. Then an easy application of the inclusion–exclusion principle shows that with the obvious proviso for set-multiplicities we have

$$C_2 = T - (T(P_1) \cup T(P_2) \cup T(P_3)) + 2T(P).$$

Since $f(x)$ is the probability density and $\int_0^\infty f(x)\, dx = 1$, we get from the preceding the transformation of the multiple integral over C_2:

$$\iiint_{C_2} f(x_1) f(x_2) f(x_3)\, dx_1\, dx_2\, dx_3 = 1 - 3\left(\int_v^\infty f(x)\, dx\right)^2 + 2\left(\int_v^\infty f(x)\, dx\right)^3.$$

The above illustrates the procedure for the branching number $k = 3$; the extension to arbitrary k may be numerically or combinatorially less simple, but it is relatively straightforward.

Certain integrodifferential equations for the pathology of branching

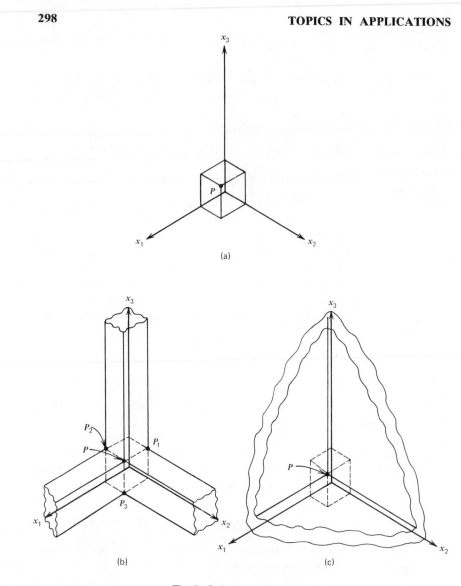

Fig. 3. Order-statistic regions.

arise out of the preceding formulas. We mention one example relevant to neuropathology [117], generalizing slightly the original version (from 2 to general k). Suppose that we have a large number of branching fibers; each one has its root at the origin and runs out along the positive x-axis as a main trunk with a single branching out into k subsidiary branches. The lengths of the main trunk and of each branch are independent identically distributed random variables with an unknown density f. Suppose that cuts can be made at $u \geq 0$, where u is at the experimenter's choice. In each cut an observable effect spreads back to the roots of the cut fibers. We can describe it by saying that the root dies if and only if either its main trunk or at least s of its branches have been cut. To sum up, there is an observable effect of the cut at u and we suppose that the fraction of surviving fibers is a potentially observable quantity as a function of u. In the special case of $k = s = 2$ we have then (11) but *now* interpreted as an integrodifferential equation for the unknown F:

$$G(u) = F(u) - \int_0^u F'(u-v)F^2(v) \, dv.$$

The reader may wish to extend this to the general k and s by means of order-statistics, and to depth ≥ 1.

(**d**) In this very small subsection we consider, very briefly and superficially, the possibility of applying branching to a problem of the geometrical type. That is, we consider some features of a branching structure which are naturally invariant under rigid motions. For this purpose let the branching structure be considered in Euclidean three-space as a dendrite, like the one shown in Fig. 1. The problem we shall consider arose from an actual study of dendrites of nerve cells, when the question came up of the origin of nerve cells. Different nerve cells differ quite considerably from each other, especially in the number and type of dendrites they have. So, for instance, some neurons (in the motor cortex and in the cerebellum—both concerned with motion!) seem to have depths reaching figures as high as 6-8. Also, they are of the bushy type: At each vertex where branching can occur, it does. Elsewhere in the nervous system, of the two branches of the lower order it is often only one that will itself branch out into two. Also, dendrites differ considerably in their intervertex behavior. So, for instance, some have quite straight branches while those of others are wavy or even kinky. Again, in some cases the length of a branch diminishes very rapidly as its order increases, while in other cases it is more nearly constant.

Yet in spite of all those (and many other) differences all dendrites arise embryologically from the same parental structure, the neuroblast. Now, the linear and bifurcating geometry of a dendrite would seem to provide a fair testing ground for certain hypotheses concerning growth.

The reader might wish to postulate that the developing dendrite, especially its growing tip, is surrounded by some growth-controlling substance whose concentration $f(x, y, z, t)$ varies both in time and in space, according to laws that could be partly deterministic and partly stochastic. It might be postulated that the growth of the tip of the dendrite takes place in the direction of the gradient of f (or at any rate, depends on it). This is somewhat analogous to the motion of a particle, smooth or with friction, down a surface of some shape. It might be attempted to account for branching by postulating that f may have saddle-points and that, under suitable conditions, when the growing tip finds itself near a saddle point of f then it bifurcates into two growing tips which will then grow down the two ridges from the saddle-point. The degree of waviness or straightness of branches is, in principle, accountable for by the scales of variations in f. Presumably, one could set up the right mechanism for such artificial sort of growth, and simulate the process on a computer. Then it might be possible to compare the results with actual dendrites, and perhaps change the values of the controlling parameters, in an attempt to get a growth picture approximating the real dendrite.

(**e**) The depth induction of branching, together with some simple analysis, seems to be the central idea behind some of the renewal theory. This is a part of probability theory, referring to certain repeating or recurrent events, and characterized analytically in the continuous case by the fact that its unknown functions f satisfy characteristic integral or integrodifferential equations. Namely, f equals a known function plus a term which is the Laplace convolution of powers of f with a kernel (which may be known or may itself involve f):

$$f(x) = g(x) + \int_0^x h(x-u)f^k(u)\, du, \quad f(x) = g(x) - \int_0^x f'(x-u)f^k(u)\, du,$$

and so on. Such equations are precisely of the type we have met in Section (c). The word "renewal" itself might be taken as a package description of the depth induction (especially if we are willing to continue the branching to infinite depth).

We shall give one simple illustration of the connection between renewal and the depth induction of branching. As before, let a particle originate at the time $t = 0$ and let its duration time be a random variable with the density $f(x)$. At its expiry, with probability p_k there appear k new identical copies of it which repeat, independently, its behavior. We are interested in the probability $F(t)$ that the whole process stops by the time $\leq t$. First, by an elementary application of conditional probabilities

$$F(t) = \sum_{k=0}^{\infty} P_k Q_k \tag{14}$$

where

$P_k = $ Prob (termination at time $\leq t|$ first branching is k-tuple),
Q_k-Prob (first branching is k-tuple) $= p_k$.

By the definition of the density f

$$P_0 = \int_0^t f(x)\, dx$$

and, more generally, by the same reasoning which led to (13) we get the analogous formula

$$P_k = \int_0^t f(t-x)F^k(x)\, dx \qquad (15)$$

which is, of course, a consequence of the induction on depth. Hence, using (14) and (15) we get

$$F(t) = p_0 \int_0^t f(x)\, dx + \sum_{k=1}^{\infty} p_k \int_0^t f(t-x)F^k(x)\, dx. \qquad (16)$$

This generalizes a result in Feller [55, vol. 2, p. 387] which is called there a nonlinear renewal equation. Before leaving the subject of branching, we recall the connection between the branching process into $0, 1, 2, \ldots$ new particles with probabilities $p_0, p_1, p_2, \ldots$ and the iterates of the generating function

$$\psi(x) = p_0 + p_1 x + p_2 x^2 + \cdots,$$

as described in vol. 1, p. 73. It will be noticed that with the above generating function (16) can be simply rewritten as

$$F(t) = \int_0^t f(t-x)\psi[F(x)]\, dx.$$

4. PURSUIT AND RELATED TOPICS

In this section we consider several questions on the subject of pursuit. This name is given to the situation when one point moves on a preassigned curve while another one pursues it, aiming straight at it at all times. The subject goes back to Leonardo da Vinci (1452–1519) and in its more mathematical form to P. Bouguer (1698–1758) and P. de Maupertuis (1698–1759). The last two were geometers at the French Académie des Sciences. Maupertuis is known as the originator of the principle of least action; the name "ligne de poursuite" appears in a 1732 memoir of Bouguer (see references in [40]).

One of the instructional benefits of our subject of pursuit is that it introduces and displays in a simple and convincing manner certain important concepts. These concern differential equations, dynamical systems and celestial dynamics, as well as some more recent phenomena from mathematical ecology, centering around the notions of periodic orbits and limit cycles. Additionally, since the equations of pursuit cannot in general be solved explicitly, we have to develop some approximate methods for the examination of trajectories. Here, too, it is possible to use pursuit as a model on which to introduce certain topics of numerical analysis in a simple geometrical fashion.

(a)   Let the point p (p for pursuer) start at the time $t = 0$ from the origin in the plane. Another point e (e for evader or escapee) is then at $(b, 0)$ where $b > 0$, say. The evader e moves with constant speed k up on the straight line C_e whose equation is $x = b$. The pursuer p moves with unit speed, aiming at all times straight at the instantaneous position of e. Under these conditions we speak of linear pursuit and we wish to determine the path C_p of p, which we call the pursuit curve corresponding to the linear escape curve C_e. It is clear that the problem can be generalized in various ways, in the first instance by allowing C_e to be an arbitrary plane curve, then by generalizing to 3 or n dimensions or to a geodesic pursuit on a curved surface, or several pursuers in cyclic order, and so on.

The pursuit problem with general escape curve C_e can be formulated purely geometrically without any appeal to kinematics. Given two points A and B and a curve C_e issuing from B, the problem is to find the curve C_p starting from A so that the tangent at any point X on C_p cuts off C_e an arc of length proportional to the length of the arc AX of C_p. Briefly, pursuit transformation means arc length proportionality under tangential correspondence. The ratio of proportionality of the are lengths is here the ratio of the speeds of the moving points p and e.

Returning to the special case of linear pursuit we let $y = f(x)$ be the equation of the pursuit curve C_p, as shown in Fig. 1a. By the arc length proportionality condition we have $|Be| = ks$ where $s =$ length op. Hence, considering the triangle pqe, we get

$$(b - x)y' = ks - y. \tag{1}$$

Differentiating with respect to x we get the differential equation of the pursuit curve C_p

$$(b - x)y'' = k(1 + y'^2)^{1/2}.$$

Here the initial conditions are $y(0) = y'(0) = 0$. Since y does not enter explicitly into the equation we reduce the order by taking y' as the new dependent and y as the new independent variable. Solving for y we have

$$y = \begin{cases} \dfrac{b}{2}\left[\dfrac{(1 - x/b)^{1+k} - 1}{1 + k} - \dfrac{(1 - x/b)^{1-k} - 1}{1 - k}\right], & k \neq 1 \\[3mm] \dfrac{b}{2}\left[\dfrac{(1 - x/b)^2 - 1}{2} - \log(1 - x/b)\right], & k = 1. \end{cases} \tag{2}$$

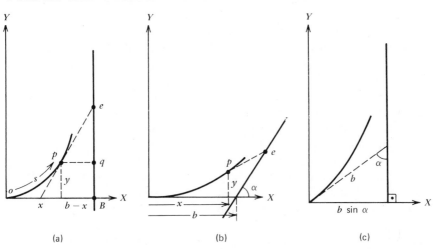

Fig. 1. Linear pursuit.

Can we similarly solve the case of oblique linear pursuit illustrated in Fig. 1b? If we proceed just as before we get instead of (1)

$$(b - x + ks \cos \alpha)y' = ks \sin \alpha - y$$

and $y(0) = y'(0) = 0$. This could be solved just as before, by a moderate expenditure of brute force. But it is easier to rotate the coordinates as shown in Fig. 1c. We have then the same problem as shown in Fig. 1a and the same differential equation as before, except that $b \sin \alpha$ replaces b and the initial conditions are $y(0) = 0$, $y'(0) = \cot \alpha$.

It is clear that the capture of e by p occurs if and only if $k < 1$. With $x = b$ in (2), the capture point is $(b, kb/(1 - k^2))$. When $k > 1$ e gets further and further away from p. On the other hand, if $k = 1$ the asymptotically attained distance between p and e is

$$\lim_{x \to b} (s - y) = \lim_{x \to b} (b - x)y' = \frac{b}{2},$$

that is, half the initial distance between p and e. The reader may wish to determine similarly the asymptotic distance for the oblique pursuit of Fig. 1b, and to extrapolate the pursuit curve to negative times t.

It is seen that capture, if it occurs at all, occurs tangentially from behind. In mechanical terms this yields smallest possible relative velocity at the moment of capture. This is of possible importance when e is an enemy missile and p is a counter-missile missile, as well as when p is a hunting animal and e its running prey (in this connection it might be of interest to determine whether phylogenetically related prey-hunting animals follow similar pursuit strategies). Clearly, there are other capture strategies besides the pursuit which is characterized by its instantaneous aiming-at. For instance, with $k < 1$ and under the conditions as in Fig. 1a, we start p from the origin and wish to have it capture e in minimum time. How should p aim now?

The reader may wish to show that the optimal strategy for p is simple interception: p forecasts e's position at the time of capture and runs straight at this extrapolated position. If e should use evasive tactics by suddenly doubling back on its line C_e (though moving afterwards still with speed k) then p need only change its motion symmetrically with respect to the horizontal line through the point of doubling back. If e is allowed to vary its velocity between $-k$ and k then p still aims at the instantaneously extrapolated capture point. A much simpler way of expressing p's strategy is to say that it moves so as to keep the same y-coordinate as e at all times. The reader may wish to compare this with reflections and images of vol. 1, p. 26. The difference between the x-coordinates of p and e decreases steadily till the capture which occurs here at a glancing angle rather than tangentially from behind.

An alternative way of expressing the intercept strategy is obtained by borrowing some simple ideas from pp. 238–241 of vol. 1. This might be of use in other pursuit problems, for instance, when the evader is not constrained to stay on a curve but in some other manner (this may be compared with the prediction problems of pp. 35–36. Let $P(t, T)$ be the exact region where p must be at time T, as computed at time t. So, $P(t, T)$ is the circular disk of radius $T-t$ about the position of p at time t. Similarly we define $E(t, T)$; since e is constrained to move on its line $E(t, T)$ is just a point. Now p moves so as to have its region $P(t, T)$ just cover $E(t, T)$; the time of capture is the smallest value of T for which this covering of regions can be done.

(**b**) Let us consider for the moment the case of arbitrary plane escape curve C_e. If its equation is known we can apply the proportional arc-length condition, to obtain eventually a nonlinear differential equation for the pursuit curve C_p. In general this cannot be solved explicitly, even in cases as simple as that of circular C_e. Indeed, it almost appears that the linear pursuit is the only directly solvable case. We say "directly solvable" because one might attempt to work backwards: A curve is taken as C_p, we construct its tangents, and apply the arc-length proportionality condition, to re-create C_e in reverse, so to say. A related matter will occupy us in the last subsection, in connection with a model for the social sciences.

Luckily, even though we cannot solve the differential equations governing pursuit, a simple graphical procedure will produce our approximation to C_p, given C_e. Moreover, we can do some error analysis, and we even have some means of effective error control. The approximation starts with replacing C_e by a polygonal line $B_0 B_1 B_2 \cdots$ of straight segments of length $k\Delta$. B_0 is the starting point of C_e; all other B_i's lie on C_e as well. From the origin O, which is the starting point of p, we draw the straight segment OO_1 aiming at B_0 and of length Δ, then from O_1 we continue with the next segment $O_1 O_2$ aiming at B_1 and also of length Δ, and so on. Now the segmented line $OO_1 O_2 \cdots$ provides an approximation to the pursuit curve C_p. Several questions could be asked now, the two most important ones being perhaps: How can we estimate the approximation error? How can we improve the approximation, other than by the obvious means of decreasing the mesh Δ?

As to the first question, a simple modification of our procedure will be helpful. Besides the approximation $OO_1 O_2 \cdots$ consider also $Oq_1 q_2 \ldots$, which is constructed as follows: We use again segments Oq_1 or $q_i q_{i+1}$, of length Δ, the start is from O but Oq_1 aims at B_1 (rather than at B_0 as before), $q_1 q_2$ aims at B_2, and so on. A simple analysis suggests that the true pursuit curve C_p lies, initially at least, between the two approximations $OO_1 O_2 \cdots$ and $Oq_1 q_2 \cdots$. This, however, is complicated by the fact that those approximations might intersect, though in this case the points of intersection correspond to different times on each approximation. It is possible to produce continually new "upper" and "lower" estimates $OO_1 O_2 \cdots$ and $Oq_1 q_2 \cdots$; by taking suitable envelopes the absolute region confining C_p can be found.

Since C_p might be reasonably expected to lie about half way between $OO_1 O_2 \cdots$ and $Oq_1 q_2 \cdots$, a simple and direct device suggests itself for improving the fit by averaging; this device in more sophisticated forms is frequently used in the numerical solution of ordinary differential equations. Namely, we draw the third polygonal line $Or_1 r_2 \cdots$, again with all segments of length Δ, and with Or_1 pointing at the midpoint of $B_0 B_1$, $r_1 r_2$ at the midpoint of $B_1 B_2$, and so on. A sample of various approximations is given in Fig. 2 which shows the true pursuit curve C_p, the two extreme approximations $OO_1 O_2 \cdots$ and $Oq_1 q_2 \cdots$, and also the averaged-out approximation $Or_1 r_2 \cdots$; the step length Δ is purposely chosen rather large for better display.

Our next device for better approximation depends on observing that it is not necessary to have the points B_0, B_1, B_2, ... equispaced on C_e so long as the ratio of the lengths of the nth steps on C_e and on C_p is always k. We exploit this by introducing what is known in numerical analysis as a variable step-length procedure. That is, we space B_i and B_{i+1} further apart where the pursuit curve C_p, as evidenced by the last few links of its approximation, curves slowly. When the curving is faster and consequently there is need for improving accuracy by decreasing Δ, we space

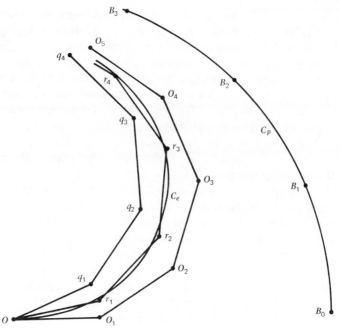

Fig. 2. Pursuit curve and its approximations.

B_i's closer to each other on C_e. As a rough measure, we could compute an approximation to the radius of curvature of C_p, and have $|B_i B_{i+1}|$ proportional to it. An example is given in Fig. 3 of a pursuit situation that calls for such a variable step-length procedure. We have here the square $KLMO$, the pursuer p starts from O and the escapee e from K; p's speed is 0.3 times that of e. The escape curve C_e is from K to L, then to M, then back to L, to K, and so on. The points A, B, C, D, ... of successive curvature maxima show that a fixed step-length procedure would be wasteful here.

(c) We consider now the circular pursuit; the escape curve C_e is a circle, of radius a. The differential equations of the pursuit curve are somewhat easier to get in the geometry of Fig. 4 than in the polar coordinates; we find that with $\rho = \rho(\theta)$, $\phi = \phi(\theta)$ C_p is given by the system of differential equations

$$\phi' = \frac{a}{\rho} \cos \phi - 1, \qquad \rho' = a \sin \phi - \frac{a}{k}.$$

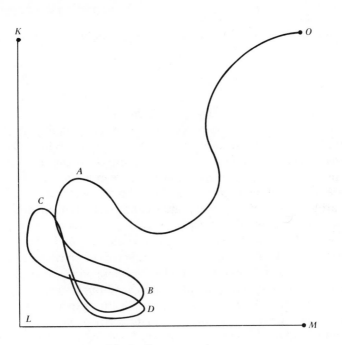

Fig. 3. Corner pursuit.

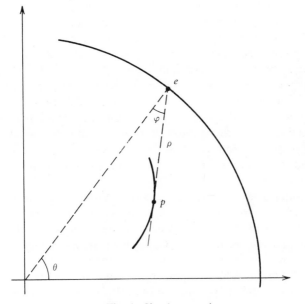

Fig. 4. Circular pursuit.

By dividing the equations, we find that the path is given by the single equation

$$\frac{d\rho}{d\phi} = \rho \frac{a(\sin \phi - 1/k)}{a \cos \phi - \rho}.$$

This equation is not explicitly solvable for ρ in terms of ϕ. However, we shall be able to find out some properties of the pursuit trajectory. Two questions we might ask are: When does capture occur? What sort of limiting situation (if any) obtains when there is no capture?

With respect to the first question it has been maintained [40, 71], that capture occurs if and only if $k < 1$. This is not strictly correct: For arbitrarily large k there exist initial configurations from which capture occurs even though the escapee moves faster than the pursuer. We demonstrate this by the simple symmetry device of time reversal: Start the system from capture position and reverse, to arrive at the necessary initial conditions. Let then $k > 1$ and somewhere on C_e let us place p and e together. Supposing that they have met head-on, let us reverse the time and so p moves in the reverse sense on C_e with its speed k while p moves in the plane aiming at all times straight *away* from the instantaneous position of e. During this motion p will trace out a certain locus C. It is now clear that we can start p from any position on C, and e from its correct place on C_e, and the resulting pursuit terminates in capture. This occurs tangentially head-on instead of tangentially from behind; the reader may wish to show that no other capture positions are possible for pursuit. The head-on capture of a faster escapee by the slower pursuer corresponds in mechanical terms to unstable equilibrium whereas the other capture position (tangentially from behind, with $k < 1$) corresponds to the stable case. The reader may also wish to examine in what essential rather than descriptive ways stability is involved here—it is not hard to show that with the head-on capture arbitrarily small perturbations of initial positions prevent capture. It must be added that the statement, "capture occurs if and only if $k < 1$," is true if "capture" is here interpreted to mean capture from arbitrary initial configuration, or "stable" capture.

We consider now the second question: What sort of limiting situation do we get when there is no capture? Suppose that e moves on its circle C_e of radius b. Recalling that $k > 1$ is the speed ratio, we let p move on the concentric circle K_p of radius b/k. If p starts its motion aiming at e then the same condition is maintained for all time. Thus we get a periodic trajectory K_p for p. What happens for other initial configurations? Using the previously outlined graphical method for several initial positions of p, both inside and outside C_e, we verify that the graph of C_p spirals into the

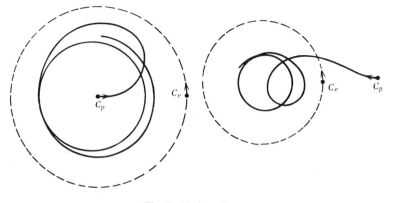

Fig. 5. Limit cycles.

circle K_p though it may intersect itself, and K_p as well, in this spiralling. As a matter of fact, it is easy to show, from the arc-length proportionality condition, that such spiralling could not occur purely from inside K_p or purely from outside. Roughly, the arc-length condition tells us that every "turn" must be of the same length. On the other hand, a closed convex curve is longer than any other closed convex curve inside it. Two examples of such spiralling are given in Fig. 5.

This exhibits something very close to a periodic motion, a phenomenon discovered by H. Poincaré, in the astronomical context of the trajectories of celestial bodies, and called by him a limit-cycle. We can describe it as asymptotically periodic motion, or a motion that differs from a true periodic motion by a vector quantity approaching 0 as $t \to \infty$.

We shall show now that with $k > 1$ for every closed convex escape curve C_e there exists a periodic pursuit curve C_p. First, we fix the matter of initial conditions. Since the motion of e is periodic we suppose without loss of generality that e starts at some fixed point A on C_e. Therefore the whole initial configuration is completely specified by the initial position B of p (this is so because C_e is convex, otherwise there could be some ambiguities). We observe now that C_p is smooth, that is, with continuously turning unit tangent vector, even though C_e may fail to be smooth. Therefore C_p, which always turns in one direction, has on it a point, say B_1, at which the unit tangent is a positive multiple of the vector BA. Loosely speaking, we could say that C_p turns in one direction and at B_1 it has completed just one turn, starting from B. We have thus defined a transformation ϕ given by $B_1 = \phi(B)$. There is no difficulty with the case when $A = B$—the motion here starts in the direction of a one-sided tangent to C_e. Now ϕ is a continuous transformation defined inside and on C_e. Hence by the Brouwer fixed-point theorem there is B_0 such that

$B_0 = \phi(B_0)$. It is clear from the construction that starting p from B_0 we get a closed periodic orbit C_p.

The reader may wish to construct a proof which exploits some mean-value property and avoids the use of Brouwer's theorem, to examine the case of nonconvex C_e (replacing, perhaps, C_e by the boundary of its convex hull), and to apply the arc-length proportionality condition to show that the periodic orbit C_p is unique.

We know now that there exists a closed periodic pursuit orbit C_p but this does not yet show that this C_p is a limit cycle; i.e. that even when the pursuit starts with p *off* C_p, there is a spiralling into C_p. We might suppose that this approach to C_p takes place in some monotonic fashion. However, the graphical methods show that a trajectory spiralling to C_p can cut it as well as itself. Therefore the monotonicity is not with respect to distance alone. Both the distance and the angle of approach matter. Generalizing the capture conditions, we might say that limit cycle approach, like capture, occurs at "0° angle." That is, we have the consecutive turns of the orbit spiralling at ever shallower angles.

(**d**) In this subsection we shall compare several quite widely differing types of motion of points, with a view to showing that in a variety of cases there are only two types of behavior. We start by assembling some examples:

(A)   motion of a point p pursuing a point e which moves on a plane closed convex curve (we suppose that there is no capture),

(B)   motion along the geodesic of a torus or, more generally, along a trajectory of a sufficiently regular differential equation on a torus ([34], cf. also p. 33 of vol. 1),

(C)   motion of a point on the periphery of a circle of radius b which rolls without slipping on a circle of radius a,

(D)   motion of a point along the path of a ray of light traveling inside a plane convex enclosure C with complete and perfect internal reflection.

It turns out that there are just two types of motion: Periodic and ergodic, each with two variants: Simple and complex. We schematize this in the table

	(1) simple	(2) complex
(1) periodic	exact periodicity	asymptotic periodicity
(2) ergodic	direct ergodicity	ergodicity w.r.t correct density

As has been claimed, pursuit motion (A) is always of type 1, and variant 1 or 2 depending on the initial conditions. Here, of course, by asymptotic periodicity we understand the presence of a limit cycle. In (B) we are interested in trajectories on a torus. Here it can be shown [34] that either the trajectory closes up and so is periodic—type 1, variant 1—or the trajectory is dense on the whole torus. In this case there is a density ρ with respect to which the trajectory is ergodic, that is, in a subset A of the torus T the trajectory spends the fraction $\int_A d\rho$ of the total time; we assume here the normalization $\int_T d\rho = 1$. To sum up, either we have type 1, variant 1 or type 2, variant 2.

In the case (C) we have an ordinary epicycloid if a/b is an integer or a still closed though self-intersecting epicycloid if a/b is rational (and $b \nmid a$). Both these belong under type 1, variant 1. If a/b is irrational then the motion is dense in a certain circular annulus, under type 2, variant 2; the reader may wish to find here the correct density ρ.

In the case (D) we have many possibilities. As was indicated in vol. 1, pp. 31–32, if C is a rectangular enclosure we have either type 1, variant 1 or type 2, variant 1 depending on the initial conditions. The same is true for a circular enclosure C. Suppose next that C is an ellipse and that the ray of light passes through a focus. If it also passes through the other focus on the same sweep then we have type 1, variant 1: The trajectory is just the focal chord described back and forth. If the ray does not pass through the other focus on its first sweep then it does so on its second sweep. Thereafter alternate sweeps cross the same focus. Moreover, the trajectory collapses *eventually* toward the focal chord, which is thus a limit-cycle: type 1, variant 2. The reader may wish to determine what happens if the trajectory never crosses a focus. We observe that it must cross the focal chord and it does so in either of the two ways: always between the foci, never between the foci.

The reader may also recall somewhat similar dichotomies (completely regular vs. completely irregular) in connection with power series of Subsection (c) of Section 5 of Chapter 2 (on recalling, additionally, certain theorems from probability and some further matters, we might be tempted to formulate a conjecture that a mathematical entity whose "complexity" is below a certain critical level, can only be "very good" or "very bad").

Let us consider the variants of type 1. The variant 1, simple periodicity, is characteristic of problems described by linear homogeneous differential equations while the variant 2, limit cycles, is proper to nonlinear problems and *cannot* occur in a problem described by a linear differential system. The difference appears to be this: Linear problems certainly admit periodic solutions but then the amplitudes of these periodic solutions depend on the initial conditions. Also, we note here the phenomenon of "indifferent equilibrium": on small perturbations a periodic orbit settles into another nearby periodic orbit. On the other hand, in the nonlinear case the (asymptotically) periodic solution has "amplitude" independent of the initial conditions; the type of equilibrium exhibited here is either stable (on small perturbations the trajectory settles back to the same limit cycle) or unstable (small perturbations are sufficient to throw the trajectory off the limit cycle it was in). Incidentally, we may recall here two other distinctions between linear and nonlinear behavior. The first one comes from the very definition: presence or absence of the superposition principle. The second one is obscurer but quite important, especially in numerical work (though also in existence proofs): Linear processes lead to the iteration of linear expressions, which we can perform explicitly, nonlinear processes lead to the iteration of nonlinear expressions which, as a rule, cannot be done explicitly.

(e)   There is a variant of pursuit which can be called the drag problem. We suppose again that a point e moves on a preassigned curve C_e in the plane, now in any manner, uniform or otherwise. It drags another point p by an inextensible rope of fixed length, say d, and the motion is with friction. Since the rope can exert only a pull but no push or lateral thrust, p will be dragged along a curve C_p so that its motion is always straight toward the instantaneous position of e. C_p is called the tractrix corresponding to the drag curve C_e (from Latin *traho* (*trahere, traxi, tractum*)—I pull). We shall first consider the relation of tractrices and pursuit curves.

Let us define a tangent correspondence between two sufficiently smooth curves C_1 and C_2; this assigns to a point x on C_1 the point of C_2 where the tangent to C_1 at x cuts C_2 (with some reasonable provision for

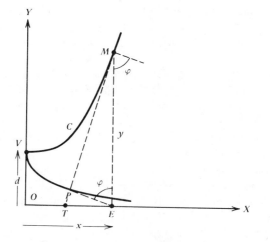

Fig. 6. Catenary and tractrix.

multiplicity and singular points). Now we can formulate a problem that generalizes both the pursuit and the drag: given a curve C_e through a point A and given a point B off C_e, to find the curve C_p through B in terms of a given property Φ which relates the tangentially corresponding points of the curves. We get the tractrices when the property Φ is the constancy of the distance. Pursuit curves result when Φ is the proportionality of the arc lengths from the initial points A and B up to any tangentially corresponding pair of points. The reader may wish to search for some other significant properties Φ. Of course, it will be clear that pursuit curves, tractrices, and general Φ-curves are considered here mainly as devices for generating interesting motions, particularly with reference to periodic orbits and limit cycles.

Returning now to tractrices we let the drag curve C_e be the straight line. In this case the tractrix C_p can be determined as follows [63]. As in Fig. 6 let C be the catenary with the equation

$$y = d \cosh \frac{x}{d}. \tag{3}$$

Let a point M on C have coordinates x and y, a simple calculation shows that the length of the arc of C from the vertex V to M is

$$s = d \sinh \frac{x}{d} = y \sin \phi.$$

But it is verified that $|MP|$ is also $y \sin \phi$ so that length $\widehat{VM} = |MP|$. We now find the locus of P as M varies. This is called the evolute (unfortunately, also the involute) of C and can be visualized by placing on C thin inextensible string along its right half, whose end is at V, then unwinding it keeping it taut; the end of the string traces out the locus of P. Referring to Fig. 6 we calculate $PE = y \cos \phi = d$; also, PE is tangent to the arc $\widehat{VP}$ at P. But this means that the tractrix conditions are satisfied. Hence the curve C_p traced out by P is the tractrix corresponding to the x-axis as the drag curve. It may be verified that the parametric equations of the tractrix are

$$x = d\left(\cos t + \log \tan \frac{t}{2}\right), \qquad y = d \sin t \tag{4}$$

and that

$$\frac{dy}{dx} = \frac{y}{\sqrt{d^2 - y^2}}. \tag{5}$$

We observe that for the circle

$$x^2 + y^2 = d^2 \tag{6}$$

we would have had

$$\frac{dy}{dx} = -\frac{y}{\sqrt{d^2 - y^2}},$$

which is similar to (5). Possibly for this reason the surface obtained by rotating the tractrix (4) about the x-axis, is called a pseudosphere. Just as the sphere obtained by rotating the circle of (6) about the x-axis happens to have constant positive Gaussian curvature $1/d^2$, so the pseudosphere has constant negative Gaussian curvature $-1/d^2$. It is noted that this pseudosphere has a singular rim—this may be regarded as a necessary consequence of a theorem of Hilbert (a complete surface of constant negative curvature cannot be smoothly embedded in the Euclidean space of three dimensions). We note that if "plane" is interpreted as "pseudospherical surface," "point in the plane" as "point on that surface," and "line" as "geodesic of that surface" then the pseudosphere provides a model of the hyperbolic plane. It is known that the "opposite" surface, the sphere, provides similarly a model for the elliptic plane.

In the original definition a tractrix is a curve along which a point p is dragged by a fixed rope of length d the other end of which moves on the drag curve C_e. Since the rope must be taut at all times it is clear that not every curve C_e admits a tractrix. For instance, no tractrix exists if C_e is a closed convex curve with a cusp of angle $\leq \pi/2$, for instance, if C_e is a square (however, a tractrix exists if C_e is the regular pentagon). Also, the length of the rope d enters into the admissibility of tractrices.

The matter of periodic orbits and limit cycles for tractrix trajectories is rather similar to the pursuit case. For instance, if the drag curve C_e is a sufficiently smooth closed convex curve and d is sufficiently small, then there exists a periodic tractrix. We can suppose that the motion starts with e at an arbitrary point A of C_e. Let the angle of the rope to the tangent (or support line) to C_e at e be $\alpha(t)$ at time t. Let T be the time for e to pass once round C_e. Then simple continuity shows that for suitable α_0 we have $\alpha(0) = \alpha_0 = \alpha(T)$; hence we get a periodic tractrix.

(f) Finally, we return briefly to plane pursuit with arbitrary escape curve C_e. If C_e were allowed to be discontinuous, that is, if e could jump across space, then the pursuit curve C_p would still be continuous though not necessarily smooth. If C_e were continuous but not smooth, for instance, if it were a closed convex curve with cusps, then at the points of C_p corresponding to the cusps on C_e the pursuit curve C_p would be smooth though not necessarily with continuously changing curvature. Generally, if the escape curve C_e is of the differentiability class $C^{(n)}$ then the pursuit curve C_p is of the class $C^{(n+1)}$ at least. Thus, the pursuit as a transformation on curves (C_e to C_p) is smoothing. Another sense in which this is also true is seen by taking C_e to be a smooth curve with a small-amplitude ripple of high frequency superimposed on it. C_p is then a smooth curve with a ripple which is both smoother and smaller than that of C_e. We have now our first premiss: Pursuit is smoothing.

Our second premiss is the Jacobi motto "one must always invert"; for this and for some applications see p. 89 of vol. 1, or the Appendix. In the present instance we intend to invert pursuit so as to obtain antipursuit. Here there are also two points, L (for leader) and H (for herd). The point H is moving on its preassigned curve C_H; L is moving ahead of H and its object is to move on a curve C_L so that C_H is the pursuit curve for the escape curve C_L. In other words, as is usual with leaders, the object of L is to keep on appearing to be a leader, with relatively less attention as to where the leadership leads. When the herd-trajectory C_H is nearly straight and very smooth, then L has no trouble staying in the lead in the above sense. But if C_H oscillates then it is observed that the antipursuit is unsmoothing—the path of L may have to undergo such violent changes that L can no longer stay a leader. The reader may wish to compare this with the limits on maneuverability in terms of curvature bounds (p. 238 of vol. 1, and p. 35, this volume). A possibly interesting variant occurs when several points $L_1, \ldots, L_k$ contend, under properly defined conditions, for the leadership, there being just one herd H.

Our third and last premiss deals with the extension problem: how to enlarge the domain of definition of a quantity, an operation, or a function. This is with special reference to extending to reals certain operations or functions given originally only for natural numbers. Examples have been given in vol. 1, pp. 59–65 and in Appendix; they include nonintegral order differentiation and iteration, factorials and the Γ-function, and so on. A useful tool in the extension problem is the conjugacy principle: If the extension problem is difficult for an entity f, it may be easier for a conjugate entity $\psi f \psi^{-1}$. For instance, it is not obvious how to iterate to real order the function $f(x) = 2x^2 - 1$. But we have $f = \psi g \psi^{-1}$ where $\psi(x) = \cos x$ and $g(x) = 2x$; hence the iteration of f is reduced to that of g. The sth iterate of g being $2^s x$, we get the sth iterate of f:

$$f_s(x) = \cos 2^s \arc \cos x.$$

We have now all the three premisses and we fomulate our problem: to construct a sufficiently continuous one-parameter family of transformations T_s, with $-1 \le s \le 1$, so that T_1 corresponds to our pursuit and T_{-1} to our antipursuit. Presumably T_0 is the identity transformation and, what is the essential point, the bigger s the more smoothing T_s. There are two principal interests attaching to such a family T_s. First, it provides a possibly fresh approach to the stability of motion. Second, it might serve as a geometrizing model in various social sciences. Thus, the antipursuit might be claimed in some measure to be a model for trend setting, trendy salesmanship, and perhaps also political leadership. This could be further augmented by feedback-type mechanisms which shift the degree of stability or instability, that is, changes in the index s of T_s. With respect to our primitive social science model, such changes might be perhaps interpreted as modeling certain historical shifts. Finally, the pursuit–antipursuit anomaly may be observed in the domain of arts when one begins with the critics in the stable pursuit of the artists and one ends up with the artists in the unstable antipursuit of the critics.

5. GENERALIZATIONS OF STEINER'S PROBLEM

Steiner's problem was considered and solved in vol. 1, pp. 139–145. Briefly stated, the problem is to construct the shortest network joining n given points in the plane. Clearly we may here consider only networks of straight segments but it is understood that vertices other than the given n points might appear. It is this feature that distinguishes Steiner's problem from the problem of the shortest spanning subtree: to construct the shortest network joining n given points and having no other vertices than these n ones.

The shortest spanning subtree problem is *discrete* since there are by Cayley's theorem (vol. 1, pp. 161–163) n^{n-2} trees on the given n vertices. Thus, finding an *effective* solution is not the question here; this could be achieved by an examination of the n^{n-2} candidate trees in turn, if no better method offers. What we do want is an *efficient* solution. While this notion is necessarily loose, it is becoming clear that a very good candidate for the efficiency criterion is the following. First, we assign a natural size to the problem in hand; for Steiner's problem we would take the number n of initial points as that size. Then we admit certain more or less specified constructions or procedures as natural steps, for example, looking up the distance between two points, comparing two such distances, drawing a line through two known points, finding the intersection of two such lines. Now, the problem has an efficient solution if for every size n the solution involves $\leq P(n)$ steps where P is a fixed polynomial. Briefly, efficiency of solution means polynomially bounded complexity in terms of the number of steps.

On the other hand, Steiner's problem is *continuous*, not discrete like that of the shortest spanning subtree. This is so because of the additional vertices each of which implies two continuous parameters. The solution of Steiner's problem we have given depended first on showing in advance that there are $\leq n-2$ of such additional vertices. Then, we have reduced the continuous problem to a large number of special cases. Thus, our solution was effective but highly inefficient (since the number of cases was a function of n growing much faster than a polynomial).

Another feature of distinction may be noted. The Steiner problem requires the geometrical structure of the plane or, minimally, some similar metric structure. It is a *geometrical* problem for n points *embedded* in a simple geometrical structure, the plane. On the other hand, the shortest spanning subtree problem may be viewed as *abstract*: here the n points need not belong to a plane nor to any other metric space; they can be considered as *unembedded*. All that we need is a set of $\binom{n}{2}$ numbers attached to the unordered pairs of these points. Moreover, the $\binom{n}{2}$ numbers, so far from being distances, need not even be positive.

In this section we consider some aspects of the Steiner and the shortest spanning subtree problems, and certain of their generalizations. These include replacing n points by n sets, considering the Minkowski rather than the Euclidean plane, and formulating a generalized plane geometrical graph minimization problem which includes many others as special cases.

(**a**)   The original problem of Steiner calls for the shortest network joining n given points in the plane; here we consider the effect of replacing the n points by n sets. The idea is that our minimal network may be joining n lakes by means of canals or n metropolitan areas by means of roads. In either case we have a Steiner-like minimization problem with the n vertex points $v_1, \ldots, v_n$ replaced by n vertex sets $V_1, \ldots, V_n$. Since our interest is in the method, not in the most general result, we shall avoid some slight complications by assuming that each V_i is a strictly convex compact set (however, there is not much difficulty under considerably milder assumptions.)

It turns out that the minimal net for n vertex sets $V_1, \ldots, V_n$ is constructed in almost the same way it was done for n vertex points. The only new feature is a suitable replacement for the fundamental lemma of Fig. 1a: If the triangle $v_1 v_2 v_3$ has no angle $\geq 120°$ and if v_{ij} is the third vertex of the equilateral triangle two of whose vertices are v_i and v_j, then (1) the three segments $v_i v_{jk}$ intersect in a point s, (2) each side of the triangle $v_1 v_2 v_3$ subtends at s $120°$, (3) the minimal net is obtained by joining v_i to s, and its length is $|v_i v_{jk}|$. Let us define the equilateral sum of two vertex sets, for instance of V_3 and V_1 in Fig. 1b, to be the set of all third vertices v_{31} of equilateral triangles two of whose vertices are v_1 and v_3, as shown in Fig. 1a, but with the proviso $v_1 \in V_1$, $v_3 \in V_3$. We construct V_{12} and V_{23} in the same way. It is not hard to show that the equilateral sum of two sets inherits a number of properties from its

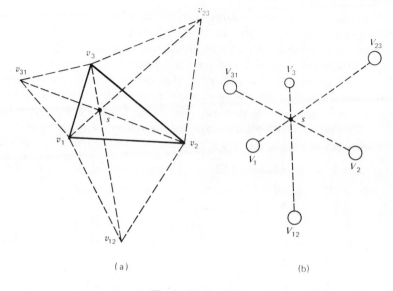

Fig. 1. Equilaterality.

summand sets; for example, it is compact and strictly convex if they are such. There is therefore a unique shortest segment joining V_1 to V_{23}, V_2 to V_{31}, and V_3 to V_{12}. It is now easy to show by applying the previous lemma, that the three segments are concurrent in s which is the Steiner point for the minimal net joining the three sets V_1, V_2, and V_3 with the same properties as before. Once the fundamental lemma has been extended to cover the case of vertex sets, the solution of Steiner's problem for n sets $V_1, \ldots, V_n$ proceeds just as for n points. For details the reader is referred to [33].

(**b**)   A brief sketch of the algorithm due to Prim [139] for solving the shortest spanning subtree problem has been given in vol. 1, p. 139. We consider this problem abstractly, that is, with n points $v_1, \ldots, v_n$ unembedded in the plane. There is a cost c_{ij} associated with the edge $v_i v_j$ and our object is to find the tree spanning the n points so as to minimize the total cost. We shall use the phrase "vertex closest to to X" to mean that vertex among all those sending branches to X, whose edge to X is cheapest. Here X can be a vertex or a fragment (= tree on some of the n vertices). This is actually an abuse of language for the costs c_{ij} do not necessarily satisfy the metric axioms; for example, $c_{ij} < 0$ is admissible. For simplicity, it may be assumed that the nearest vertices are unique; otherwise, a simple modification of our construction will yield all the trees corresponding to the same minimal total cost.

Prim's algorithm consists of choosing a vertex, joining it to the vertex closest to it to form the initial fragment, joining this fragment to the vertex closest to the fragment, and so on until all vertices have been exhausted. If desired, the work could be programmed for parallel computation by choosing several initial vertices and working out of them by successive adjunctions. Taking as unit steps the operations such as look-up of the "distance" of two points or finding the minimum of two numbers, we show without difficulty that Prim's algorithm is efficient, with the number of steps needed to solve the n point problem being bounded from above by a fixed cubic polynomial in n.

Two interesting dual observations are made by Prim in [139]:

(a)   The shortest spanning subtree also minimizes all increasing symmetric functions of the costs c_{ij} and maximizes all decreasing symmetric functions;

(b)   the longest spanning subtree does the reverse.

The following application of this is pointed out in [139]. A message is to be passed to all members $m_1, \ldots, m_n$ of an underground organization; it is supposed that when m_i and m_j communicate there is probability p_{ij} of being "found out." Any specific way of distributing the message from member to member is represented by a tree T on the n vertices $m_1, \ldots, m_n$ and the probability of a leak is

$$1 - \prod_T (1 - p_{ij}).$$

This being an increasing symmetric function of the probabilities p_{ij}, we find the optimal tree T by the shortest spanning subtree technique as above.

(c)   We formulate now a general geometrical problem concerning graph minimization in the plane. Let us recall that if v is a vertex of a (geometrical) tree U in the plane then the valency or weight $w(v)$ is the number of edges incident on v and the length $L(U)$ of the tree is the sum of the lengths of all edges in it. Our problem is the following: Letting α, β, γ be three nonnegative real numbers, and n points $a_1, \ldots, a_n$ be given in the plane, to find the integer k and k additional points $p_1, \ldots, p_k$ and then to construct a tree U with the $n + k$ vertices $a_1, \ldots, a_n; p_1, \ldots, p_k$ so as to minimize the sum

$$L(U) + \alpha \sum_{i=1}^{n} w(a_i) + \beta \sum_{i=1}^{k} w(p_i) + \gamma k. \qquad (1)$$

The following interpretation of our problem is terms of cost can be offered. Let the points $a_1, \ldots, a_n$ represent the terminals of a communication net to be built, for example, n cities to be joined by a road system. A point where s roads meet will be called an s-junction. Suppose that the cost of building one unit length of the road is 1 in some monetary units, that $s\alpha$ is the cost of a city s-junction, $s\beta$ the cost of any other s-junction, and finally that there is a fixed charge γ for each junction outside a city (such a junction must be necessarily an s-junction with $s \geq 3$). In view of the possibility of traffic lights, under- and overpasses, labor costs in and out of cities, and so on, it is possible to support the claim that equation (1) represents a simple but realistic model of pricing the network of roads. Our problem is now simply to construct the cheapest road network joining the n cities. We note the following special cases.

CASE 1. If $\alpha = \beta = \gamma = 0$ we get just the Steiner problem.

CASE 2. Suppose that $\alpha = 0$ and that max (β, γ) is very large. This puts a premium on avoiding any junction additional to the n cities. Therefore no new vertices will appear in the minimal tree: $k = 0$. Since the length is still to be minimized we get simply the problem of the shortest spanning subtree.

CASE 3. Suppose now that $\beta = 0$, γ is big and $\alpha > \gamma$. Since α is big we try to avoid city junctions but since each city must be in the network $w(a_i) \geq 1$ for all i, thus $w(a_i) = 1$. There is also a premium on having possibly few junctions outside cities, but many roads can meet since there is no premium on having β low. Hence we get $k = 1$: There is exactly one extra-city junction p, and its weight is n. Our problem is accordingly to find that junction, that is, to minimize the sum of the n distances

$$\sum_{i=1}^{n} |pa_i|.$$

The reader may wish to find other special cases. Also, one might ask whether a specific graph-minimization problem, for example, the geometrical case of the traveling salesman problem, can occur as a special instance, for suitable α, β, and γ. Further, in the general problem one may ask for the maximum on k in terms of n, α, β, and γ.

(d) In this final subsection we collect some odds and ends on Steiner's problem and related ones. First, we may consider Steiner's problem in planes other than the Euclidean one. A fairly extensive generalization are the Minkowski planes; these can be shown to be the same thing as two-dimensional Banach spaces. Here we have a closed convex curve C with the center of symmetry o, functioning as the unit sphere (or circle). Referring to Fig. 2a we determine the distance from x to y by translating

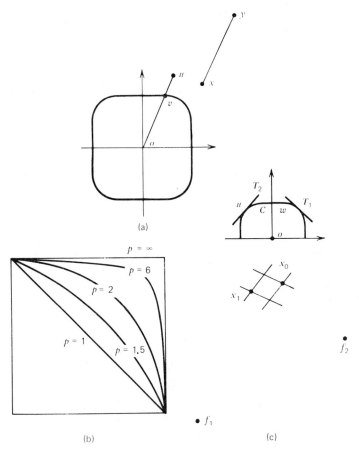

Fig. 2. M-Circles and M-ellipses.

the vector xy to the position ou; if v is the point where ou (or possibly its continuation beyond u) cuts C then the Minkowski distance $d(x, y)$ is the ratio of the Euclidean lengths $|ou|/|ov|$. A Minkowski plane is, therefore, a metric space with a nonisotropic metric: one which behaves differently in different directions. Consequently, we cannot define angles unless the M-circle C reduces to the ordinary Euclidean circle. This gives us an elementary analog to the distinction of Hilbert spaces among all other Banach spaces, as being the ones which admit inner product. The analogy is furthered by observing that the Minkowski planes include as a special subclass the L^p-planes. These are M-planes where the M-circle has the (ordinary Cartesian) equation $|x|^p + |y|^p = 1$, $p \geq 1$ (for $p < 1$ we do not get a convex curve). Examples of L^p-circles are shown in Fig. 2b; on account of symmetry only the first quadrants are shown. In particular, the L^p-planes include the two extreme cases $p = 1$ and $p = \infty$. The L^1-plane carries the metric

$$d((x_1, y_1), (x_2, y_2)) = |x_2 - x_1| + |y_2 - y_1|$$

known also as the Manhattan reactangular grid metric, and the L^∞-plane has the metric

$$d((x_1, y_1), (x_2, y_2)) = \max (|x_2 - x_1|, |y_2 - y_1|).$$

Steiner problem (as other related ones) in the L^1-metric has several applications to rectangular grid communication networks. An example arises in printed circuitry when n points on an insulated plate are to be electrically connected. For important technical reasons the nozzle that sprays thin metal lines onto the plate moves only up–down or right–left but not both at the same time. Now the L^1-metric Steiner problem is formally identical with the problem of the shortest sprayed-on connecting net for the n points; for further details see [66a].

We consider next very briefly the Steiner problem for the general Minkowski plane. First, we observe the Minkowskian ellipses, or briefly M-ellipses; these are the exact counterpart of the Euclidean ones: loci of points the sum of whose M-distances to two fixed points is constant. An approximate graphical construction of an M-ellipse is shown in Fig. 2c. The coordinate system with the curve C that functions as the M-circle is shown in the small sketch above Fig. 2c, to remind us of the nonisotropy of an M-geometry. Let f_1 and f_2 be the foci of the M-ellipse passing through the point x_0. On C we locate points w and u such that ow and ou are parallel to $f_1 x_0$ and $f_2 x_0$. Then we draw through x_0 two straight lines parallel to the tangents of C at w and at u. One of those two lines through x_0 is moved by a small amount away from f_2, and the other one by the same amount toward f_1, by parallel motions. The two new straight lines intersect at x_1 and now the same procedure is repeated with x_1 in place of x_0. In this way a polygonal approximation $x_0 x_1 x_2 \cdots$ to the M-ellipse will be obtained.

We turn now to the Steiner problem of the shortest network S joining three points a, b, c in a Minkowski plane. Like in the Euclidean case, S is either composed of two segments joining one of the three points to the other two, or S consists of three segments joining some point x to a, b, and c. The reader may wish to use the last possibility to introduce into the M-plane something like the 120° angle.

To determine x in the second case we observe that it is the intersection of two curves K_a and K_b defined as follows. K_a is the locus in which the M-circles, centered at a, touch M-ellipses with the foci b and c; K_b is similarly defined. Those curves can be found by the following fast-converging algorithm. Let x_0 be arbitrary and draw through x_0 the M-circle centered at a and the M-ellipse with the foci b and c. The intersection is a spindle-shaped convex region with two cusps, at x_0 and at y_0. Now let x_1 be the midpoint of $x_0 y_0$ and repeat the procedure replacing x_0 by x_1. The sequence $x_0, x_1, x_2, \ldots$ converges very fast to a point on K_a.

Finally, we observe on an example the differences that may exist between the geometrical and the abstractly given forms of the same minimization problem. Our example is the traveling salesman problem: how to make the shortest round tour that visits n sites $v_1, \ldots, v_n$. In the geometrical form the vertices v_i are points in the Euclidean plane and the distance from v_i to v_j is the ordinary Euclidean distance (the reader may wish to consider the case of Minkowski plane). In the abstract form of the problem there is a table of $n(n-1)$ numbers c_{ij} corresponding to the $n(n-1)$ ordered pairs (i, j), and c_{ij} is taken as the "distance" from v_i to v_j. We can use the name "cost" instead of "distance" since none of the three metric axioms need hold: c_{ij} is not always c_{ji}, $c_{ij} < 0$ is possible, and so is $c_{ij} + c_{jk} < c_{ik}$. Hence in this form of the problem we are looking for the permutation $\pi = (\pi(1)\pi(2) \cdots \pi(n))$, $\pi(1) = 1$, which minimizes the sum

$$\sum_{i=1}^{n} c_{\pi(i)\pi(i+1)} \tag{2}$$

and is *cyclic* (i.e., does not permutate any smaller number than all n objects among themselves).

As is well known, it is the requirement of cyclic permutation that causes difficulty. If this is dropped, we get the problem of minimizing (2) over all permutations π, which are allowed now to have cycles of length $< n$. This is the so-called assignment problem and is known to have an efficient solution (with the number of steps to solve it bounded by a quartic polynomial in n). On the other hand, for the minimizing of (2) by cyclic permutations one needs about $2^n n^2$ steps, and it is even conjectured that no real improvement is possible.

In the geometrical case, as was noted in vol. 1, pp. 138–139, we show easily that the minimal path cannot be self-crossing; this enables us to solve the problem in certain special cases. For instance, if the points $v_1, \ldots, v_n$ are the vertices of a convex n-gon P then by the non-self-crossing condition P itself is the minimal path. For a possible generalization let us recall the concept of layers for the n points $v_1, \ldots, v_n$. The first layer consists of those v_i's that are the vertices of the convex hull of all v_i's (in more graphical terms, the first layer of pins $v_1, \ldots, v_n$ on which an elastic rubber band rests). These are removed and the second layer consists then of points v_i that are the vertices of the convex hull of the remainder, and so on. We note that $v_1, \ldots, v_n$ are vertices of a convex n-gon if and only if there is exactly one layer. Hence comes the possibility that an efficient solution can be produced for n points which have $\leq k$ layers where k is fixed (while n, of course, is not). The reader might wish to attempt a transfer of the layer concept to the abstract form of the problem.

Next, instead of minimizing the quantity (2) we could maximize it; this might correspond to the traveling salesman who gets paid for his travel rather than for his sales. In the abstract form of the problem there is no difference, since the quantities c_{ij} of (2) can be negative and to minimize S is to maximize $-S$. For the geometrical case of the problem the non-self-intersection condition on the minimum path suggests that the longest path is one with many self-intersections. We could also formulate a separate problem: that of producing the piecewise rectilinear path $v_1 v_{n_1} v_{n_2} \cdots v_1$ which maximizes the number of self-intersections. It may be conjectured that this maximum, $n(n-3)/2$, is attained if and only if n is odd and the n points v_i are the vertices of a convex n-gon (in which no three diagonals can be concurrent, except at a point v_i).

6

FUNCTIONAL EQUATIONS AND MATHEMATICAL MODELING

This chapter deals with some functional equations, principally as they arise in applications. We shall not be particularly concerned with the existence of solutions or with the methods of solving, whether analytically or numerically. Rather, our interest will be in the classification of functional equations and in deriving them. Presumably, the ability to derive the equations governing the phenomenon in question is an important component of the art of mathematical modeling. Since not just books but whole libraries have been written on our subjects, it is clear that not even a general introduction could be given within our compass and at our elementary level; we shall be even more arbitrarily selective here than elsewhere.

1. CERTAIN TYPES OF FUNCTIONAL EQUATIONS

(a) The phrase *functional equation* appears in mathematics in several quite different senses. First, it may be used to denote an equation that is, loosely speaking, of the form: a known function of the unknown function of a known function of (the same) unknown function of $\cdots$ is 0. For instance, we have the Cauchy equation

$$f(x+y) = f(x) + f(y), \tag{1}$$

325

which could also be written as

$$h[f(g(x, y)), g(f(x), f(y))] = 0$$

with $h(u, v) = u - v$ and $g(u, v) = u + v$. If f is continuous then the solution of (1) is $f(x) = f(1)x$: First, (1) implies that $f(n) = f(1)n$ for every integer n, and similar argument shows that

$$f\left(\frac{p}{q}\right) = \frac{f(1)p}{q}, \qquad q \neq 0,$$

for all integers p and q. In other words, $f(x) = f(1)x$ for rational x; the assumed continuity allows us to pass to real x. But the same conclusion holds under the much weaker hypothesis of measurability of f. We give a proof due to M. Kac [90] that uses only the fact that a bounded measurable function is integrable. Let the number a be such that

$$\int_0^a e^{if(y)} \, dy \neq 0;$$

then from (1) we gather that

$$e^{if(x)} = \int_x^{a+x} e^{if(y)} \, dy \Big/ \int_0^a e^{if(y)} \, dy.$$

It follows by a very simple estimate that $e^{if(x)}$ must be continuous and therefore, by an application of the previous result, we have

$$f(x) = f(1)x + 2\pi K(x),$$

where $K(x)$ is a *continuous integer-valued* function. Hence $K \equiv 0$ and we are finished. The reader may wish to compare this with the proof of Rouché's theorem (where similar use is made of a continuous integer-valued function) and of Liouville's theorem for harmonic functions (where also one passes from $f(x)$ to $e^{if(x)}$) in Section 6 of Chapter 2.

The equations

$$f(x + y) = f(x)f(y) \qquad \text{and} \qquad f(xy) = f(x) + f(y),$$

which are to the exponential and to the logarithm what (1) is to cx, can be similarly treated.

A generalization of (1) is the Sinzow equation

$$F(x, y) + F(y, z) = F(x, z) \tag{2}$$

[160, 1] which reduces to (1) with $F(x, y) = f(x - y)$ since $x - z = x - y + y - z$. While (1) characterizes the linear function cx, (2) could be said to characterize subtraction as a particular binary composition. It is therefore

not surprising that the general solution of (2) is $F(x, y) = h(x) - h(y)$, h arbitrary. These simple remarks might perhaps help to show how functional equations, in this sense of the phrase, arise from characterizing functions and operations. So, for instance, we could observe a specific functional form of the addition theorem for a trigonometric function, and we could ask whether that form characterizes the function. Similarly, we have seen how (2) is related to the operation of subtracting.

We solve such functional equations sometimes by reducing them to differential equations. As a simple example, with $f(x + y) = f(x)f(y)$ we have

$$f'(x) = \lim_{h \to 0} \frac{f(x+h) - f(x)}{h} = f(x) \lim_{h \to 0} \frac{f(h) - 1}{h} = f'(0)f(x),$$

supposing that the derivative is defined at $x = 0$ (it must then be defined for all x, cf. pp. 64–66). The equation of d'Alembert

$$f(x + y) + f(x - y) = 2f(x)f(y) \tag{3}$$

can be solved by the same device of exploiting the definition of the derivative, here the *second* derivative. From (3) one gets

$$\frac{f(x+h) - 2f(x) + f(x-h)}{h^2} = 2f(x) \frac{f(h) - 1}{h^2},$$

and already the R.H.S. may suggest the cosine function. Letting $h \to 0$ we get the differential equation $f'' = cf$. Hence

$$f(x) = \cosh kx \quad \text{if} \quad c = k^2,$$
$$f(x) = \cos kx \quad \text{if} \quad c = -k^2.$$

Occasionally, we can solve a functional equation by using telescoping cancellation (vol. 1, p. 81). Consider, for instance,

$$f(x + y) = f(x)f(y)e^{P(x,y)}, \tag{4}$$

where P is a suitable polynomial. To illustrate our method we use the symmetric cubic

$$P(x, y) = c_1 xy(x + y + c_2);$$

equation (4) may then be written as

$$f(x + y) = f(x)f(y)a^{xy}b^{xy(x+y)}. \tag{5}$$

With $x = y = 0$ we find that $f(0) = f^2(0)$ so that $f(0) = 0$ or 1. The first possibility leads to $f(x) \equiv 0$, at any rate for sufficiently smooth functions, for now $f^{(n)}(0) = 0$ for all n; we take therefore $f(0) = 1$. In (5) put $y = x$, then $2x$, $3x$, $\ldots$, nx; multiplying the resulting equations and cancelling we get

$$f((n+1)x) = f^{n+1}(x) a^{x^2 \sum\limits_{j=1}^{n} j} b^{x^3 \sum\limits_{j=1}^{n} j(j+1)}.$$

Next, let $nx = u$ be fixed while $x \to 0$ and $n \to \infty$. Then, since $f(0) = 1$,

$$\lim_{n \to \infty} f^{n+1}\left(\frac{u}{n}\right) = \lim_{n \to \infty} \left[1 + f'(0)\frac{u}{n} + O\left(\frac{1}{n^2}\right)\right]^{n+1} = e^{ku}$$

while

$$\lim_{n \to \infty} \frac{u^2}{n^2} \sum_{j=1}^{n} j = \frac{u^2}{2}, \qquad \lim_{n \to \infty} \frac{u^3}{n^3} \sum_{j=1}^{n} j(j+1) = \frac{u^3}{3}.$$

Therefore the solution is

$$f(u) = e^{ku} a^{u^2/2} b^{u^3/3};$$

however, it would be simpler to solve (4) by guessing the solution and then verifying.

In the second place, a functional equation may signify just an equation involving functionals. Here by a functional we mean a real- or complex-valued function of functions, that is, a numerical-valued function defined on some function space. For instance, let $U(x)$ and $P(x)$ be fixed suitable functions, then

$$R(K) = \int_0^\infty U[P(K(t)) - K'(t)] \, dt$$

is a functional of the function $K(t)$. This, the so-called Ramsey functional, has a significance in economics. Letting $P(K)$ be the production expressed as a function of the capital $K(t)$ at the time t, it is supposed that all of this production is either consumed or reinvested. This reinvestment is K'-rate of change of the capital, so that consumption is the difference $P(K) - K'$. $U(x)$ is the instantaneous "satisfaction" or "utility," expressed as a function of the consumption x. Thus the Ramsey functional is the total utility. For the problem of maximizing R, that is, of prescribing the whole future course of investment and capital programs for best total utility, see [140] and [65]. The problem of maximizing $R(K)$ is historically the first variational problem considered in mathematical economics. It was formulated by F. P. Ramsey (of the Ramsey theorem in combinatorics, see Section 4 of Chapter 3).

Third, there is a minor loose usage of "functional equations" in such phrases as "functional–differential equations." This is intended to convey the idea that here the derivatives occur in some nonlocal sense. For instance,

$$f'(t) = F(t, f(t))$$

is an ordinary "local" differential equation. It is local because different values of the independent variable t do not get mixed in the equation. Alternatively, the equation connects $f(t)$ with f at the "infinitesimally near" value $t + dt$. On the other hand, the values of t do get mixed in such equations as

$$f'(t) = F(t, f(t-1)), \qquad f'(t) = F(t - h(t)), \qquad f'(t) = F\left(f(t), \ f\left(\frac{t}{2}\right)\right).$$

These "nonlocal" equations are sometimes called functional–differential. The first one could be more exactly described as a fixed delay difference–differential equation; the second one represents then a case of variable delay.

(b) Finally, in a fourth and even looser sense, a functional equation is any relation referring to an unknown function. We give a small list of some of the commoner special types. It is supposed that $y = f(x, t)$ is the unknown function; t is a scalar time variable, x is a scalar or vector space variable, and y itself can be scalar or vector. If there is no dependence on x we have $y = y(t)$.

REMARKS.

1. Examples 1–5 are local equations: Here the values of the independent variables are either taken at one point (x, t) or in an "infinitesimal" neighborhood of one point. Examples 6–10 are nonlocal. The local–nonlocal division refers to the equations, not to their solutions. It has no relation to the difference of behavior of certain partial differential equations where in one type (hyperbolic) a localized change in the initial data $f(x, 0)$ near some $x = x_0$ leads only to a localized perturbation in the later values $f(x, t)$.

2. A system of first-order equations yields a single higher-order equation. Conversely, one higher-order equation can be represented as a first-order system (we co-opt sufficiently many derivatives as auxiliary variables).

3. In several places of our table we distinguish between explicit relations (with F) and implicit ones (with G).

4. No higher-order equivalents have been given for difference, difference–differential, and integrodifferential equations. Certain further mixed types have been omitted.

	Function	Depends on	Form of relation	Name of equation
1	$y(t)$	$y(t)$	$y = F(y)$ or $G(y) = 0$	ordinary algebraic or transcendental
2	$y(t)$	$y(t), y(t-dt)$	$y' = F(y)$ or $G(y, y') = 0$	first-order autonomous O.D.
3	$y(t)$	$y(t), y(t-dt), \ldots, y(t-n\,dt)$	$y^{(n)} = F(y, y', \ldots, y^{(n-1)})$ or $G(y, y', \ldots, y^{(n)}) = 0$	nth order autonomous O.D.
4	$y(t)$	$t, y(t), y(t-dt),$ $\ldots, y(t-n\,dt)$	$y^{(n)} = F(t, y, y', \ldots, y^{(n-1)})$	general nth order O.D.
5	$y(t, x)$	$t, x, y(t, x), y(t-dt, x),$ $y(t, x-dx), \ldots$	$G(t, x, y, y_x, y_t, y_{xx}, \ldots) = 0$	general P.D.
6	$y(t)$	$y(t), y(t-a)$	$G(y(t), y(t-a)) = 0$	first-order difference
7	$y(t)$	$t, y(t), y(t-a),$ $y(t-dt)$	$G(t, y(t), y'(t), y(t-a)) = 0$	first-order difference-differential
8	$y(t)$	$y(\tau), -\infty < \tau \leq t,$	$G\left(y(t), \displaystyle\int_{-\infty}^{t} h(\tau, y(\tau))\, d\tau\right) = 0$	integral
9	$y(t)$	$t, y(t), y(t-dt), y(\tau),$ $y(\tau-d\tau), 0 \leq \tau \leq t,$	$G\left(t, y, y', \displaystyle\int_{0}^{t} h(\tau, y(\tau), y'(\tau))\, d\tau\right) = 0$	integrodifferential
10	$y(t, x)$	$t, x, y(t, x), y(t-dt, x),$ $y(t, x-dx), \ldots, y(t, x-k(x, t))$	$G(t, x, y, y_t, y_x, \ldots, y(t, x-k(x, t))) = 0$	nonlocal P.D.

5. The integral and integrodifferential equations (8) and (9) could be called *causal* with finite and with infinite memories respectively, because in either case the present state $y(t)$ of the system does not depend on a future state. On the other hand, the integral equation

$$y(t) = y_0(t) + \int_{-\infty}^{\infty} h(\tau, t, y(\tau)) \, d\tau$$

could be called *acausal* because here the present state depends on the future.

6. Autonomous versus nonautonomous is an important distinction for ordinary differential equations. The name comes from two Greek words: *autos*—referring to oneself, and *nomos*—law; the literal etymology yields "subject to own law."

7. In some functional equations involving time derivatives $y, y', \ldots, y_t,$ $y_{tt}, \ldots, y_{tx}, \ldots$ we speak of the steady state as the limiting condition for large t when there are no time changes. This amounts to setting all time derivatives to 0, and often yields a much simpler functional equation. The opposite behavior, referring to the phenomena that die out for large t, is referred to as the transient state.

One of the main divisions for functional equations is the local–nonlocal ·dichotomy. When the space variable x is absent and the time t alone remains, the local type is best exemplified by the differential equations of *classical mechanics*. The nonlocal type was introduced under the contrasting name of *hereditary mechanics* by E. Picard in 1907, [133, 40]. We quote his remarks at some length: "In all this study of classical mechanics the laws which express our ideas on motion have been condensed into differential equations, that is to say, relations between variables and their derivatives. We must not forget that we have, in fact, formulated a principle of *nonheredity*, when we suppose that the future of a system depends at a given moment only on its present state, or in a more general manner, if we regard the forces as depending also on velocities, that the future depends on the present state and the infinitely neighboring state which precedes. This is a restrictive hypothesis, and one which, in appearance at least, is contradicted by the facts. Examples are numerous where the future of a system seems to depend upon former states. Here we have *heredity*. In some complex cases one sees that it is necessary, perhaps, to abandon differential equations and consider functional equations in which there appear integrals taken from a distant time to the present, integrals which will be, in fact, this hereditary part. The proponents of classical mechanics, however, are able to pretend that the heredity is only apparent and that it amounts merely to this: that we have fixed our attentions upon too small a number of variables."

This is illustrated by Volterra, [40, 176a] who considers the dependence of the angle T of torsion of a metal bar, on the torque M affecting it. In the first approximation there is a linear dependence $T = kM$. Volterra goes on to say that the elastic substance of the metal bar exhibits the phenomenon of fatigue, which results in a sort of heredity: the past influences the present via an integral in which the remoteness of the past is taken into account; T and M become functions of time and we have

$$T(t) = kM(t) + \int_{t_0}^{t} K(t, s)M(s) \, ds,$$

The kernel $K(t, s)$ is called by Volterra the coefficient of heredity. A special case arises when the degree of influence of the past on the present depends only on the time lapsed: $K(t, s) = K(t - s)$.

What Picard and Volterra call heredity could also be called memory ("heredity" sounds perhaps more species-oriented but it is a trifle "undemocratic"; "memory" is more flattering to the individual). If the "memory" thus exhibited by the mechanical, electromagnetic, or optical system is "reversible" then it may be considered as a basis for building an actual computer storage unit, that is, a computer memory.

Emile Picard (1856–1941) was a French mathematician who worked principally in analysis. The well-known method of solving functional equations by successive approximations and iteration is named after him ($y = Ay$: let y_0 be a suitable initial approximation and let $y_{n+1} = Ay_n$, hopefully $y_n \to y$), as well as a theorem on entire functions (they assume every complex value infinitely many times, with one possible exception; an example is e^z which is never 0).

Vito Volterra (1860–1940) was an Italian mathematician (the name Volterra comes from the old Etruscan city Volterra, in Etruscan form *Velathri*; strangely enough, Picard too was named after a place name, *Picardy*). Volterra's interest was in analysis with special reference to integral and related functional equations. His work on "hereditary mechanics" helped to pioneer developments both theoretical (functional analysis) and applied (rheology, that is, the science of flow of nonstandard fluids with long "memories," prestressed concrete, mathematical ecology).

Young Volterra discovered integral calculus by himself at the age of fourteen; shortly after, his parents wished him to be practical and take up commerce while Volterra himself wanted to be a scientist; the case was submitted to the arbitration of a very rich and successful distant cousin who, however, came out on the boy's side (and became, a number of years later, his father-in-law). In 1931, at the age of 71, Volterra refused to take the oath of allegiance to the fascist government of Italy, and was deprived of his professorship as well as of memberships in Italian scientific societies. He continued his work in his country house in Ariccia, not far from the location of the sacred grove of Diana by the lake Nemi.

It may be difficult for us nowadays to appreciate Picard's and Volterra's insistence on "hereditary mechanics." But, then, we have largely liberated ourselves from the classical mechanico–mathematical Newton–Laplace view of the universe that runs on autonomous ordinary differential equations. What Picard calls hereditary, as distinguished from classical, mechanics has an exact counterpart in probability: Markovian versus non-Markovian stochastic processes. Probability, being a newer discipline, has perhaps avoided such sharp dichotomizing and partisanship as mechanics.

Outside of mathematics, physics, and biology further examples of "hereditary" versus "classical" treatment could be given in mathematical economics. Here we meet many differential or difference equations, connecting such quantities as production, investment, and consumption, when these are considered as functions of time. This corresponds to classical mechanics with its local equations expressing the Newton laws. On the other hand, it is not unreasonable to attempt to study the effect of habit on those quantities. For instance, it seems natural to suppose that the history of one's previous consumption of a commodity will affect the price one is willing to pay for it presently (hence advertising campaigns that depend on low initial prices and on habituation). Now the differential equations change to integrodifferential ones, and so they are no longer "local" or familiar. Since the integrodifferential equations are so much more difficult and forbidding-looking than the ordinary differential ones, it is understandable why the economists (and others) shy away from them. Besides the influence of habit on, say, consumption one might want to study the opposite effect of "novelty." Bringing in habit changes differential equations to integrodifferential ones—one may speculate on the type of functional equations that might appear in a mathematical treatment of novelty.

2. DERIVATION OF CERTAIN SIMPLE FUNCTIONAL EQUATIONS

(a) For our first example we consider in the k-dimensional Euclidean space a large number of point-particles distributed independently and at random so that the average density is b particles per unit volume. It is desired to find the probability density $w(r)$ for the distance of a particle to its n.n. (= nearest neighbor). Following Hertz [74] we develop a simple functional relation for $w(r)$, based on the elements of probability. Let any one of the particles be chosen as the origin O and let $K_r(O)$ be the solid ball of radius r about O. We recall that the volume of $K_r(O)$ is

$$\pi^{k/2}\frac{r^k}{\Gamma(1+k/2)} = v_k r^k \qquad \text{say,} \qquad (1)$$

as shown in Chapter 2. By our hypothesis the events occurring in disjoint volumes are independent, and so

> Prob(n.n. to O is at distance r to $r+dr$) =
> Prob($K_r(O)$ contains O but no other particles).
> Prob(the spherical shell between $K_{r+dr}(O)$ and $K_r(O)$ is not empty).

From (1) the volume of the spherical shell is $kv_k r^{k-1}\,dr$, and so the last probability is $bkv_k r^{k-1}\,dr$. The probability which multiplies it is, by the definition of $w(r)$,

$$\int_r^\infty w(r)\,dr$$

while the term on the other side of the equation is $w(r)\,dr$. Hence

$$w(r) = bkv_k r^{k-1}\int_r^\infty w(r)\,dr. \qquad (2)$$

This is converted to a simple differential equation by multiplying with r^{1-k} and differentiating:

$$w' + [(1-k)r^{-1} + bkv_k r^{k-1}]w = 0;$$

we solve the above and get

$$w(r) = Cr^{k-1}\,e^{-bv_k r^k}.$$

The constant C is obtained from (2) by letting $r \to 0$ since

$$\int_0^\infty w(r)\,dr = 1$$

so that

$$w(r) \sim bkv_k r^{k-1}, \qquad r \text{ small.}$$

Hence $C = bkv_k$ and we get the nearest-neighbor probability density

$$w(r) = bkv_k r^{k-1}\,e^{-bv_k r^k}. \qquad (3)$$

Without loss of generality we normalize to density $b = 1$; this is equivalent to a change of unit of volume. Now the moments of w can be found in terms of Γ-functions:

$$\overline{r_k^m} = \int_0^\infty r^m w(r)\,dr = \pi^{-m/2}\Gamma^{m/k}\left(1+\frac{k}{2}\right)\Gamma\left(1+\frac{m}{k}\right);$$

in particular we have the expected nearest-neighbor distance

$$\overline{r_k} = \pi^{-1/2}\Gamma\left(1+\frac{1}{k}\right)\Gamma^{1/k}\left(1+\frac{k}{2}\right).$$

The behavior of $\overline{r_k}$ as a function of the dimension k is shown in the table below

k	1	2	3	4	$\cdots$	k large
$\overline{r_k}$	1/2	1/2	0.5540	0.6082	$\cdots$	$\sim k^{1/2}(2\pi e)^{-1/2}$.

We note that $\overline{r_k}$, in terms of a real variable k, attains its minimum for k between 1 and 2 (while here there appears to be no obvious way of assigning meaning to the fractional dimension extremum, there are, occasionally, situations where such results may correspond to some reality). Next, let us compare in k dimensions two arrangements of point-particles: random arrangement with density 1, and the unit cubic lattice (with the same density). Then the asymptotic result for k large shows that near dimension $k = 17$ both arrangements have the same expected nearest neighbor distance. The reader may wish to improve the asymptotic result for k large to more terms; it will be found that we have here a situation in which all the three constants π, e, γ of analysis appear together.

There are some obvious possibilities of applying the nearest neighbor statistics for $k = 2$ and $k = 3$. So, in pure statistics we may consider tests for deciding whether a given distribution is random or not. In astronomy we have the possibility of applying the n.n. to stellar statistics and dynamics, in botany (with $k = 2$) to random seeding, and in physical chemistry to ionic calculations.

In neurocytoarchitectonics the nearest neighbor statistics might turn up in connection with some selective nerve-cell staining methods. These methods go by the name of Golgi or Golgi–Cox; they use some compounds of heavy metals: silver, osmium, and mercury. These compounds are used to impregnate the nerve-tissue specimens so as to stain nerve-cells that are ordinarily transparent. As a result, the stained nerve cells show up in black or other dark color, against clear background, and so they are observable under a microscope. One of the features of the Golgi–Cox family of stains is their strange selectivity. Simplifying the picture we can put it thus: Only one out of every few hundred cells gets stained, and then it gets stained completely, with all its parts. This selectivity is in some respects a very fortunate circumstance; without it (i.e., if all cells stained) the microscope picture would be a black forest in which it might be hard to distinguish much. However, we should still like to know: Why this selectivity? At least two classes of mechanisms might be responsible; let us call them specific and stochastic. Roughly, a specific mechanism explains the selective "taking" of the stain by some chemical, mechanical, or electrobiochemical peculiarity of the cells that stain. A stochastic mechanism is best explained by assuming that the same block of tissue could be stained, then "unstained," then stained again, and so on. On the stochastic view, it is no longer true that the same cells will stain each time, but only the same number of them, and with approximately the same spatial distribution.

To explain a possible stochastic model of selective staining, let us describe the tissue block as a sandbox and the nerve cells as crumpled-up paper balls, coated with a semi-impervious layer of some kind. Now a container of thick ink is poured into the box and slowly diffuses through the sand. Supposing that the ink density is fairly variable, we hypothesize that if the ink concentration at the surface of a ball is high enough then the protective layer may break down and as a result, a large amount of ink is locally absorbed by the paper interior of the ball. We should then expect that the close neighbors of a cell that stains are highly unlikely to get stained themselves. This observation indeed appears to be true with actual Golgi–Cox stained cells, from the work of E. Ramon-Moliner [139a]. We have omitted several complicating factors; for example, pouring more ink will not produce more stained cells because the protective layer of a cell gets very rapidly impervious as the post-mortem time grows. Thus the probability of a breakdown in that layer decreases very fast and so, on this model, the selective stain phenomenon arises from a race between the diffusion of staining substance (and fluctuations in its density) and the imperviousness of the protective layer of the cell.

Now we ask: How could one possibly distinguish between a specific and a stochastic model of selective staining? It is clear that precisely the nearest-stained-neighbor distribution may serve as the first distinguishing feature.

(b) In this subsection we derive a simple linear partial differential equation, the telegraphy equation, so called from its application to the study of signals in cables. It exhibits the following phenomenon: Even

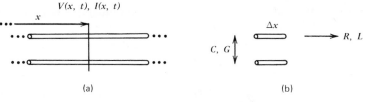

$V(x, t), I(x, t)$

(a)   (b)

Fig. 1. Voltage and current relations.

though we may wish to know just one function describing our system, it is best to introduce two such functions, the one we want and another, an auxiliary one. We then express each function in terms of the other, and we eliminate. It often happens under these conditions that both functions satisfy the same functional equation even though the two component equations expressing each function by the other one, are different. Under the paradigm $u = F(v)$, $v = G(u)$ so that $u = F(G(u))$, $v = G(F(v))$, it might be suggested that we deal here with a commuting pair of operators F and G.

Suppose that we want to study the signal propagation in a pair of long parallel conductors as shown in Fig. 1a. The following factors must be considered:

(a) The system is described by the voltage $V(x, t)$ and the current $I(x, t)$, which are functions of time t and distance x, measured along the cable;
(b) there are series resistance R and series inductance L, *along* the cable;
(c) there are parallel capacitance C and leak conductance G, *across* the cable;
(d) the four quantities R, L, C, G are uniform and refer to a unit length of the cable (of course, new and more difficult problems result if some of the four parameters are not constant but, say, periodic in x on account of some faults in cable winding on a drum, or slowly varying with t).

We set up now two relations that amount to expressing V by I and I by V. These are obtained by using a conservation law: What goes in $-$ what goes out $=$ what stays in. (The reader may wish to compare this with the derivation of transport equations on p. 271). Consider a short length Δx of the cable, as in Fig. 1b; we argue as follows. The voltage change ΔV along the element Δx of line length must be accounted for by the sum of voltage drops due to the series phenomena (b), and the current change ΔI must be similarly balanced by the parallel phenomena (c) across the line.

When (d) is taken into account we can express our balances as

$$\Delta V = \left(-L\frac{\partial I}{\partial t} - RI\right)\Delta x,$$

$$\Delta I = \left(-C\frac{\partial V}{\partial t} - GV\right)\Delta x. \tag{6}$$

The minus signs indicate drops, not rises. Dividing by Δx and going to the limit $\Delta x \to 0$ we get the pair of equations

$$\frac{\partial V}{\partial x} + L\frac{\partial I}{\partial t} + RI = 0,$$

$$\frac{\partial I}{\partial x} + C\frac{\partial V}{\partial t} + GV = 0. \tag{7}$$

As could be expected much earlier in our formulation, our equations turn out to be local. Suppose now that we wish to obtain an equation for V alone. Then we eliminate I out of (7) and this amounts to a thinly disguised use of the principle of computing one thing in two different ways: we express $(\partial/\partial x)(\partial I/\partial t)$ from the first equation of (7) and equate it to $(\partial/\partial t)(\partial I/\partial x)$ computed from the second equation. This yields

$$V_{xx} = LCV_{tt} + (RC + GL)V_t + RGV,$$

repeating the same process for I we find that it satisfies the same equation. Thus, both V and I satisfy the telegraphy equation

$$f_{xx} = LCf_{tt} + (RC + GL)f_t + RGf. \tag{8}$$

It turns out that simplification results if we introduce new constants

$$LC = \frac{1}{c^2}, \qquad a = \frac{G}{C}, \qquad b = \frac{R}{L};$$

(8) may now be written as

$$c^2 f_{xx} = f_{tt} + (a+b)f_t + abf. \tag{9}$$

This may be simplified further by finding an integrating factor: If we write $h(x, t) = e^{kt}f(x, t)$ where k is a suitable constant $(=(a+b)/2)$ then h satisfies the simpler equation

$$c^2 h_{xx} = h_{tt} - \left(\frac{a-b}{2}\right)^2 h.$$

In the "resonance" case of $a = b$ h satisfies the wave equation

$$h_{xx} = \frac{1}{c^2} h_{tt};$$

this has the well-known general solutions $F(x \pm ct)$ representing left-traveling and right-traveling waves on the cable. It follows that when $a = b$ the telegraphy equation has solutions of the form

$$f(x, t) = e^{-(a+b)t/2} F(x \pm ct).$$

This has the important consequence that the signal is transmitted along the cable in a form that keeps constant shape (though not constant size).

(c) The same phenomenon of two different first-order equations for two related physical quantities, both of which satisfy the same second-order equation, occurs also in the following nonlocal case. Consider a helical coil in the shape of very tightly wound helix of uniform uninsulated wire; let the length of one turn of the helix be 1. Let dc voltage be applied across the coil terminals. The system (in its steady state) is given by three functions $V(x)$, $J(x)$, $I(x)$. Here x is the distance measured along the helix, $V(x)$ and $J(x)$ are the voltage and current at x while $I(x)$ is the total sideways current leak up to x. That is, we have the current $\Delta I(x)$ flowing from the "element" $x - x + \Delta x$ to the corresponding "element" $x + 1 - x + 1 + \Delta x$ on the next turn of the coil.

We could derive now the three equations connecting V, J, and I but to simplify such derivation we shall not proceed directly. Rather, we apply the technique of lumping the circuit parameters. This transforms the continuous distributions into discrete ones: The (continuous) wire of the coil is thought of as composed of a large number of equal short segments. Each such segment is connected to four others: one immediately preceding and one immediately following, and to two others that are its neighbors on the preceding and the following turns of the coil (and whose distance to it is therefore 1, measured along the coil wire). The interconnection scheme suggests in advance that our equations will be nonlocal, and difference–differential, to be precise: the "differential" part will be due to the two immediate neighbors and the "difference" part to the two neighbors at distance 1. The corresponding circuit diagram is shown in Fig. 2a, we have a periodic structure of which just one repeating part is shown. We suppose that the resistance of the coil wire is a ohms per unit length and the sideways conductance, from turn to turn, is b mhos per unit length. Referring to Fig. 2b we consider one mesh of our equivalent circuit and we get three equations as follows. First,

$$b \Delta x [V(x+1) - V(x)] = \Delta I(x) \tag{10}$$

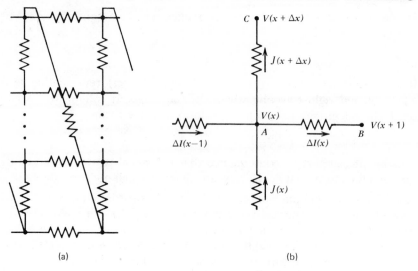

Fig. 2. Equivalent circuit with lumped parameters.

because the conductance of the AB link, like that of any other horizontal link, is $b \, \Delta x$ while the voltages of A and B and the current from A to B are as shown in Fig. 2b. Next,

$$J(x+\Delta x) - J(x) + \Delta I(x-1) - \Delta I(x) = 0 \tag{11}$$

because the sum of all currents at the node A is 0. Finally

$$V(x+\Delta x) - V(x) = a \, \Delta x \, J(x+\Delta x) \tag{12}$$

by considering the circuit link AC with voltages and current as shown. Passing to the limit $(\Delta x \to 0)$ and integrating, we get from equations (10), (11), and (12)

$$b[V(x+1) - V(x)] = I'(x), \qquad J(x) = I(x) - I(x-1) + \text{const.},$$
$$V'(x) = aJ(x).$$

The second equation is used to eliminate J from the third one, yielding the pair

$$b[V(x+1) - V(x)] = I'(x), \qquad V'(x) = a[I(x) - I(x-1)] + \text{const.} \tag{13}$$

We can now eliminate either V or I from (13) and it turns out that *both* satisfy the same nonlocal second-order difference–differential equation

$$f''(x) = ab[f(x+1) - 2f(x) + f(x-1)]. \tag{14}$$

It must be added this is only an approximation since (14) is obviously incorrect with respect to the first and the last turn of the coil.

Equation (14) is a very special example of a linear difference–differential equation with constant coefficients

$$\sum_{j=0}^{J} \sum_{k=0}^{K} A_{kj} f^{(k)}(x + a_j) = g(x), \tag{15}$$

where A_{kj} and a_j are constants. We recall the operatorial form of Taylor's theorem (vol. 1, p. 109): The shift operator S_h given by $S_h f(x) = f(x + h)$ and the differential operator D given by $Df(x) = f'(x)$ are connected by

$$S_h = e^{hD}.$$

With this (15) can be written simply as

$$E(D)f(x) = g(x),$$

where $E(D)$ is an exponential polynomial in D. On account of its symmetric behavior with respect to shift we can use Fourier transform here: If

$$\int_{-\infty}^{\infty} e^{iux} f(x)\, dx = F(u), \int_{-\infty}^{\infty} e^{iux} g(x)\, dx = G(u)$$

then (15) becomes

$$F(u) \sum_{j=0}^{J} \sum_{k=0}^{K} A_{kj} (-iu)^k e^{-iua_j} = G(u).$$

Hence, under suitable conditions $F(u)$ can be found and by inverting the transform we get the solution $f(x)$. For sets of such conditions and for further details see [17]. For instance, it is shown there that the equation

$$(1+c)f(x) - \frac{c}{2}[f(x+1) - 2f(x) + f(x-1)] - f''(x) = g(x), \qquad |c| < 1,$$

has the solution

$$f(x) = \frac{1}{2\pi} \int_{-\infty}^{\infty} K(y) f(x-y)\, dy,$$

where

$$K(x) = \int_{-\infty}^{\infty} \frac{e^{ixt}}{1 + t^2 - c \cos t}\, dt.$$

If the functions $f(x)$ or $g(x)$ do not behave sufficiently well for large x, so that the Fourier transforms F or G are not defined, it may still be possible to use finite integral transforms, for instance that of Euler–Laplace

$$\int_a^b e^{-ux} f(x)\, dx = F(u), \int_a^b e^{-ux} g(x)\, dx = G(u).$$

For this, and for other methods of solution, see [133a].

As a much simpler technique, we can use the method of trial solution just as with linear differential or difference equations, with constant coefficients. For instance, letting $f(x) = e^{mx}$ in (14) we get the characteristic equation involving exponential polynomials

$$m^2 = ab(e^m - 2 + e^{-m}), \tag{16}$$

which can also be written as

$$m^2 = 4ab \sinh^2 \frac{m}{2}.$$

If we put $ab = k^2$ and separate m into real and imaginary parts, $m/2 = u + iv$, we obtain the system

$$u = k \cos v \sinh u, \qquad v = k \sin v \cosh u. \tag{17}$$

For an nth order linear differential equation with constant coefficients, the trial solution e^{mx} leads to the characteristic equation $P(m) = 0$ where P is an nth degree polynomial. Hence we construct n linearly independent solutions and so, eventually, the general solution. Here, for linear difference–differential equations with constant coefficients, we meet further complications because it may happen that the exponential–polynomial characteristic equation has infinitely many roots. We show this for the equation (14) for which the characteristic equation is (16) or, equivalently, (17).

Our object, then, is to prove that system (17) has infinitely many (real) roots u, v. This system is of the general form

$$f_1(u) = g_1(v), \qquad f_2(u) = g_2(v). \tag{18}$$

We start with a simple graphical method that sometimes will show the roots of systems like (18). With reference to Fig. 3, we suppose that all four curves are continuous, we start with two values u_1, u_2 on the first graph and continue round the properly aligned graphs, as shown in Fig. 3, getting eventually two values U_1, U_2. From the basic properties of continuity it follows that system (18) represented graphically in Fig. 3, has roots u, v satisfying $u_1 < u < u_2$, $v_1 < v < v_2$. A proportionality argument even suggests that we might try

$$\frac{u_1 U_2 + u_2 U_1}{U_2 + U_1}, \quad g_1^{-1}\left(f_1\left(\frac{u_1 U_2 + u_2 U_1}{U_2 + U_1}\right)\right)$$

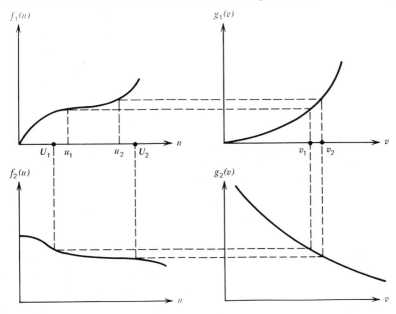

Fig. 3. Real roots of systems.

as the approximation to those roots. Combining perhaps the Volterra method of locus plotting (Section 2, Chapter 5) and some ideas from the graphics of iteration (vol. 1, Chapter 2), the reader may wish to extend the above to a method for the roots of the system

$$f(u, v) = 0, \quad g(u, v) = 0$$

which is more general than (18).

The system (17) is of the form (18) where

$$f_1(u) = \frac{u}{\sinh u}, \quad g_1(v) = k \cos v, \quad f_2(u) = k \cosh u, \quad g_2(v) = \frac{v}{\sin v}.$$

We suppose that $k \geq 1$ and we use the above method, as it is illustrated in Fig. 3, taking n to be an integer >0 and the graphs of our four functions as follows:

$$f_1 \quad \text{on} \quad 0 < u < \infty$$

$$g_1 \quad \text{on} \quad 2n\pi < v < (2n + 1/2)\pi$$

$$f_2 \quad \text{on} \quad 0 < u < \infty$$

$$g_2 \quad \text{on} \quad 2n\pi < v < (2n + 1)\pi.$$

The result is that a pair of roots u, v is shown to exist, such that $2n\pi < v < (2n + 1/2)\pi$. Since n is arbitrary there are infinitely many roots.

3. GEODESIC FOCUSING AND THE MOSEVICH FUNCTIONAL EQUATION

Some of the examples of functional equations shown in the table on p. 330 may seem quite involved. However, the classification of that table is very incomplete, and in this section we give an example of an implicit nonlocal functional equation far more complex than those of the table of p. 330. Our aim is to instance mathematical modeling, not to outdo in functional complexity. The point here is that the functional equation discussed below arose from a simple and practical problem in radio engineering. The problem was: how to construct the parallel-plate equivalent of a parabola. Since this uses terminology which may be unfamiliar, we remark preliminarily that the problem asks for an equivalent of a parabolic reflector: to do for microwaves under certain simple conditions what the parabolic reflector does for visible light, namely to transform the radiation from a point source into a plane beam. Briefly: to focus. We shall begin with some elementary questions in what is known as geometrical optics, in order to build up to the statement of our problem and the formulation of our equation. However, some of those elementary questions may be of interest in their own right.

(**a**) The simple and well-known focusing properties of the conic sections are shown in Fig. 1. In Fig. 1a we have the internal reflection and point-to-point focusing in an ellipse. Figure 1b shows the external reflection and the so-called virtual focusing by the hyperbola. The type of focusing we want, point source to plane beam, is exhibited by the parabola in Fig. 1c. In each case the reflection is described by the usual law: The incident and

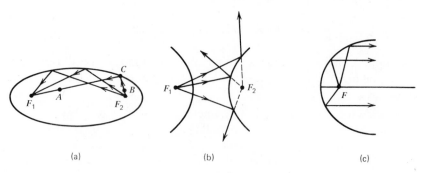

Fig. 1. Focusing and conic sections.

the reflected ray make equal angles with the normal at the point of incidence. Alternatively and equivalently, the reflection is described by a simple minimum property: for the ellipse, for instance, ACB of Fig. 1a is the shortest path from A to B which touches the ellipse. Since in each case in Fig. 1 the x-axis can act as the axis of revolution, we get all three types of focusing in three dimensions just by generating suitable surfaces of revolution. (One of our difficulties will be that our parallel-plate parabolic equivalent is *not* a system of revolution.) We describe next a different focusing arrangement, based on refraction rather than reflection.

The Luneburg lens [104] is a hypothetical variable-index device which performs its focusing as follows. First, let us define a radial lens to be the unit disk about the origin O in which there is a variable index of refraction $n(r)$ that is a function of the distance r from O only. We suppose that $n(r)$ is sufficiently smooth and sufficiently monotonic function of r. In particular, $n(1) = 1$ so that on its boundary our radial lens matches exactly the index 1 of the surrounding medium. Under these conditions we may ask whether the function $n(r)$ can be found so as to perform point-to-point focusing. That is, as shown in Fig. 2b, all rays from a point P_0 on the negative x-axis, which pass through the lens, will focus at a point P_1 on the positive x-axis. On account of rotational symmetry we could revolve our plane system about the x-axis, getting a three-dimensional spherical lens rather than a plane circular one.

The Luneburg lens is a special case of a radial lens in which

$$n(r) = \sqrt{2 - r^2} \qquad (1)$$

and the focus P_0 is on the rim of the lens boundary while P_1 recedes to infinity. In other words, it does just the focusing of point-source to plane-beam.

To prove it, we must first derive the condition that governs the motion of light rays in a medium with radially varying index of refraction. In analogy to it, we have the condition (3), p. 95 which describes the optical path in the plane when the index of refraction is a function of y alone. Here, the condition we want is more complex but it is not hard to get at, by a suitable reduction of our new situation to one we have met. Namely, we observe that, optically speaking, our plane is not really the geometrical Euclidean flat, but a surface of revolution. Now, we might remember the Clairaut theorem, which states that a geodesic G on a surface of revolution cuts all parallel circles at the angle ψ so that we have: Along G, $r \sin \psi$ is constant. In our system we are interested in the *optical* distance and our ray paths are geodesics which minimize that optical distance. Therefore r must be replaced by $rn(r)$ and now we get

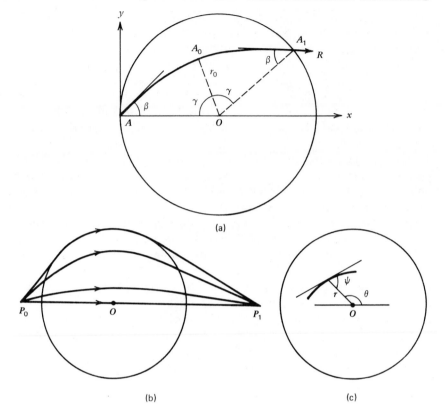

Fig. 2. The Luneburg lens.

our condition at once: In a radial lens the path of a ray of light is given by

$$rn(r) \sin \psi = c \qquad c = \text{const.} \qquad (2)$$

here r and ψ are as in Fig. 2c. Let us now prove the focusing property of the Luneburg lens. Taking $n(r) = \sqrt{2 - r^2}$ and referring to Fig. 2a, we let a ray R of light be projected from A at an angle β, $0 < \beta < \pi/2$. Since $n(r)$ is monotone it is not difficult to see that R gets closer and closer to O, attains a minimum distance r_0 at A_0 and then moves away from O again. The line OA_0 acts as the axis of symmetry for R, hence the angle at A_1 between the tangent to R and the radius vector OA_1 is β again. To find out just how R emerges, relative to the x-axis, we must determine the angle γ. Recalling that for the polar coordinates the angle ψ is given by

$\tan \psi = r/r'$, we get from (2)

$$d\phi = \frac{c\,dr}{r\sqrt{r^2(2-r^2)-c^2}}$$

and by integrating

$$\gamma = c \int_{r_0}^{1} \frac{dr}{r\sqrt{r^2(2-r^2)-c^2}}. \tag{3}$$

Putting $r = 1$ and then $r = r_0$ in (2) we relate the constants of our problem:

$$\sin \beta = c = r_0\sqrt{2-r_0^2}.$$

With these, the elementary integral in (3) is evaluated (for instance, by substituting $r^2 = 1/v$ and then completing the square) and we get

$$\gamma = \frac{\pi - \beta}{2}. \tag{4}$$

Therefore $\sphericalangle AOA_1 = 2\gamma = \pi - \beta$, which means that R emerges parallel to the x-axis. Since R was an arbitrary ray from A, we have proved the focusing property of the Luneburg lens.

Suppose now that the function $rn(r)$ has suitable monotone and boundary behavior, let r_0 be as before the distance of closest approach to the origin, and let R_0 and R_1 be the distances of P_0 and of P_1 to the origin. With some attention to the monotonicity and symmetry the reader may wish to show that the function $n(r)$ which performs the focusing of Fig. 1b, satisfies the Luneburg integral equation

$$\int_{r_0}^{1} \frac{c\,dr}{r\sqrt{r^2n^2(r)-c^2}} = \frac{1}{2}\left(\pi + \arcsin\frac{c}{R_1} + \arcsin\frac{c}{R_0} - 2\arcsin c\right). \tag{5}$$

In the special case of Luneburg lens $R_0 = 1$, $R_1 = \infty$ and $n(r) = \sqrt{2-r^2}$; (5) is then equivalent to (3) and (4).

(b) Before taking up a yet different method of focusing we consider what happens to a parallel beam incident on a curve that does not have the point-focus property of the parabola. The simplest nontrivial case is that of a circle. Consider a circle K centered at C and a concentric circle K_1 of radius half that of K. Further, let the circle K_2 whose radius is a quarter that of K roll without slipping on K_1 starting so that the initial tangency point was at I, where the x-axis cuts K_1. This tangency point, if tied rigidly to K_1 traces out the arc E of an epicycloid. Suppose now that the tangency point between K_2 and K_1 is at T as shown in Fig. 3a. Since

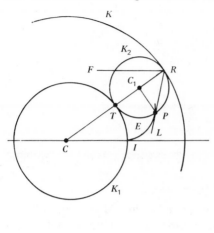

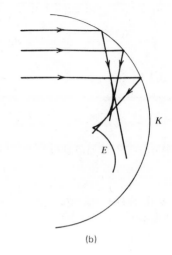

(a)

(b)

Fig. 3. Circle caustic.

K_1 is twice the size of K_2 we have

$$\sphericalangle ICT = 1/2 \sphericalangle PC_1 T = \sphericalangle PRT.$$

Also, $\sphericalangle FRT = \sphericalangle ICT$. Therefore the straight line CR, normal to K at R, bisects the angle $\sphericalangle FRP$. Also, the tangent L to the epicycloid E at P passes through R because T is the instantaneous center of rotation of K_2. It follows now that if an incoming bundle of rays, all parallel to the x-axis, gets internally reflected from the circle K, as in Fig. 3b, then the reflected rays do not focus to a point but they envelop a curve E, which plays the role of the focus of a parabola. This curve, as we have shown, is here an epicycloid of two cusps. Its generic name is the caustic; here, for instance, we say that the two-cusped epicycloid is the caustic of the circle. The name comes from Greek *kaustikos* = burning; it may be noted here that *focus* is a Latin word meaning hearth. With a certain empirical bent it is not hard to produce caustics of circles and other curves by reflecting natural or artificial light inside shining cans or tinfoil sheets.

Let us return now to a parabola with its ideal focusing of parallel rays to a point focus F. What happens if the rays, while remaining parallel, are now incident off-axially? Or, conversely, if the point-source of radiation at F, giving a parallel emergent beam, is now displaced off F? Before considering the answers let us examine the questions. Why displace the beam or the point source? For the passive case, where the parabolic mirror collects the incoming rays, as for instance in a telescope, we are interested in what happens to rays coming at an angle θ to the axis. If

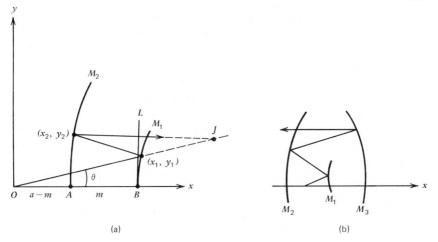

Fig. 4. The Schwarzschild system.

there is an image distortion growing fast with θ, this means that our instrument can photograph only a small piece of the sky. If we can find a way to control this distortion then the telescope will be useful over a bigger area. For the active case, when a point source of radiation sits at the focus F, we may wish to perform a scan, that is, sweep the parallel beam across an angle, back and forth. This could be accomplished by moving the whole apparatus, mirror and all. However, it is far preferable to keep the mirror stationary and just move back and forth the usually small and light point source.

Now, it happens that the parabola while perfect at focusing, is very bad at such scanning: If the point source is displaced from the focus the parallelism of the beam deteriorates very badly even for small displacements. Somewhat surprisingly, the circle, which does not have the ideal focusing, is better in this way. So, certain optical systems that have to perform at off-axis angles will sometimes use a spherical mirror with an added corrector device, rather than a parabolic mirror.

It is this off-axis error control that suggests a yet different method of focusing, developed by the astronomer K. Schwarzschild [155]. The point-source-to-parallel-beam is done here by reflections, but with two mirrors rather than one. The idea is that with two curves at our disposal we can achieve not only the perfect point focus as with parabola, but also a much smaller off-axis error. The Schwarzschild system is shown in Fig. 4a. There are two reflecting curves M_1 and M_2 (really, two surfaces of revolution about the x-axis), positioned as shown in the figure. A ray projected from O at an angle θ passes *through* M_2, gets reflected from

M_1, then from M_2, and emerges parallel to the x-axis, crossing M_1, as shown. An equivalent demand is that the optic path length along the ray, from O up to the point where the path cuts L, is constant for all rays:

$$(x_1^2 + y_1^2)^{1/2} + [(x_2 - x_1)^2 + (y_2 - y_1)^2]^{1/2} + a - x_2 = a + 2m, \qquad (6)$$

since $a + 2m$ is the length of the path $OBAB$. The other condition is still at our disposal, to ensure improved off-axis performance.

Formally, we could proceed as follows. Let the point-source be displaced from O by a small amount ε, positive or negative, along the y-axis. We then introduce the quantity Q, which measures how the beam gets distorted away from parallelism due to the shift ε. When the quantity Q is developed in a power series in ε symmetry considerations suggest that $Q(\varepsilon) = -Q(-\varepsilon)$; thus only odd powers are present in the expansion

$$Q(\varepsilon) = A_1 \varepsilon + A_3 \varepsilon^3 + A_5 \varepsilon^5 + \cdots. \qquad (7)$$

The coefficients $A_1, A_3, A_5, \ldots$, are called the (successive) aberrations. A_1 is the primary one and our focusing condition (6) makes $A_1 = 0$ (also for a parabola $A_1 = 0$; however, then A_3 is big). For good off-axis performance we should like to have $A_3 = 0$ as well. It is proved in the works on optics [21] that this is assured by the so-called Abbe sine condition: The point J of Fig. 4a lies on a circle about O. Now, taking this circle to be of radius 1 and introducing the parameter t by $t = \tan \theta/2$ we express the Abbe condition as

$$y_2 = \sin \theta = \frac{2t}{1 + t^2}. \qquad (8)$$

The definition of t gives us

$$\frac{y_1}{x_1} = \frac{2t}{1 - t^2} \qquad (9)$$

and equations (6), (8), (9) yield eventually the parametric representation of G. E. Roberts [144]:

$$x_1 = \frac{m(1 - t^2)}{t^2 + \dfrac{m}{a}\left(1 + \dfrac{m-1}{m}\right)^{1/(1-m)}}, \qquad y_1 = \frac{2mt}{t^2 + \dfrac{m}{a}\left(1 + \dfrac{m-1}{m}\right)^{1/(1-m)}},$$

$$x_2 = \frac{a\left(1 + \dfrac{m-1}{m}t^2\right)^{(2m-1)/(m-1)} + t^2/m}{(1 + t^2)^2} - m, \qquad y_2 = \frac{2t}{1 + t^2}.$$

Of course, we could also add a third mirror, as in Fig. 4b, and cancel with it the fifth-order aberration, making $A_1 = A_3 = A_5 = 0$ in (7), and so on.

(c)   We describe now parallel-plate systems as focusing devices. Such a system is the region between two metal plates that are not too sharply curved and, without being necessarily plane, are parallel. That is, every normal to either surface is also normal to the other one. It follows that the normal separation of the plates is constant. Let M be the mean surface consisting of all midpoints of those double normals. Mechanically, we notice that a ball of diameter equal to the plate separation could roll around between the plates keeping contact with both, and its center would then trace out M. Suppose now that microwave radiation is fed in between the plates. Then to a sufficiently good approximation we are in the domain of geometrical optics, instead of wave fronts we can speak of rays (which are normal to them) and these rays are geodesics of M. Such parallel-plate devices are important as optical systems for at least two good reasons. The first one arises from the difficulty of producing variable index of refraction, the second one is due to the self-obstruction of the Schwarzschild system.

Both difficulties are overcome in a parallel-plate system. Here we achieve the effect of a variable index of refraction by geodesic focusing, that is, designing the mean surface M so that its geodesics have the desired optical properties. Also, by connecting three flat pieces with two curved bends we obtain an equivalent of the Schwarzschild system without any self-obstruction. So long as we deal with axially symmetric systems no new mathematics appears. For instance, the geodesic equivalent of the Luneburg lens may be obtained by solving integral equations of the same type as the Luneburg integral equation (5). The situation is radically changed if we ask an even simpler, indeed basic, question: What is the parallel-plate equivalent of the parabola? The question is, how to generate the proper bend B, connecting two flat parts P_1 and P_2 in Fig. 5a, so that all geodesics starting from a point O in P_1 as straight rays, which move over B, emerge in P_2 as parallel straight lines. Thus we are asking for a yet another method of point-to-parallel beam focusing, and there does not appear to be anything in our simple and practical problem to suggest the complexity of the functional equation emerging from it.

Our question is not completely specified yet: We now add the requirement that the focusing bend B should be a part of a tubular surface T. Such a surface T is most easily described as the envelope of the family of spheres of equal radius whose centers are on a directrix curve D. We suppose, for regularity, that the radius of curvature of D is sufficiently large, as compared to the sphere radius. When D is a straight line then the tube T is a cylinder, when D is a circle T is a torus, and so on; the general case is sometimes called a canal surface as well as a tubular surface.

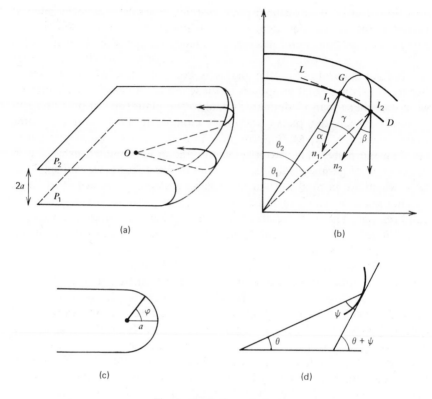

Fig. 5. Parabolic equivalent.

Why demand that our bend B should be a part of a tube, smoothly joining P_1 and P_2? The two dictating reasons unite here: Both for mathematical tractability and for ease of mechanical generation of parallel metal plates, tubular surfaces are best. The directrix D of our focusing bend is clearly a plane curve, exactly halfway between the two copies of it which form the rims of the flat parts P_1 and P_2 in Fig. 5a. Looking at the system from above we have the diagram of Fig. 5b. D is the directrix whose polar equation $r = r(\theta)$ we want, and the whole course G of a geodesic from O is shown. The normals at the points I_1 and I_2 to D are n_1 and n_2, γ is the angle between them; the separation between P_1 and P_2 is $2a$; for vertical positioning of a point on G we have the latitude angle ϕ, shown in Fig. 5c; this drawing is in a plane *normal* to D.

In terms of the angles of Fig. 5b the focusing condition we want is

$$\theta_1 + \gamma = \alpha + \beta. \tag{10}$$

We observe that when $a = 0$, then the whole of G shrinks to one point, I_1 and I_2 coincide, and so $\gamma = 0$ and $\alpha = \beta$; equation (10) defines then the parabola. The Mosevich functional equation is obtained by expressing γ, α, β as functions of θ_1 and functionals in $r = r(\theta)$, and substituting these expressions into (10). The result is a functional equation for $r(\theta)$ which, in principle, determines D.

There is no difficulty with α; exploiting the standard elements of polar coordinates we have

$$\tan \alpha = \frac{r'(\theta_1)}{r(\theta_1)}. \tag{11}$$

We shall obviously need the knowledge of the full course of the geodesic G over B; taking it as given by a relation $\theta = \theta(\phi)$, we apply standard classical differential geometry of surfaces [165], and get the nonlinear second-order differential equation

$$\theta'' = a^{-1}R^{-2}(r'^2 + r^2)(R + a \cos \phi) \sin \phi \theta'^3$$
$$+ \left[\frac{R'a \cos \phi}{R(R + a \cos \phi)} - \frac{rr' + r'r''}{r'^2 + r^2} \right] \theta'^2 + \frac{2a \sin \phi}{R + a \cos \phi} \theta', \tag{12}$$

where R is the radius of curvature of D in polar coordinates:

$$R = \frac{(r'^2 + r^2)^{3/2}}{(r^2 + 2r'^2 - rr'')}, \quad r' = \frac{dr}{d\theta}.$$

We must now connect the angles θ_2 and β with θ_1. For this, we solve (12) subject to two initial conditions that tell where and at what angle the geodesic G starts at I_1:

$$\theta\left(\frac{-\pi}{2}\right) = \theta_1, \quad \frac{d\theta(-\pi/2)}{d\phi} = (r'^2 + r^2)^{-1/2} a \tan \alpha \tag{13}$$

with $r = r(\theta_1)$, $r' = r'(\theta_1)$, and α as in (11). Solving (12) subject to (13), and in the solution $\theta(\phi)$ putting $\phi = \pi/2$, we get our expressions

$$\theta_2 = F(r(\theta), \theta_1), \quad \beta = H(r(\theta), \theta_1), \tag{14}$$

which are functions of θ_1 and functionals in $r = r(\theta)$. Finally, γ is obtained once we know θ_2; for the angle between normals is also the angle between tangents; using Fig. 5d we get

$$\gamma = \theta_2 + \psi_2 - \theta_1 - \psi_1, \quad \tan \psi_i = \frac{r(\theta_i)}{r'(\theta_i)}, \quad i = 1, 2. \tag{15}$$

When equations (11)–(15) are used to express γ, α, β as functions of θ_1 and functionals of $r(\theta)$, and substitutions are made into (10), we get the

Mosevich functional equation for $r(\theta)$. The reader may wonder perhaps what makes this equation so brutally involved, how does this equation relate to our classification of p. 330, and last but not least, how to go about solving it.

Concerning the first two points, we observe that the nonlocal nature of the Mosevich equation is worse than the simple nonlocalness of the various difference and integral equations of p. 330. We could perhaps describe it by saying that the Mosevich equation is "functionally-implicitly" nonlocal. The mathematical–optical reason for it is that only the surfaces of revolution and one other class, the Liouville surfaces [165], have mathematically tractable geodesics.

As to a practicable method of solving the Mosevich equation numeri-cally, this is based on the osculation principle, commonly used in geometry, analysis, and especially, in differential geometry. However, the beginnings of this principle we observe already in simple calculus. The idea is that we replace a complicated structure locally by a simpler one of best fit. So, the curve near a point is replaced by the tangent to the curve at that point; more generally and abstractly, a manifold is locally approxi-mated by its tangent space. If such lowest-order approximation is insuffi-cient, we set up a higher-order one. At the same time, we speak of corresponding higher-order properties. For instance, tangents to curves give us the first-order properties of directions, rates, and angles (of intersection of two curves). But for curvature properties they are insuffi-cient and we have to go to the higher-order osculation, getting local circle and local radius of curvature, and so on.

This technique is very well exemplified by one method of a numerical treatment of the Mosevich equation. We go back to Fig. 5b and the directrix D. If this curve is replaced at the entry point I_1 by its tangent L there, then L is the directrix of the simple tubular surface—the circular cylinder C of radius a. This circular cylinder C approximates locally to our curved bend B, and it has the great advantage of possessing geodesics that are easily tractable mathematically. However, the order of approxi-mation is too low. Hence, we approximate to D at I_1 by its circle of best fit—the curvature circle K of D at I_1. This circle K is the directrix of another tubular surface, the torus, which fits B better than C did. We call it the osculating torus of the tubular surface B. Now, the geodesics of a torus are not as easily handled as those of a cylinder. However, they are *explicitly* expressible, by elliptic functions, and so we shall avoid the implicit nonlocalities of the full Mosevich equation. With the use of osculating tori, we can set up a numerical scheme for a step-by-step determination of the directrix D. For further details and results, the reader is referred to the work of Mosevich, [128, 129].

4. THE HARTLINE–RATLIFF EQUATIONS AND PATTERN PERCEPTION

Schematic background to the nonlocalness of some functional equations is provided by looking at the systems described by those equations as though they were simple control devices. In certain temporal processes, for instance, we may monitor the input rate with a part of the output. To put it differently, a sample of the "after" is sent to the "before" to be mixed with it. This is called feedback and usually it leads to some sort of autonomous differential equations. If the system has lag then the input and the output are for different values of t and nonlocalness appears: difference–differential equations instead of ordinary differential ones. In this section we consider a spatial analog of the temporal feedback process; here we have spatial mixing in which samples of local output are sent to other places. This spatially nonlocal process is called lateral inhibition, and it arises from the anatomy and physiology of vision in a simple visual system. After a discussion of the biological background we introduce the Hartline–Ratliff equations, which express mathematically the lateral inhibition. It turns out that the spatial nonlocalness of the lateral inhibition gives rise to a certain type of pattern perception.

(a) A simple traditional picture of the activity of the nervous system is as follows. First, events take place in the environment and the corresponding disturbances reach the sense organ. There the information is coded into trains of discrete uniform impulses in the sensory nerve fibers. These transmit the information from the sense organ to a ganglion or a brain center where it is received, "decoded," compared perhaps with past experience or correlated with simultaneous reception from other senses, and then acted upon. That is, messages are sent out to effector organs (muscles, glands, vocal apparatus). From this point of view the role of sense organs is to detect, and that of sensory nerves is to dump the detection message into the right slot in the brain. The anatomical image which this picture evokes is that of a receptor mosaic of the sense organ; each receptor has its separate fiber and these fibers run together, forming the sensory nerve, but they remain separate, somewhat after the manner of distinct insulated strands of a cable. Aside from possible temporal effects we would be inclined to suppose that the equations describing the process ought to be local since there are no point-to-point interactions on the receptor mosaic.

In the case of vision (and for other senses too, most likely) this picture is incorrect in at least three important and intimately related respects:

(1) The nerve fibers from the sense receptors *are* in fact interconnected by a lateral network ("network" in Latin is *rete*; the diminutive form

is *retina*; hence this interconnecting network gives the name to an important part of the eye);
(2)   the processing of visual information begins in the eye itself, not in some higher brain center (such as the lateral geniculate body or the visual cortex);
(3)   the equations describing the process are nonlocal.

These three respects provide an anatomical–physiological–mathematical correlation that is perhaps better known for vision than for other senses but probably is no different there.

The receptor apparatus transforms light intensity incident on different visual elements into a pattern of varying frequencies: If the intensity of illumination received by a receptor goes up or down then so does the output frequency in the receptor's output fiber, while the amplitude of the response remains constant. Detailed researches on this process in the horshoe-crab *Limulus* have been carried out by Hartline, Ratliff, and their collaborators, [69, 70]. The eye of *Limulus* is compound and consists of about a thousand separate visual elements called the ommatidia. When the ith ommatidium alone is illuminated, with the light of intensity I_i, its output fiber fires off at some frequency e_i. When several ommatidia are illuminated and the illumination of the ith one is still I_i, it is found that the ith fiber fires off at some frequency x_i generally lower than e_i. That is, each ommatidium inhibits the activity of its neighbors by lowering their firing frequencies (provided that both it and the neighbors are active). Since that inhibition takes place at "right angles" to the "line of propagation" (which is receptor to fiber to brain), it is called *lateral inhibition*.

Empirically, the lateral inhibition is found to be approximately linear, and additive over different inhibitors but it does not set in below a certain threshold. All this is summed up in the Hartline–Ratliff equations

$$x_i = e_i - \sum_{\substack{j=1 \\ j \neq i}}^{n} k_{ij} \max (0, x_j - t_{ij}), \quad i = 1, 2, \ldots, n. \qquad (1)$$

Here all quantities are nonnegative, k_{ij} is the coefficient of inhibition of the jth element on the ith one while t_{ij} is the corresponding inhibition threshold. It is not always true that $k_{ij} = k_{ji}$ or $t_{ij} = t_{ji}$. As a rule, the further the ith and jth elements lie on the retina the higher t_{ij} and the lower k_{ij}; a rough order-of-magnitude guide is that $t_{ij}k_{ij}$ is nearly constant. To allow for the possibility that the sum term in (1) exceeds e_i and to prevent the consequent impossible negative frequencies, we should really have had instead of (1) the full system

$$x_i = \max\left[0, e_i - \sum_{\substack{j=1 \\ j \neq 1}}^{n} k_{ij} \max(0, x_j - t_{ij})\right], \quad i = 1, 2, \ldots, n.$$

However, there is a simple correspondence between the two systems and between their solutions, and we shall continue to use (1).

Would one be justified in calling system (1) nonlocal? To exhibit the matter more clearly we apply a familiar technique: Starting with a finite system we idealize it to an infinite one by a passage to the limit. This is done as follows. The domain of the discrete index $i(=1, 2, \ldots, n)$ is replaced by the two-dimensional plane or curved region R, corresponding to the retina; the variables u and v (which replace the indices i and j of (1)) will be vector variables ranging over R. Instead of the input e_i and output x_i we have now the known input density $e(u)$ and the unknown output density $x(u)$. The two-index quantities t_{ij} and k_{ij} are replaced by $T(u, v)$ and $K(u, v)$, which take over their functions. Like all the quantities in (1), all the preceding functions are nonnegative scalars. Now (1) generalizes naturally to the nonlocal nonlinear integral equation

$$x(u) = e(u) - \int_R K(u, v) \max[0, x(v) - T(u, v)] \, dv. \tag{2}$$

Recalling the title of the present chapter, we observe how a very simple and natural generalization of our model (1) has taken us to a rather nonobvious relation (2). With respect to the classification of the degrees of nonlocalness by means of kernels such as that of (2), the reader may wish to recall some terminology of nonlinear functional analysis. In particular, we have the Urysohn and the Hammerstein operators defined by:

$$Ux(u) = \int f(u, v, x(v)) \, dv,$$

$$Hx(u) = \int f(u, v) g(v, x(v)) \, dv.$$

We note that (2) contains a special case of Urysohn operator but that it is "too nonlocal" to be expressible with a Hammerstein operator. Equation (2) suggests that there may be classes of operators intermediate between the types U and H.

Both the discrete form (1) and the continuous form (2) could be said to describe a nonlocally interactive phenomenon. This nonlocal interaction exhibits in spatial terms what Picard and Volterra have called hereditary mechanics for time processes.

(b) We take up briefly the question of uniqueness of the solutions of equation (1) (their existence is guaranteed by a simple argument based on the intermediate value property). This uniqueness of solutions is a rather natural question since it guarantees that when the animal looks at one thing (i.e., with given e_i's) it sees one thing (unique x_i's). However, we consider the uniqueness partly to indicate some perhaps unexpected developments, principally in matters of convexity. In fact, we have already seen some of them: the apertures and inner apertures of vol. 1, p. 45 as well as the 2^n-rule, generalizing the *regula falsi*, of p. 122, arose precisely from the Hartline–Ratliff equations and their generalizations. While these are matters of no particular moment, they do illustrate in their small way the indebtedness of mathematics to other sciences, in this case to neurophysiology. Indeed, we may recall here a quotation from J. Fourier (1768–1830), the discoverer of Fourier series: "L'étude approfondie de la nature est la source la plus féconde des découverts mathématiques." (A thorough study of nature is the most fertile source of mathematical discovery.)

Returning to our question of the solutions of system (1), let us consider graphically the system for $n = 2$, as shown in Fig. 1. The equations are here

$$x_1 = e_1 - k_{12} \max (0, x_2 - t_{12})$$
$$x_2 = e_2 - k_{21} \max (0, x_1 - t_{21}). \tag{3}$$

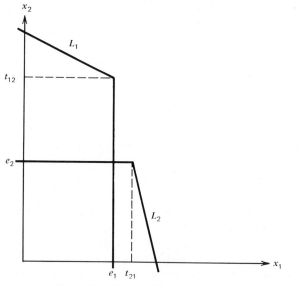

Fig. 1. Hartline–Ratliff system for $n = 2$.

We observe that the slopes of L_1 and L_2 in Fig. 1 are $-k_{21}$ and $-1/k_{12}$. Also, we refer for future use to the various coordinates noted on the axes in Fig. 1; in particular, we observe that as the parameters e_1, e_2, t_{12}, t_{21} vary the two graphs get translated, and conversely. Considering the graphs $x_2 = f(x_1)$ and $x_1 = g(x_2)$ of Fig. 1 we find three possibilities:

(a) If $k_{12}k_{21} < 1$ then (3) has only one solution, irrespective of the values of the other parameters;

(b) if $k_{12}k_{21} > 1$ then there is either one solution or three, depending on the e's and the t's;

(c) in the dividing case of $k_{12}k_{21} = 1$ there is either one solution or a whole continuum of them

The case illustrated in Fig. 1 corresponds to (b): The two piecewise linear graphs intersect in one point only, but a suitable translation of one of them will yield three solutions.

This leads us to the following problem: for what values of k_{ij}'s in (1) will there be a unique solution for x_i's no matter what the e_i's and t_{ij}'s may be?

The assumption of variability of e_i's is quite natural since these parameters describe the illumination incident upon the eye and so are variable within wide limits. The assumption of variability of t_{ij}'s is not natural since these are constant internal parameters of the eye itself. Hence we'll have unnecessarily stringent conditions on k_{ij}'s. However, in our formulation the problem has greater symmetry and possibly wider applicability.

We note that with both the e_i's and the t_{ij}'s variable, instead of (1) we could consider the system

$$x_i = -\sum_{\substack{j=1 \\ j \neq i}}^{n} k_{ij} \max(0, x_j), \qquad i = 1, 2, \ldots, n. \tag{4}$$

Thus our problem is that of translational uniqueness for intersection of convex surfaces in E^n. Indeed, let S_i be the surface in the n-dimensional Euclidean space E^n, defined by the ith equation of (4). The quantities

$$t_{1i}, t_{2i}, \ldots, t_{i-1i}, e_i, t_{i+1i}, \ldots, t_{ni}$$

are n positioning parameters, in fact they are merely the components of a vector such that the ith equation of (1) represents S_i translated by just that vector. In geometrical formulation our problem is then: Determine the conditions on the k_{ij}'s in (4) so that the n surfaces $S_1, \ldots, S_n$ should always intersect in one point, no matter how we translate them. An attempt to investigate system (1) or (4) for $n = 3$ shows that the direct inspection method which worked for $n = 2$ is not likely to work for $n \geq 3$.

To produce a workable indirect approach we note that the surfaces S_i are convex and they are boundaries of convex polyhedral cones C_i with the origin O as vertex. It is therefore natural, as is often the case with convex cones, to consider here the dual cones N_i. Roughly, N_i is the generalized spherical representation of S_i: the set of all directions normal to S_i. More precisely, we have

$$N_i = \{x : x \in E^n, x \cdot y \leq 0 \text{ for any } y \in C_i\}$$

where, as usual, we identify points in E^n and the corresponding vectors from the origin O.

It turns out [118, 100] that S_i's have translationally unique intersection if and only if no plane through O has interior points in common with all n cones N_i. Equivalently, any m interiors of N_i's should be separable from the interiors of the remaining $n - m$ N_i's by a plane through O. A proof is given in Lewis [100] of the following criterion for translational uniqueness. Let K be the $n \times n$ matrix with the diagonal elements 1 and off-diagonal elements k_{ij}. Let $\mathcal{T}_1, \mathcal{T}_2$ be two nonempty disjoint subsets of $\{1, 2, \ldots, n\}$ and define

$$K(\mathcal{T}_1, \mathcal{T}_2) = (a_{ij}),$$
$$= k_{ij} \quad \text{if} \quad i \in \mathcal{T}_1 \quad \text{and} \quad j \in \mathcal{T}_2 \text{ or vice versa,}$$
$$a_{ij} = 1 \quad \text{if} \quad i = j,$$
$$= 0 \quad \text{otherwise.}$$

Now the necessary and sufficient condition of Lewis for the uniqueness of solutions of (1) for arbitrary e_i's and t_{ij}'s is:

$$\det K(\mathcal{T}_1, \mathcal{T}_2) > 0 \quad \text{for every } \mathcal{T}_1, \mathcal{T}_2.$$

The reader may wish to produce a corresponding uniqueness criterion for the continuous analog (2) of (1).

(c) We shall examine in this subsection the possibilities of lateral inhibition as a pattern detecting mechanism. For a simple illustration we start with the problem of recognizing a plane convex p-gon P as such, $p \geq 3$, irrespective of size, shape, orientation and location. Imagine P to be projected upon the simple Limulus retina as a uniformly lit shape against dark background. We recall here that k_{ij} gets larger and t_{ij} smaller, the closer the ith and jth visual elements are situated on the retina.

Consider now a visual element lying well within P. It is active but so are all of its near neighbors; hence the total effective inhibition is high and the output x_i of our element will be low. Next, a receptor inside P and close to a side of the polygon P but not too near a vertex will also be inhibited but less strongly since only about half of its near neighbors are active and inhibiting. Finally, a receptor inside P and close to a vertex of it is least inhibited since here the least fraction of its neighbors will inhibit. Therefore, among all the output fibers we should observe p fibers, or perhaps p small groups of fibers, firing at maximum frequencies. The number of maxima corresponds exactly to the sole pattern invariant p of the polygon P, and by postulating a hypothetical local comparator to detect the maxima we get the number p. It may be of interest to see what happens when the convexity restriction is lifted. If P is allowed to be nonconvex, that is, re-entrant, while remaining simple, that is, not self-crossing, maxima will appear only at the receptors close to the p_0 vertices with interior angles $<180°$, and $p = p_0$ if and only if P is convex. A simple though artificial modification enables us to count p. We use the trick of reversal: After projecting P as a light pattern against dark background we project it as a dark pattern against light background. If P is convex there are no maxima under the dark image, and in the general case we shall have exactly $p - p_0$ maxima. Thus the combined number of maxima is the number we want: p. As an additional bonus we get here the possibility of automatic pattern detection of convexity of polygons. A more refined data-processing technique would not only count the numbers p_0 and $p - p_0$ of "light" and "dark" maxima but it would note also their relative positions and sizes. Hence we get the possibility of automatic perception of star-shaped polygons and of measures of roundess for polygons.

Further considerations of this kind indicate that lateral inhibition might provide a means of detecting local variations in the curvature of the light–dark interface. Briefly, it fastens onto the corners of visual patterns. Now, it is well known from the work of psychologists, that with any figures or patterns the points on their contours corresponding to maximum contour curvature are among the principal clues to pattern recognition; see, for example, Attneave [6]. An excellent example is the Attneave cat of Fig. 2. This was obtained by starting from an actual photograph of a sleeping cat, drawing the animal's contour, marking on that contour the 38 points of maximum curvature, then correctly joining these points by straight segments. It will be noted that here the pattern detection (i.e., recognition of "cat") is broken up into several parts. One of these, the first in the series, is the determination of points of maximum curvature; this is precisely the job to which lateral inhibition is well adapted. The second job is to connect those points *correctly*, that is, according to the proper connectivity, and already this is harder than the

Fig. 2. The Attneave cat. (Copyright 1954 by the American Psychological Association. Reprinted by permission.)

previous task. This second job must be followed by several more before the final aim is attained—identifying the figure as that of a sleeping cat. It is likely that several of these tasks are carried out fairly "high up" in the nervous system. However, the beginning of the series, the pre-processing of the visual information so to say, can very well begin in the eye itself, even as simple an eye as that of *Limulus*, with its lateral inhibition.

Another phenomenon explainable by the simple lateral inhibition mechanism is provided by the Mach bands [141]. These are narrow successively dark and light rings surrounding a bright seen object V; they turn out to be just the successive crests and dips of the inhibition–disinhibition wave. So, V is bright and therefore the receptors outside and immediately bordering it are strongly inhibited, giving us the first, dark, ring about V. The neighbors of *those* neighbors of V are therefore not inhibited strongly, hence their outputs are relatively high and so we get the second, bright, ring about V, and so on.

There is yet another way in which lateral inhibition could be used toward pattern perception. It is known from the work of Hubel and Wiesel [80] on the vision in cats that certain units in the visual cortex respond to moving spots of light in a manner which depends on the direction of the movement. For instance, a unit may show brisk response when a spot of light sweeps horizontally across some point on the retina. Other units may respond similarly to vertical sweeps or to sweeps at other angles. It is remarked in [80] that such responses must result from complex integrative mechanisms. We shall show here that similar nonisotropic response may occur in the static case, due to a suitable retinal lateral inhibition mechanism.

As was remarked before, the inhibition on a receptor depends principally on the activity of its near neighbors. It was tacitly assumed that the inhibition is isotropic: k_{ij} and t_{ij} depend only on the distance between the ith and the jth receptors but not on their orientation, one relative to the other. Let us now suppose that the ith receptor r_i has an associated inhibition field F_i which is not necessarily circular. For simplicity we might suppose that a receptor r_j can inhibit r_i if and only if it lies in F_i. This is achieved by putting $k_{ij} = 0$, or $t_{ij} = \infty$, whenever $r_j \notin F_i$. Suppose now that all inhibition fields are thin ellipses stretched out in, say, the horizontal direction. If a bright narrow segment L is projected on the retina so that its center falls on the receptor r_i, it will be found that the output x_i varies as L rotates about r_i: There is a maximum for L vertical and a minimum for L horizontal (i.e., coinciding with the direction of the inhibition field F_i). Assuming that the elliptical inhibition fields of various receptors, instead of being parallel, are oriented in different directions one gets a situation analogous to that in Hubel's experiments. Namely, we get an orientational preference in response from receptor to receptor. The origin of such differently oriented inhibition fields might perhaps be due to sharply varying local growth-gradients in the embryonic cells. As a result, certain cells might exhibit preferential directions of growth, leading to richer cellular interconnections in those directions.

The use of nonisotropic inhibition fields offers some possibilities for automatic pattern-detection. For instance, we may be interested in automatically correct lining up of printed or written text or of similar material. Since this exhibits strong directionality, suitably shaped and aligned inhibition fields might be considered for detecting "which way is up" or at least "which way is vertical." It will be observed that for an efficient detection of a pattern by means of a nonisotropic inhibition field we ought to make that field similar (in some sense) to the pattern to be detected. This might be of some possible application in biological laboratory procedures where hundreds of thousands of microscope photographs of cells have to be examined and one wants to pick out the few cells with some specific abnormality of shape.

The preceding material amply justifies a detailed mathematical study of the Hartline–Ratliff system (1) and, more especially, its continuous analog (2). If the functions K and T of (2) are such that unique solution $x(u)$ exists for every given $e(u)$, then we are interested in the properties of the solution operator B given by

$$x(u) = B(e(u)).$$

In particular, we are interested in what happens under B to inputs $e(u)$, which are characteristic functions of simple and sufficiently regular plane sets S. This corresponds to having those shapes S projected on the retina as uniformly lit against dark background. Some preliminary numerical work indicates that B is in some sense a de-averaging operator (this compares with the previously described Mach bands). From the biological point of view this makes excellent sense: The processing or pre-processing of the visual information by the eye should be directed to emphasizing contrasts, not to smoothing them out by an averaging process. The reader may wish to formulate this de-averaging more precisely, to find out what local curvature invariants of the light–dark interface play a role, and to determine whether the precise form of the kernel in (2) has some extremizing property from the point of view of information transforming and extraction.

(d) In the previous subsections we spoke repeatedly of patterns and pattern detection, abstracting, or perception. We consider now the problem of defining patterns. In a simple prototype version it is the problem of set membership of elements of disjoint unions: Given a collection $\{X_i : i \in I\}$ of pairwise disjoint sets X_i indexed by I, and given an element x of the union $\bigcup_{i \in I} X_i$, to find the unique value $i = i(x)$ of the index, such that $x \in X_i$. The index set I is the set of patterns and $i(x)$ is the pattern of x.

For instance, in one of our previous examples we have had $I = \{3, 4, \ldots\}$, with X_i as the set of all convex i-gons in the plane. In another example I might be the finite set consisting of the letters of the alphabet and the punctuation marks (including the blank) and $\bigcup_{i \in I} X_i$ could then be the collection of all printed marks occurring in a book. It is not hard to set up in the same way the problems of recognition of speech or hand script, or some remoter examples such as being able to assign to a previously unknown piece of art the label of African sculpture, Bach, or early Mozart (or late Haydn).

Several sources of difficulty arise at once from our simplistic formulation:

(a) lack of an operational assignment of x to its correct set X_i (ours above is purely existential in nature),

(b) stationarity of the problem, and

(c) the assumption of pairwise disjoint X_i's.

As to (a), we have had a sample of what is meant by an operational and effective assignment $x \in X_i$ in our simple problem where lateral inhibition was used for the assignment of a convex i-gon to its correct index i. In other cases such effective and operational assignment of x to X_i may be a complicated matter needing such tools as, for instance, statistical decision theory [57]. Under (b) we have mentioned the stationarity of the problem; the following is meant. Taking a dynamical and hierarchical point of view and aiming at something like a theory of learning, we might prefer a growing, or at least a changing, model. For instance, the sets X_i

may grow or shrink. Also, the elements of I could form new sets Y_j with a second-order index set $\mathcal{T}$, these would be patterns of patterns, and so on. Again, once the $x \in X_i$ assignment of x to its correct X_i is effectively instrumented we may suppose that the assignment method itself is variable and depends on certain "success" criteria. That is, some parameters governing the x-to-X_i assignment change, depending on past degree of success in correct assignment. Briefly, something like training or learning is aimed at, in the sense of modeling at least [146, 124].

Under (c) we had the disjointness of X_i's; this will be discussed at greater length. On the grounds of threshold phenomena, Heisenberg uncertainty principle, looseness of our judgement etc., the sets X_i may be not disjoint, or not exactly disjoint. The "etc." here stands, of course, for something like "nonrecognizability of arbitrarily small differences" or "indistinguishability of arbitrarily close objects." One might be interested in something statistical or, at any rate, intermediate between complete certainty and uncertainty: Not the surety that $x \in X_i$ or $x \notin X_i$ but perhaps a probability or a truth value intermediate between T and F; something like $f(x, i) = \mathrm{Prob}(x \in X_i)$. More radically, we may want something like a generalized topology (to be described) that would allow a comparison of the events $x \in A$ and $y \in B$.

One mathematically radical approach was suggested by the work of Menger, 1942 [120]. According to his own words he was guided by the following suggestion of Poincaré: In the actual physical world we meet triples, p, q, r such that

$$p = q, \qquad q = r, \qquad p \neq r. \tag{5}$$

Here the equality might be taken as physical identity or as operational indistinguishability. Extending (5) to longer chains we have n-tuples

$$p_1 = p_2, \qquad p_2 = p_3, \ldots, \qquad p_{n-1} = p_n, \qquad p_1 \neq p_n. \tag{6}$$

This has considerable bearing on the question of disjointness of the sets X_i. The connection is seen by looking at (6) from the point of view of "intermediate forms": p_1 might stand for a triangle, p_2 for a very slightly distorted triangle, p_3 for a slightly more distorted one and so on, finishing with p_n which might be a quadrilateral rather than a triangle.

Menger takes care of such threshold phenomena and measurement difficulties and transfers then into the underlying geometry, by introducing statistical metric spaces. In such a space there is associated with every two points p, q a function $f(x; p, q)$ of a real variable x such that $0 \leq f(x; p, q) \leq 1$. The intended interpretation is: $f(x; p, q)$ is the probability that the distance from p to q is $\leq x$. The axioms correspond to the usual metric space axioms:

$$f(0; p, p) = 1, \qquad f(0; p, q) < 1 \quad \text{if} \quad p \neq q; \qquad f(x; p, q) = f(x; q, p).$$

In addition we have an equivalent of the triangle axiom

$$T[f(x; p, q), f(y; q, r)] \leq f(x + y; p, r), \tag{7}$$

where $T(u, v)$ is a real-variable function defined on the unit square, such that

$$0 \leq T(u, v) = T(v, u) \leq T(1, 1) \leq 1, \qquad T(u, 1) > 0 \quad \text{if} \quad u > 0,$$

$T(u, v)$ nondecreasing in u and v. There are some difficulties with the triangle axiom (7); an alternative form has been suggested here (the convolution form of A. Wald). As with ordinary metric spaces it is possible to introduce into the statistical metric spaces the apparatus of geodesics, betweenness, convexity, and so on. Menger's work is continued by Schweitzer, Sklar, and Thorp [122, 156, 169], both as a general investigation and with specific reference to the problem of metrizing a statistical metric space S. This is an embedding of S into a larger, metric, space, with some preservation of the structure of S.

Menger's work, though never lost, has been often rediscovered. For instance, a 1962 paper [188] by E. C. Zeeman introduces "tolerance spaces" in order to investigate the "topology of brain." Those turn out to be just sets on which there is defined a symmetric and reflexive but not necessarily transitive relation. This is clearly related to the Poincaré–Menger chains (5) and (6). In a 1964 paper [187] the so-called "fuzzy sets" make their appearance. Here we have the underlying space X and for $A \subseteq X$ the set membership $x \in A$ is replaced by the map

$$f: X \times 2^X \rightarrow [0, 1]; \tag{8}$$

in particular, $f(x, A)$ is interpreted as "grade of membership" rather than probability, and hypothetical connections with multivalued logic are stressed. Such work on continuum-valued logics replacing the ordinary two-valued one has been done by Keisler and Chang [92]; the objects are known as logics with truth values in a Hausdorff space, or continuous models. Yet this aspect, too, is related to an early paper of Menger [121] on a probabilistic theory of relations.

What seems to be a possible object of study is something like a probabilistic analog of a Fréchet space (or pre-topological space). Such spaces are described in the book [159] of Sierpinski; our probabilistic analog is a set X with the function f of (8), subject to the sole "dominance axiom": $f(x, A) \leq f(x, B)$ if $A \subseteq B$. The object of such a study would be presumably a direct "statisticizing" or "smearing out." Instead of the "infinite precision" of usual mathematics, in which we postulate the ordinary real number system and then introduce random variables, distributions, and so on, we now wish to achieve the same effect directly, so to speak. Hence (8) still suffers from the shortcoming of taking for granted the ordinary real numbers with their "infinite precision"; this enters with the interval $[0, 1]$ in (8). For that reason one might perhaps replace (8) by a map

$$f: X \times 2^X \to A,$$

where A is a structure with something like partial order or uniformity. Presumably, one might study sequences of such maps in which the limit map gives the idealized relation of ordinary set membership:

$$F: X \times 2^X \to \{0, 1\}$$

or even the ordinary topology.

(e) The appearance in the foregoing of set theory in its foundational rather than finite and combinatorial aspect may be surprising, especially in a book of the type of ours. Yet the subject has wide and sometimes perhaps unsuspected relevance in applied and very concrete problems. We shall offer an illustration to finish this book with: Set theory and the logical status of usury.

For this purpose let us begin with one of the historically most influential paradoxes, the so-called Russell paradox. This is apparently due to Zermelo but it has been popularized by Russell and used by him to controvert an early axiomatic system of Gottlob Frege. The paradox is as follows. Let Z be the class of all, and only those, sets which are not members of themselves

$$Z = \{x: x \notin x\}. \tag{9}$$

Is Z a member of itself or not? If it is not then by putting $x = Z$ in (9) we find that it is; but if it is then by the very definition it cannot be. A somewhat vulgarized version is known as the barber paradox: in a certain place the barber shaves all, and only those, men who do not shave themselves; does the barber shave himself?

We run very briefly over a few other paradoxes. First, there is the ancient classic: "This statement is false,"—is it true or false? Simple analysis lands us in a circularity: If it is true then what it asserts is indeed the case, that is, it is false, and conversely. This is related to "X, who is a Cretan, says that all Cretans are liars,"—are all Cretans liars or not? Next, let us call an adjective *autological* if it could be said to apply to itself, and *heterological* otherwise. For instance, "short" is short and "English" is English, thus both are autological. But "long" is not long and "Finnish" is not Finnish, thus both are heterological. Now we ask: is "heterological" itself autological or heterological?

Limiting ourselves to positive integers we find that certain numbers can be defined by sentences or phrases without any symbols, just by English words. For instance, the phrase "least integer which is a sum of two cubes of integers in two different ways" uniquely defines the Ramanujan number

$$1729 = 1^3 + 12^3 = 9^3 + 10^3.$$

A simpler phrase is "seventeen hundred twenty nine." Since the number of sentences or phrases of exactly n words is finite, it is clear that there are numbers requiring arbitrarily long phrases. Consider now the phrase "smallest number not defined in English by less than twelve words." It defines a unique number, yet it uses itself only *eleven* words.

What do all those, as well as many other similar, paradoxes have in common? It strikes us at once that they share a direct self-referential quality; something is reflex in its action: It names itself, defines itself, contains (or does not contain) itself, even shaves itself (or not). The reader may wish to refer here to parts of Chapter 4 where we speak of Gödel's theorem and its relation to oblique or indirect self-reference.

Since we obviously wish to avoid paradoxes, at least professionally, we legislate them out of existence by various axiomatic procedures. In the Zermelo–Frankel set theory, for instance, we do not allow unbridled set formation: If $P(x)$ stands for a property or predicate applicable to sets then

$$\{x: P(x)\}$$

is forbidden. In its place we allow set formation with an upper bound: If u is a set then

$$\{x: x \in u \quad \text{and} \quad P(x)\}$$

is a legitimate set. Thus the Russell paradox is eliminated, not because we forbid $x \notin x$, but because we do not allow $\{x: P(x)\}$ where $P(x)$, in particular, could stand for $x \notin x$. If the Russell paradox is attempted in a rigorous manner with $Z = \{x: x \notin x\}$ replaced by

$$Z = \{x: x \in u \quad \text{and} \quad x \notin x\}$$

we only get the strange but not at all paradoxical conclusion that there is no universe. For, all that we can now infer is "for every u there is some x outside u."

The Zermelo–Frankel theory is homogeneous, in the sense of dealing with one kind of objects only: Sets. In the nonhomogeneous, or two-object, theories, such as those of von Neumann, Bernays, Gödel, and Mostowski, we have two kinds of entities: Sets and classes. The classes may even include the universal class, which contains *every set* as a member. But even though here there is a universe, unlike in the Zermelo–Frankel theory, it is a puny universe, and it works for sets only, not for classes. Indeed, the Russell paradox is avoided because, while the construction

$$\{x: P(x)\}, \qquad P(x) \text{ a property of } x,$$

is allowed, its result is a *class*, not a set. Further, classes are not members of anything: $x \in A$ for some A implies that x is a set. Thus here we can form the Russell class

$$Z = \{x: x \notin x\},$$

but since it is a proper class, that is, not a set, $Z \in Z$ makes no sense in these theories.

In the Russell–Whitehead theories of types, simple types, branched types, and so on, we escape the self-reference, or impredicativity as Russell prefers to call it, by a stratification procedure. All entities considered here are not of one kind as in Zermelo–Frankel theory, or of two kinds as in the von Neumann–Bernays–Gödel theory. They are divided into types: Those of type 0, of type 1,..., usually with a transfinite induction allowing for any ordinal number type. Now, it is decreed that a set of type $\alpha + 1$ must consist of entities whose type is α, and of no others. Hence self-reference in general, and the Russell paradox in particular, will be excluded here by legislating out the reflex action: The type stratification implies that nothing can contain itself.

It is perhaps unexpected that such matters should have any remote connection with political economics; with money, credit, interest, and so on; briefly, with unearned income; in one word: usury. Many passionate attacks have been made on usury, and they have been countered by many defenses, equally passionate. However, both were based on religious, ethical, or at least philosophico–psychological grounds. We shall now apply the axiomatics of set theory to indicate a possibility of a purely logical rather than morally prescriptive treatment.

Suppose that a definition of money is to be given. Taking goods, or rather commodities (to use a more inclusive term) for granted, we could say that one monetary unit is a collection of commodities which are equivalent or exchangeable: x pounds of salt, y pounds of iron, z minutes of a particular service etc. Thus the logical status of money, in its aspect of exchange, is that of a set of commodities. Yet, money is "put to work," used, it is desired, it is bought and sold. Briefly, it behaves like a commodity. We have therefore arrived at the classical self-referential situation characterizing paradoxes. Here money is both a commodity and a set of commodities.

Some difficulties could be listed, but they do not appear insurmountable. For instance, if money is an equivalence class of commodities rather than just a set of them, we may recall that relations are expressible by means of sets, and there appear slight technical complications only. It could be objected that a pound of salt here and now is not the same as a pound salt here and last year, or a pound of salt now but far away from here. This might be answered by taking the equivalence relation on larger sets, perhaps involving Cartesian products of commodities, space and time.

Since economics sometimes considers itself as referring to real life it might perhaps claim exemption from logical rigor. After all, even the paradoxes did not have much impact outside the foundations of mathematics. But what if the paradoxical two-level self-referential role of money turns out to be involved in crimes which start from countless murders and culminate in planetary exhaustion by uncontrolled growth?

APPENDIX

Here we collect some "principles" that have appeared in vols. 1 and 2, explicitly or implicitly. Without taking them too seriously, we may use them at our convenience, either for help in understanding or as a guide in problem solving and formulation. Some of our illustrations are new but mostly they draw upon previous material. Also, in order to exhibit the operation of some of those principles we make references that go perhaps beyond the level of the book itself.

1. COMPLEXITY STRATIFICATION

This is a principle of wide spread and deep applicability. The idea is, roughly, when dealing with objects of a certain kind, to stratify them into layers L_0, L_1, L_2, ... which are disjoint and exhaustive, on the basis of increasing complexity. Here the simplest, or initial, objects make up L_0, those next to them in simplicity, or perhaps those generated out of them, make up L_1, and so on. Let F be a property that may or may not hold for our objects. Then we note the following "complexity schema" for proving that no object satisfies F. If such an object x existed then it would belong to some L_a where a is smallest possible; however, the nature of the property F may then force x to belong to an even simpler layer L_{a-1}.

This is the method of Abel's proof of the impossibility of solving the general quintic equation by radicals. The proof goes like this: Let a root of the quintic be expressible by radicals and let it be expressed as a function (of the equation-coefficients), which is most economical in the use of radicals. Equivalently, consider the root as a function of the coefficients and let its algebraic complexity be lowest possible. Then it turns out that the root in question has complexity even lower than its hypothetical minimum.

The reader may wish to refer here to vol. 1, pp. 205–212, for a brief discussion of complexity stratification, and for relations of Liouville's work on functional complexity to Abel's proof. There the algebraic functions form the simplest layer L_0 and each next layer L_{a+1} consists of functions that depend algebraically on a finite number of exponentials and logarithms of functions in L_a. The totality of functions of all levels L_a forms the class E of elementary functions. Now, the complexity schema is used by Liouville to derive necessary and sufficient conditions for an algebraic function, that is, one in L_0, to have an integral which is elementary, that is, is a function in E. Similarly, Liouville proves that certain differential equations (for instance, the Bessel equation) have no elementary solutions, except when the coefficients of these differential equations have certain special values. Related ideas occur in the work of Ritt [143] and Siegel [158] on the transcendency of values of the solutions of linear differential equations.

We consider next the Hölder theorem on the Γ-function [78a] : $y = \Gamma(x)$ satisfies no polynomial differential equation

$$P(x, y, y', \ldots, y^{(n)}) = 0.$$

It must be observed that this theorem goes back to the pre-Cantorian days when there was no agreement as to what constituted a function. Partly under the impact of Fourier series, it was debated whether

$$f(x) = \begin{matrix} x^2 & x \geq 0, \\ 0 & x \leq 0, \end{matrix}$$

was one function or two. Solutions of differential equations were admitted as bona fide functions; hence the background of Hölder's theorem, and its role in enlarging our notion of function. Now, every proof of Hölder's theorem proceeds by a direct application of the complexity schema. Thus, if there exists such a polynomial differential equation for $\Gamma(x)$ then there must exist a simplest one. But then, the basic recurrence relation $x\Gamma(x) = \Gamma(x+1)$ forces a further simplification of even that "simplest" equation. Here the complexity refers in a simple way to the "size," degree, "weight," and so on, of the polynomial.

The foregoing examples have in common a certain implicitness: In their usual form they do not refer to complexity, complexity scale, and complexity measures by name. The case is quite different with Gödel's result [60] on the consistency of the G.C.H. (= generalized continuum hypothesis). Gödel's theorem is not so much a consistency result as an equi-consistency result. It states that if a set theory S without the G.C.H. is consistent then it remains consistent when the G.C.H. is adjoined as an additional axiom to S. The proof of Gödel's theorem is by an appeal to model theory: If a theory has a model then it is consistent. It is therefore necessary to form a model for $S+$G.C.H. *within S.* This is done much as in geometry when we prove, for instance, that the hyperbolic geometry is consistent relative to the Euclidean one. This is accomplished by showing that with suitable interpretation of lines, points, and so on, the interior of a circle *in the Euclidean plane* is a model for the hyperbolic geometry.

With set theory we perform a similar restriction of the "large" universe that is, the class of all sets (corresponding to the Euclidean plane) to a "small" universe (corresponding to the interior of the circle as above). This is done in detail precisely by applying a complexity measure. But unlike in the other examples where we had the complexity layers L_0, L_1, $L_2, \ldots$, with $\omega = \{0, 1, 2, \ldots\}$ as the indexing set, here we continue the complexity stratification into the transfinite domain and introduce layers L_a for every ordinal number a. In more detail, we call a set X constructible over the set Y if X is obtained by correctly framed formulas of the theory S, in which all bound variables are restricted to Y. We also put

$$X' = X \cup \text{ all sets constructible over } X$$

and we apply the usual transfinite induction by iterating the priming operation for nonlimit ordinals, and by collecting together the results of all previous stages at the limit ordinals. Thus we get the scale of constructible sets, where L_0 is the empty set and

$$L_a = (L_{a-1})' \qquad a \text{ not a limit ordinal,}$$

$$L_a = \bigcup_{b<a} L_b \qquad a \text{ is a limit ordinal.}$$

Strictly speaking, the above is an *inclusive* scale with each L_a containing all the previous ones. An *exclusive* scale is obtained by replacing L_a with $L_a - L_{a-1}$. The class L which is the union of all L_a's is called the class of constructible sets. A longer name that would be perhaps more descriptive is: the class of all sets of ordinally classifiable constructional complexity. This L is now taken as the "small" universe and we get our desired model: *Within L* the G.C.H. turns out to be a *provable theorem.* Why? The detailed answer, of course, is not easy to give and a certain description of the G.C.H. itself is necessary. What is it about? It concerns infinite sets and in particular, it concerns the comparison of the two canonical ways of producing larger sets out of smaller ones. So, given an infinite set X we have the cardinal increase:

X is replaced by its power set 2^X, that is, by the collection of all subsets of X,

and also the ordinal increase:

X is replaced by the set $H(X)$ of all ordinals of the same cardinality as X.

The ordinal increase turns out to be the slowest possible one: We have

$$\text{if} \quad |X| = \aleph_a \quad \text{then} \quad |H(X)| = \aleph_{a+1}.$$

Thus certainly

$$|2^X| \geq |H(X)|,$$

and the G.C.H. is just the simplest possible statement on the relative position or size of the cardinal and ordinal increases; it states that they are equal:

(G.C.H.) $$\qquad\qquad |2^X| = |H(X)|.$$

Now, we could say that *inside* the class L the G.C.H. is a provable theorem because sufficient hold is obtained in L on the power sets 2^X—the indefiniteness of the phrase "all subsets of (an infinite set) X" has been adequately muzzled by means of the constructivity.

2. CONJUGACY

This simple but pervasive and profound principle may be schematized by the following image: To cross an impassable wall we sink down a shaft S, then we make a horizontal tunnel T, finally we dig up the reverse shaft S^{-1}; the whole trip is schematized by $A = STS^{-1}$. The relation of A and T is called conjugacy and it is an equivalence relation. The range of applicability is enormous. In abstract algebra it is the basis of conjugacy classes and hence a cornerstone of group theory. In linear algebra it expresses the similarity of matrices. In analysis it is the very basis of solving functional equations by means of integral transforms. Also, it appears in an essential, if transparent, way in iteration (vol. 1, pp. 54, 59, 75).

Perhaps one of its most important roles is in suggesting approaches to the central "extension problem"—how to extend an operation originally defined for integers to a larger set, for instance, to the set of all real numbers. An elementary example is supplied by the power x^a of a number x. This is defined in the obvious way when a is an integer. To extend powers to noninteger values of a, we introduce exponentials and logarithms and put

$$x^a = e^{a \log x},$$

which is

$$x^a = F(aF^{-1}(x)), \qquad \text{where} \qquad F(u) = e^u.$$

We have in effect conjugated raising to a power to the simpler operation of just multiplication.

This is quite similar to the iteration paradigm: If O is conjugate to T so that $O = STS^{-1}$, and if further T happens to be easily iterable, to any real order a, then so is O and we have $O_a = ST_aS^{-1}$. So, for instance, one of the solutions of the functional equation

$$f(f(x)) = 2x^2 - 1$$

is $f(x) = \cos \sqrt{2} \text{ arc cos } x$. We see this by putting $2x^2 - 1 = g(x)$ and observing that

$$\cos 2x = g(\cos x) \quad \text{or} \quad g(x) = \cos 2 \cos^{-1} x$$

which establishes the conjugacy of $g(x)$ and the simply and easily iterable function $y = 2x$. Hence the half-order iterate as solution: $\cos \sqrt{2} \text{ arc cos } x$. We recall that on a deeper level a similar use is made of conjugacy in the solution of Hilbert's fifth problem (vol. 1, pp. 61–62). It will be noted that in both these problems, as well as in the matter of extending the factorial $n!$ to noninteger values, we run into the uniqueness problem: the question of unique extension arises, apart from the question of existence of extensions. This is illustrated by the multiplicity of square or other roots, of fractional order iterates, of would-be Γ-functions, and so on (vol. 1, pp. 59–63).

Generally, we might say that conjugacy enables us to reduce doing something new to doing something we already know. A particularly illustrative example is from Section 1 of Chapter 2, where we extend the notion of a geometrical mean from a finite set of positive numbers to a positive function defined over an interval $[a, b]$. First, we observe that there is no difficulty in defining the arithmetic mean for the continuous as well as the discrete case:

$$A(f) = \frac{1}{b-a} \int_a^b f(x) \, dx.$$

Next, we have the conjugacy of arithmetic and geometric means under logarithms:

$$\text{G.M.}(x_1, x_2, \ldots, x_n) = F^{-1}(\text{A.M.}(F(x_1), F(x_2), \ldots, F(x_n))),$$

where $F(x) = \log x$. Therefore we extend the geometric mean to the continuous case and define

$$G(f) = \exp A(\log f).$$

Similar use of conjugacy occurs in vol. 1, p. 214, in the derivation of the Faa di Bruno formula for the nth derivative of a compound function $f(g(x))$. An entirely practical use is made of conjugation in the Cauer method of filter design, in the last part of Section 1 of Chapter 5. At the other end of the spectrum we have the theory of categories where the conjugacy principle would appear to be connected with the so-called adjoint functors.

3. JACOBI INVERSION

The best statement of this principle is the original German quotation from Carl Gustav Jacob Jacobi (1804–1851): "*Man muss immer umkehren*"—one must always invert. We note at once a connection with the previous principle. What Jacobi meant was most likely the inversion problem for elliptic integrals which may be summarized thus: While the elliptic integrals are what we meet first and naturally, it is the inverse object—the elliptic functions—that is more natural. This is illustrated by the elementary example of

$$x = \int_0^y \frac{dy}{\sqrt{1-y^2}} = \text{arc sin } y, \text{ as against } y = \sin x.$$

However, the usefulness of the Jacobi principle extends far down, sideways and (perhaps) up. On the more elementary level we meet the inversion principle in the telescoping cancellation for series and products. That is, in order to *add*, in particular to add a_j with j ranging over a finite or infinite set, we invert and represent a_j as the *difference* $b_{j+1} - b_j$. Then the telescoping cancellation gives us

$$a_1 + a_2 + \cdots = b_2 - b_1 + b_3 - b_2 + \cdots = b - b_1,$$

where b is either the last b_j or, in the infinite case, $\lim b_j$. Similarly, we work with products, replacing sums by products and their inverse, the differences, by quotients. So, for instance, we can recognize at once the *feasibility* of expressing the product

$$\prod_{n=1}^{\infty} \frac{(n-a_1) \cdots (n-a_k)}{(n-b_1) \cdots (n-b_k)}$$

as a finite product of Γ-functions (see vol. 1, p. 101) precisely because the basic recursion $\Gamma(x+1) = x\Gamma(x)$ suggests the correct multiplicative cancellation in the above infinite product. For refinements, related techniques, and illustrative examples the reader may wish to refer to vol. 1, pp. 81–95. What we have called the device of telescoping mirroring is a related technique often useful in deriving combinatorial and other identities; this is described and illustrated in the section on Gaussian binomial coefficients and other q-analogs in Chapter 3.

From the point of view of general usefulness probably the most important example of inversion is provided by (the principal theorem of) calculus. Here our object is to integrate $f(x)$; while this operation is defined by a certain limiting procedure, this is not the way we usually obtain the integral

$$F = \int f(x)\, dx.$$

What we do is precisely to invert: Given f we look for its antiderivative, that is, the function whose derivative is f. In other words, we have set out to integrate and we have passed to the inverse operation of differentiating. For this and for the close parallel between the role of logarithms in integration and the role of the digamma function in summation the reader is referred to vol. 1, p. 89.

Finally, we recognize at once the operation of Jacobi's inversion principle in the classical and even more in the generalized Möbius inversion (cf. the relevant parts of Chapter 3).

4. STRUCTURE

Here we shall note and discuss several more or less related principles of quite different ranks. The first two are procedural or problem-solving devices.

(a) Structuration. This is meant to apply to cases where there is not enough structure in the problem as it is originally given or stated, and the solution or treatment depends on bringing in successfully further structure. This may often take the form of introducing extra variables or functions. For instance, to evaluate

$$\sum_{k=0}^{n} \binom{n}{k} k^2 \qquad \text{or} \qquad \prod_{n=1}^{\infty} (1 - 2^{-n})$$

we introduce the expressions

$$\sum_{k=0}^{n} \binom{n}{k} x^k \qquad \text{or} \qquad \prod_{n=1}^{\infty} (1 - x^n)$$

and then proceed as was done in vol. 1, pp. 95–96, 220. Similar use is noted in proving the Euler theorem on the function $f(x_1, \ldots, x_n)$ homogeneous of degree a:

$$\sum_{j=1}^{n} x_j \frac{\partial f}{\partial x_j} = af$$

or the Taylor theorem for functions of several (say, two) variables:

$$f(x+h, y+k) = f(x, y) + hf_x + kf_y$$
$$+ \frac{1}{2!}(h^2 f_{xx} + 2hk f_{xy} + k^2 f_{yy}) + \cdots.$$

We get simultaneous structuration and reduction by introducing an extra variable t and operating on our functions as though they were dependent on t alone. In particular, we introduce

$$F(t) = f(tx_1, \ldots, tx_n) = t^a f(x_1, \ldots, x_n)$$

in one case and

$$F(t) = f(x + ht, y + kt)$$

in the other case. This achieves a reduction from several variables to one, by the somewhat paradoxical device of bringing in one more variable; we might call it the structuration variable. In the first case we differentiate with respect to t and then put $t = 1$; a slight extension of Euler's theorem suggests itself naturally; it is to differentiate several times and then put $t = 1$. In the second case we apply the McLaurin expansion to $F(t)$ and then put $t = 1$.

Very similar structuration occurs in the process of expanding products in series; for instance, the Euler expansions

$$\prod_{n=0}^{\infty}(1 + x^{2n+1}) = 1 + \sum_{n=1}^{\infty} x^{n^2} \bigg/ \prod_{j=1}^{n}(1 - x^{2j}),$$
$$\prod_{n=1}^{\infty}(1 + x^{2n}) = 1 + \sum_{n=1}^{\infty} x^{n(n+1)} \bigg/ \prod_{j=1}^{n}(1 - x^{2j}),$$

can be obtained in this way. We introduce the Euler parameter t (indeed, the device could almost be named after L. Euler) and consider

$$F(t) = \prod_{n=0}^{\infty}(1 + tx^{2n+1});$$

telescoping mirroring derives for us the functional equation

$$F(t) = (1 + tx)F(tx^2)$$

and this enables us to determine recursively the coefficients in the power series for $F(t)$. Then we put $t = 1$ and $t = x$ and we have our desired expansions.

In the discrete domain we use structuration in problems on counting where, for instance, the number $f(n)$ of ways of doing something, relative to a set of n elements, is desired. To enable us to take a hold we bring in additional structure and complicate our problem, for instance, by introducing the function $f(n, k)$ such that

$$\sum_k f(n, k) = f(n).$$

Usually $f(n, k)$ will be the number of ways of doing our thing relative to our n-element set so that an additional condition depending on k holds. As was shown on the example of counting the zigzag permutations in vol. 1, p. 104, the structuring parameter may enable us to perform a successful recursion on n which then leads to the solution of the counting problem. The previously mentioned Euler device of introducing an additional parameter in the expansion of a product in series has the same effect: successful recursion.

We have met the same device—of successful recursion brought about by suitable structuring—with the Ramsey theorem of Section 4, Chapter 3. The simplest way to state this theorem is in terms of graph theory and with one sole parameter: For every n there is a finite $f(n)$ such that for any graph G on n vertices either G or the complementary graph $\bar{G}$ contains a complete n-graph. However, with just one parameter there is too much parsimony and we do not get sufficient structure for an inductive hold on the problem. We therefore use the less symmetric form of stating Ramsey's theorem, with several parameters, and sufficient structure for successful induction.

A different example of structuration was given in vol. 1, p. 156 as an illustration of the method of composite estimates. We wished there to get a possibly low estimate on the number d such that every plane set P of diameter 1 is a union of three sets of diameters $\leq d$. As was shown there, P can be assumed to be of constant width 1. We introduce a continuous parameter $x = x(P)$, which is the diameter of the largest equilateral triangle inscriptible into P. This x is the structuring parameter that enables us to perform the double estimate thus. The number d is first shown to have a good lower estimate $L(x)$. That is, $d \leq L(x)$ where $L(x)$ increases with increasing x. Then, a good upper estimate $U(x)$ is developed; here "upper" refers to the fact that $d \leq U(x)$ where $U(x)$ is decreasing when x increases. Hence $d \leq x_0$ where x_0 is the sole root of the equation $L(x) = U(x)$.

(b) Specialization. This is meant to apply to the opposite sort of cases: Here we have, so to say, "de-structuring." Simply put, the problem has too much structure and we proceed by specializing: The general case is concluded from a selected special case. A fairly typical example is a plane geometrical problem that can be simplified by suitable projections. For instance, we have the Routh theorem of vol. 1, pp. 7–8, referring to the ratio of areas of a triangle T and another triangle inscribed into T. It is possible to prove Routh's theorem as in [37] by introducing some new concepts, but we can also prove it more simply. For, our ratio of areas is unchanged by orthogonal projection on another plane. Hence we can assume, without loss of generality, the "simplified structure" of having an isosceles right-angled triangle T. In this special case we have a simple direct proof. Similar use of projections will sometimes simplify plane problems dealing with ellipses: We project the configuration so that the ellipse projects onto a circle.

An even simpler collapse of structure is provided by the first step we take in evaluating such integrals as

$$\int_{-\infty}^{\infty} e^{-ax^2} \cos bx \, dx;$$

here we use a substitution and reduce the number of parameters from two to one. A slightly more sophisticated example of the same kind occurs in the following evaluation of the integral

$$I = \int_0^{\infty} \frac{\alpha y^2 + \beta y + \gamma}{Ay^4 + By^2 + C} \, dy,$$

where A and C are positive and $4AC > B^2$. We have first

$$I = \alpha K_2 + \beta K_1 + \gamma K_0$$

putting

$$K_n = K_n(A, B, C) = \int_0^{\infty} \frac{y^n \, dy}{Ay^4 + By^2 + C}, \qquad n = 0, 1, 2.$$

The integral K_1 is simply evaluated by the substitution $y^2 = x$. Also, the substitution $y = 1/x$ shows the symmetry

$$K_2(A, B, C) = K_0(C, B, A).$$

To find K_2 fastest we employ the Schlömilch formula of vol. 1, p. 200:

$$\int_0^{\infty} f\left(cx - \frac{a}{x}\right)^2 dx = \frac{1}{c} \int_0^{\infty} f(y^2) \, dy.$$

Taking $f(u) = (u + q^2)^{-1}$, with suitable q, a, and c, we get K_2. Then, collecting various terms we have

$$I = \frac{\pi}{2} \left[\left(\frac{\alpha}{\sqrt{A}} + \frac{\gamma}{\sqrt{C}} \right) U + \beta V \right] - \beta V \arctan BV,$$

with the abbreviations

$$U = (B + 2\sqrt{AC})^{-1/2}, \qquad V = (4AC - B^2)^{-1/2}.$$

Another use of specialization was made in Section 1 of Chapter 6, in vol. 1. There the identities of Leibniz and of Faa di Bruno were proved for arbitrary functions by means of specializing to the exponential function. For example, Leibniz formula for the nth derivative of a product

$$[f(x)g(x)]^{(n)} = \sum_{j=0}^{n} \binom{n}{j} f^{(j)}(x) g^{(n-j)}(x)$$

was proved by first showing that

$$(fg)^{(n)} = \sum_{j=0}^{n} A_{nj} f^{(j)} g^{(n-j)},$$

where A_{nj} are constants independent of f and g. Then we specialized by choosing $f(x) = e^{ax}$ and $g(x) = e^{bx}$ so that

$$(a+b)^n = \sum A_{nj} a^j b^{n-j}.$$

Perhaps the most characteristic example of the specialization principle is the Schubert theorem of enumerative geometry (Section 8 of Chapter 1). This deals with hyperplanes which satisfy prescribed intersection conditions with certain given hyperplanes. For instance, we ask about the number of lines in the three-dimensional Euclidean space E^3 which intersect four given lines L_1, L_2, L_3, and L_4. The essence of Schubert's theorem is just specialization: The theorem asserts that we can specialize the given hyperplanes in any way provided that the intersection number does not become thereby infinite. Here, for instance, we choose the four lines in two intersecting pairs, and we find then easily that exactly two lines intersect all four L_i's. The specialization contained in Schubert's theorem exploits effectively the simple principle that an integer-valued continuous function is a constant. The same thing was used in Section 6 of Chapter 2 to prove Rouché's theorem.

(c) A variant of the structuring and of specialization is obtained when we use natural or specially introduced symmetry. This was illustrated on a number of examples in vol. 1, pp. 26–34 and 145–148.

(d) Structure Interaction. Unlike perhaps the previous cases, this principle is one of the mathematical "heavies." We refer here to the effects when two different structures on one set reinforce each other, or otherwise interact in more or less unobvious ways. Two rather extreme cases of such structure reinforcement were mentioned in vol. 1, p. 62. One of them was the simple example concerning the fundamentals of complex variables and analytic functions: If $f(z)$ is differentiable then it is analytic. Here the interaction is between the limits implicit in differentiability, and the topology of the plane. Namely, the limit

$$\lim_{z \to z_0} \frac{f(z) - f(z)}{z - z_0}$$

must exist no matter how z approaches z_0. On a line, where the topology is essentially determined by the order, there are just two ways for such approach, from the right and from the left. In the plane the topology is incomparably richer and therefore the requirement of a unique limit under all modes of approach has much profounder implications.

At the other extreme we had one of the achievements of modern mathematics, the proof of the conjecture of Hilbert's fifth problem: that a locally Euclidean topological group is a Lie group. Here the algebraic and the topological structures of the group interact so strongly that the minimal smoothness, viz. continuity, of the coordinate functions implies their maximal smoothness, viz. analyticity.

It is a platitude to ascribe the immense difficulties of such problems as the existence of infinitely many twin primes, the Goldbach conjecture, or the existence of infinitely many primes of the form $P(x)$—where P is any polynomial of degree at least two, with integer coefficients—to the interaction of the additivity and multiplicativity structures of the integers. For, in each case we have a problem dealing with a sum, yet the summands are primes defined multiplicatively. The same interaction appears in a different form in the incompleteness theorem of Gödel, especially when we contrast it with the previously known result of Lindenbaum that if only the addition is imposed as the sole structure on integers then the system is complete: Every theorem that is true is also provable. With reference to the crucial device of Gödel encoding, and to our remarks on oblique self-reference at the end of Section 6 in Chapter 4, we observe that it is precisely the existence of both operations, addition and multiplication, that enables us to perform the Gödel encoding. This in turn leads to the "obliquity" of self-reference: the key proposition in Gödel's proof does not state "I am unprovable" directly but only via the Gödel encoding.

Finally we note the great difficulty of the generalized continuum hypothesis: The theorems of Gödel [60] and Cohen [35] show that no decent known axiom system suffices to prove either its truth or its falsity. As was remarked in the first section of this Appendix, a convenient way of expressing the generalized continuum hypothesis is that it is the simplest possible statement relating the "ordinal" and the "cardinal" ways of producing out of an infinite set a bigger set. So, it may be said that here, too, the difficulty is due to the interaction of two different structures, the ordinal one and the cardinal one.

5. COMPUTING ONE THING IN TWO DIFFERENT WAYS

This principle has a wide coverage; one is sometimes tempted to consider it as almost characterizing the combinatorial ingredient in mathematics. However, it is useful in cases involving the continuous domain as well. Examples are provided by the use of Fubini's theorem to evaluate integrals by interchanging the order of integrations in a double integral (vol. 1, pp. 175–179) or employing the Cavalieri principle (vol. 1, p. 47; end of Section 2 in Chapter 1) to compute volumes. Somewhat more remotely, the same idea is involved in computing integrals by the right choice of volume element, as in Section 1 of Chapter 5, especially with the Catalan method. In geometry the principle of computing one thing in two different ways is at the very basis of Max Dehn's solution of the Hilbert problem on inequivalent pyramids (vol. 1, p. 16). If we substitute "composing" for "computing" then our principle is also seen to be behind such pathologies as the so-called Banach–Tarski paradoxical decomposition of the sphere (due to Hausdorff).

However, many principal applications are in the finite, or at any rate discrete, domain. Several applications to combinatorial identities are given in Section 4 of Chapter 3. A fairly typical elementary application is in vol. 1, p. 140 in Steiner's problem. It is necessary there to show that the number of certain vertices is bounded in a certain way. This turns out to be a simple consequence of counting all the edges in two different ways: once by going over the edges themselves, and once by counting at the vertices all the edges issuing from them, and adding. In the process we have to divide the last sum by two to prevent multiple counting. This division by two suggests that other symmetry and multiplicity removals also are related to this principle. Very similar, though subtler, counting in two ways, once by vertices and once by edges, was also at the core of the proof of Cauchy's theorem on the rigidity of convex polyhedra in Section 6 of Chapter 1.

Extensive use is made of our principle in number theory. For instance, some of the many proofs of the Gauss law of quadratic reciprocity employ the multiple counting and rather little else [174, 67]. In connection with number theory, the reader may wish to consider the combinatorial restatement of the Lindelöf hypothesis in view of multiple counting, for this see Section 6 of Chapter 3. Counting in two different ways, or even just looking in two different ways, occurs in the theory of partitions. This is especially true where the Ferrers graphs come in, since these readily lend themselves to rearrangements and so to counting in two different ways.

In simple elementary combinatorics the principle of counting in two different ways is listed at the beginning of Section 4 of Chapter 3; however, it is also involved in other principles listed there: the $p:q$ correspondence, and the inclusion–exclusion. As a typical example of a straight combinatorial application we mention the well-known identity

$$\sum_{k=0}^{n} \binom{n}{k}^{2} = \binom{2n}{n}.$$

A simple proof is obtained by multiplying out and equating the coefficients of x^{n} in the identity

$$\left(\sum_{k=0}^{n} \binom{n}{k} x^{k}\right)^{2} = (1+x)^{2n}.$$

The equating of coefficients qualifies the above as an application of our principle. However, we note the following proof, which makes a more essential use of the double counting; it is based on the idea of interpreting the two sides of our identity as different ways of doing something. The coefficient $\binom{2n}{n}$ is plainly the number of ways of choosing, say, a committee of n persons from the total of $2n$. Supposing these to be women and men in equal numbers, we find $\binom{n}{k}\binom{n}{n-k}$ ways for choosing a committee of k women and $n-k$ men; summing over all possible values of k we get

$$\binom{2n}{n} = \sum_{k=0}^{n} \binom{n}{k}\binom{n}{n-k} = \sum_{k=0}^{n} \binom{n}{k}^{2}.$$

6. EXTREMIZATION PRINCIPLES

Several small and partly related principles are mentioned here, all of them referring to extremization or optimization, as it is also known.

(a) Minimum Perturbation. The informal Fermat–Pascal principle states that near an extremum a (smooth) function varies slowly. We can also say that at the extremum every local perturbation must exhibit the right monotonicity. Our principle is then simply that in search for extrema we consider first the effect of those perturbations which vary possibly few variables only. For, plainly, with a function of several variables these are simplest to examine. For instance, by moving just one vertex at a time this enables us to prove very simply that the n-gon inscribed into a circle has largest area when it is regular. Dually, the n-gon P circumscribing a circle C has least area when it is regular. Here we consider a perturbation in which only one tangency point rides on C; we are interested in the area of the triangle T based on the corresponding side and bounded by the extensions of the two neighboring sides. The area of T is to be *largest* and so T is isosceles. Another simple application proves the Weitzenböck inequality

$$a^2 + b^2 + c^2 \geq 4\sqrt{3}\,A$$

for a triangle of area A and sides a, b, c. Here we just minimize $b^2 + c^2$ keeping a and A constant. Hence $b = c$ and so by symmetry $a = b = c$; this is verified to be the only case in which the equality in the above is attained.

Our principle is often usefully applied in graph minimization. For instance, consider the point X that minimizes the sum of distances $|AX| + |BX| + |CX|$, supposing ABC to be a triangle with angles less than $120°$. We cannot vary the length of only one of the segments XA, XB, XC. Therefore we do the next best thing and instead of keeping $|XA|$ and $|XB|$ constant we keep their sum constant. Hence under this perturbation X stays on an ellipse with A and B as foci. Clearly XC should be normal to that ellipse; a symmetric argument enables us to show that not only the angles CXA and CXB are equal but all three angles CXA, CXB, BXA are equal and so each is $120°$. A modification of the same procedure enables us to solve in Section 8 of Chapter 1 the three-dimensional analog of minimizing the sum of distances from a variable point to the four vertices of a tetrahedron. Again, another application in vol. 1, p. 138 shows that the shortest path of the traveling salesman in the Euclidean plane cannot be self-intersecting.

In the graphical form our principle often leads to the determination of certain geometrical maxima and minima in terms of tangency points of certain curves. For instance, the rectangle of maximum area that can be inscribed into the ellipse $x^2/a^2 + y^2/b^2 = 1$ has as its vertices the tangency points of the ellipse and the hyperbolas $xy = \pm c$. Again, suppose that we wish to find a straight segment from a point p in the first quadrant to the parabola P given by $y = -x^2$, along which a particle slides smoothly in shortest possible time from p to P, with constant downward gravity. Since the points which such a sliding particle can reach at any time t when sliding down inclined planes form a circle, we solve our problem by constructing the circle tangent to P, whose highest point is p.

One must be careful with the application of this principle since a function of several variables may get increased by a perturbation of each variable alone, yet the point in question may fail to be a minimum (although it must be a higher-dimensional equivalent of a saddle-point, if the function is smooth enough).

(b)   Interdependence of Extrema. This is: Under certain conditions the existence of certain extrema implies the existence of some further extrema. A simple illustration is that between each pair of consecutive maxima of $f(x)$ there is a minimum, and conversely. A more recondite example is provided by the problem of three double normals in vol. 1, p. 44. We have there a continuous real-valued function defined on a two-dimensional sphere, with the extra condition that at the antipodal points the function takes equal values. By the standard theorems on continuity and compactness we conclude the existence of a maximum and a minimum. Hence we have a pair of maxima and a pair of minima. Under these conditions the function must have also a saddle-point (hence two saddle-points). Thus there are three antipodal pairs of stationary points, giving us the three double normals. On further extension this principle leads to the beginnings of the Morse theory of critical points on manifolds.

(c)   Transversality. This states, roughly, that if an arc C is to extremize a functional and the end points of C are free to move on certain two loci, then in the extremizing position C is at right angles to these loci. An adaptation is involved in Steiner's proof of the isoperimetric property of the circle (vol. 1, pp. 20–22). In a modified form this is that the closed curve of fixed circumference, which encloses maximum area, is a circle because it must cut at right angles every straight line which bisects its circumference (and hence also, by taking mirror reflections, the area enclosed by it).

An elementary but characteristic use of this principle is found in the following somewhat counter-intuitive solution of a well-known elementary puzzle. Let T be an isosceles triangle with the vertex angle $2\pi/n$ where n is fairly large, say $n \geq 12$. We wish to find the shortest arc C which bisects the area of T. One might be tempted to claim that such C is a straight segment parallel to the base of T. However, this is not so. First, on account of the hypotheses, C is a curve joining the two equal sides of T. Since the end points of T are free to slide on those equal sides of T, the transversality principle suggests that C cuts the two sides at right angles. Hence it could not be a straight segment though it might be guessed to be a circular arc. Which is correct: the straight segment or the circular arc?

The circular arc turns out to be the correct answer. This can be shown by using a symmetry principle and reflecting T successively in its two long sides (and reflecting the reflections) till the configuration closes up with n copies of T forming a regular n-gon P. The n copies of C form then a closed curve K which must enclose a fixed area, namely half the area of P. Also, K inherits from C the shortest length condition. Hence we have reduced our problem to the isoperimetrics of the circle: K is a circle and C is a circular arc.

(d) **Convex Inscription and Extreme Points.** This refers to inscribing a maximal polyhedron P into a convex body K. The principle asserts that the vertices of P will often be among the extreme points of K. This was shown to be the case in vol. 1, p. 149 when P was a tetrahedron of largest volume, inscribed into a circular cylinder K. However, K can be arbitrary for if a vertex v of P were not extreme, we could apply the principle (a) and perturb v so that the volume is either increased or, at any rate, not decreased.

A related form is the diameter principle: here the maximal "polyhedron" has just two vertices p and q and we want these to be furthest apart. It turns out that p and q are extreme points of K and the planes through these points, orthogonal to the connecting straight segment pq, support K from opposite sides and contain it between them.

(e) **Duality.** A simple form of this is present in the "weakest link" argument—the *largest* weight that a chain can carry is determined by the *weakest* link. The same argument is used in vol. 1, p. 21 to determine the longest straight rod which can be carried horizontally round a right-angle bend in a corridor. Another simple example of duality is seen in the extremal n-gons circumscribed and inscribed to a fixed circle: Maximal inscribed n-gon as well as the minimal circumscribed one are regular. The quantity being extremized is either the circumference or the area. In terms of a symmetrizing operation more is true: Given a general inscribed n-gon, if we move a vertex v so as to tend to equalize the two sides meeting in v, then the length or area of the n-gon increases; with the circumscribed n-gon there would be a decrease under these conditions.

The last example of duality provides a passage to the isoperimetric duality stated in vol. 1, p. 20: If an integral $I(f)$ depending on the function f and its derivatives is to be maximized while another such integral $J(f)$ is kept constant, then the same function that maximizes $I(f)$ subject to the condition $J(f) = $ const. also minimizes $J(f)$ subject to the condition $I(f) = $ const. We have used this in solving the isoperimetric problem of the helix: Of all fixed-length arcs in E^3, one turn of the circular helix (of pitch $1/\sqrt{2}$) maximizes the volume of the convex hull.

The duality is observed and used by noting that the helix is a geodesic, that is, shortest-length curve, on its projecting circular cylinder.

(f) Equiminimax. This occurs when we wish to minimize the largest or maximize the smallest one of several quantities depending on x

$$\min_i \max [f_1(x), \ldots, f_n(x)] \quad \text{or} \quad \max_i \min [f_1(x), \ldots, f_n(x)].$$

It often happens that the values of the f_i's are equal then. For instance, in the problem of sequential localization of the zero of a monotonic continuous function (vol. 1, p. 159) we cannot do better that bisect the localization interval each time, since

$$\min_x \max (x, 1-x) = \tfrac{1}{2}$$

and we have for the minimax $x = 1 - x = 1/2$. Again, if three points A, B, C are to be determined in a square S so as to be possibly far from each other, we want to find

$$\max_{A,B,C \in S} \min (|AB|, |AC|, |BC|).$$

Here the optimal configuration is the equilateral triangle ABC, which has one vertex at a corner of S and the other two on the sides of S. We note that the three distances are equal.

A much more serious occurrence of the equiminimax is in the Chebyshev criterion for minimax approximation (Section 1, Chapter 5). Loosely put, it states that a fixed function $f(x)$ is best approximated by a function $g(x)$ belonging to a certain set of functions if the difference $f(x) - g(x)$ exhibits equal ripple: possibly many maxima and minima, and with the absolute value of the difference assuming the same value for all these extrema.

7. PARITY COUNTING

This minor but useful principle applies to the cases when we wish to show that two finite sets are of equal size (without needing the size itself), or when we wish to prove that they cannot be of the same size by demonstrating that no $1:1$ correspondence exists, or when we want to count the difference $A - B$ of two large but nearly equal numbers. In the simplest application of this principle we show that the symmetric group has as many even as odd permutations since there is a simple $1:1$ correspondence by means of a transposition. Another simple use is in the solution of a well-known puzzle: if two squares at the opposite ends of a long diagonal are removed from a chessboard, and if a domino piece is a rectangle just covering two chess squares, can the mutilated chessboard be covered by dominoes exactly? No, by the parity counting, because each domino covers one black and one white square, while the two removed squares are of the same color. The same argument will show that a whole unmutilated n-by-n board can be covered exactly by the dominoes if and only if n is even. In other words, if n^2 points form a square array then a nearest-neighbor reshuffle leaving no point fixed is possible if and only if n is even. These simple examples suggest the possibility of a more sophisticated parity counting, for instance, with dominoes of other than rectangular shape.

A much subtler use of parity counting was made in the Franklin proof of Euler's identity, vol.1, pp. 221–223, when we counted the difference between partitions into even and into odd number of parts. Since this difference was either 0 or ±1, it was convenient to check, without counting, the even partitions against the odd ones, noting only those cases when no 1:1 correspondence exists.

A simple application arises in number theory, in connection with the number $R_2(n)$ of representations of n as a sum of two squares. As usual, we count in $n = n_1^2 + n_2^2$ the orders n_1, n_2 and n_2, n_1, as well as the signs $\pm n_1$ and $\pm n_2$ as distinct. The identity

$$\left(\sum_{n=-\infty}^{\infty} x^{n^2} \right)^2 = 1 + 4 \sum_{k=1}^{\infty} \frac{(-1)^{k+1} x^{2k-1}}{1 - x^{2k-1}}$$

shows, on expanding and equating the coefficients of like powers, that

$$R_2(n) = 4 \left(\sum_{\substack{d|n \\ d=4k+1}} 1 - \sum_{\substack{d|n \\ d=4k+3}} 1 \right).$$

This raises the question of an independent and simple proof that the second sum in the above never exceeds the first one. That is, we want to show that every number has no fewer divisors of the form $4k+1$ than of the form $4k+3$. First, a direct argument reduces the case of general n to the form where n is odd and all its prime divisors are of the form $4k+3$. Writing then

$$n = \prod_{i=1}^{s} p_i^{a_i}$$

we identify the divisors of n with the set of s-tuples of integers

$$V = \{(x_1, x_2, \ldots, x_s) : 0 \le x_i \le a_i, i = 1, \ldots, s\}.$$

In terms of the integer lattice in the s-dimensional Euclidean space, where each lattice point $(x_1, x_2, \ldots, x_s)$ has the signature

$$(-1)^{x_1+x_2+\cdots+x_s}$$

we wish to show that the sum of all the signatures over V is nonnegative. However, this follows by a simple parity count: It is easy to draw a curve which starts from the origin $(0, 0, \ldots, 0)$ and goes through all the lattice points of V, passing from neighbor to neighbor. Hence the sum of signatures over V is $1-1+1-\cdots\pm1$, and so it is ≥0.

8. REDUCING MULTIPLICATION TO SQUARING

In its elementary form this principle is expressed by the identity

$$xy = \frac{(x+y)^2 - x^2 - y^2}{2}.$$

That is, if the "simpler" operations of addition, subtraction and division by 2 are defined then multiplication is reducible to squaring. In a slightly more general form we have

$$xy + yx = (x+y)^2 - x^2 - y^2.$$

In the context of vector spaces this principle takes a different appearance: the inner product is given by the norm as

$$xy = \tfrac{1}{2}(\|x+y\|^2 - \|x\|^2 - \|y\|^2).$$

A simple but space-saving application was made in Section 2 of Chapter 5 where a quadratic transport operator (corresponding to the norm) was used to induce the bilinear operator corresponding to the inner product.

A more sophisticated application of a related type was in the proof of the theorem of Titchmarsh in Section 8 of Chapter 2, which is essential in the Mikusinski approach to generalized functions as members of the quotient field of a certain ring of functions. We want to show that the convolution product

$$f * g = \int_0^x f(y)g(x-y)\,dy$$

of two continuous functions is 0 if and only if f or g is 0. Here the general case

$$f * g \equiv 0 \rightarrow f \equiv 0 \quad \text{or} \quad g \equiv 0$$

is deduced from the special, or "square" case, proved first:

$$f * f = 0 \rightarrow f = 0.$$

A rather unexpected application of our principle is in the following proof of the complete multiplicativity of the Legendre symbol of quadratic residuacy [174]:

$$\left(\frac{a}{p}\right)\left(\frac{b}{p}\right) = \left(\frac{ab}{p}\right).$$

Here p is a fixed odd prime and

$$\left(\frac{a}{p}\right) = 1 \quad \text{or} \quad -1$$

depending on whether a is, or is not, a quadratic residue for p. That is, the value is 1 or -1 depending on whether the congruence

$$x^2 \equiv a \pmod{p}$$

has solutions or has none.

Using R and N to to denote the quadratic residues and nonresidues generally, we prove simply that $R_1 R_2 = R$ and $RN_1 = N_2$. To prove the remaining and hardest case, that the product of two nonresidues is a residue, that is, that $N_1 N_2 = R$ we again reduce the multiplication $N_1 N_2$ to pure squaring. For, it is observed that if N is a fixed nonresidue and

$$R_1, R_2 \ldots, R_{(p-1)/2}$$

are all the $(p-1)/2$ residues, then the multiples

$$NR_1, NR_2, \ldots, NR_{(p-1)/2}$$

are incongruent nonresidues. Since there are $(p-1)/2$ of them, they make up all the nonresidues there are. Hence each nonresidue can be represented as NR_i. Now we have

$$N_1 N_2 = NR_i NR_j = N^2 R_i R_j = R$$

since by its very definition N^2, like any perfect square, is a quadratic residue.

It may be observed that

$$xy(x+y) = \frac{(x+y)^3 - x^3 - y^3}{3},$$

giving us something like the original but for cubes. It is by far not as useful as the original form for squares, but it finds an occasional application. For instance, let I_1 be the interval $[0, 1]$ and let $x_1, x_2, \ldots$ be any infinite sequence of real numbers, dense in I_1 and with $0 < x_i < 1$. Removing x_1 from I_1 breaks it up into two intervals, then x_2 is removed and so one of these two is broken up again, and so on. Eventually, every interval

I_n which appears in the process gets broken up; the earliest time it does so let it be into parts of length u_n and v_n. Then an application of the above formula and an easy telescoping cancellation show that

$$\sum_{n=1}^{\infty} u_n v_n (u_n + v_n) = \tfrac{1}{3}.$$

9. GENERALIZATIONS

Finally, we shall indicate on a few scattered examples how to generalize some classes of widely differing problems, generating thereby new problems. One may sometimes start by asking whether the mathematical situation in hand represents "the whole truth," and if it does not, making it do so more nearly. For instance, let us consider the following three problems.

(A) Since π is transcendental, for every polynomial

$$P(x) = \sum_{j=0}^{n} a_j x^j, \qquad \text{all } a\text{'s integers,}$$

we have $P(\pi) \neq 0$. Does this represent "the whole truth" or can we say more?

(B) Let C be a simple plane closed rectifiable curve of length $L(C)$, enclosing the area $A(C)$. Then the isoperimetric inequality states that

$$L^2(C) - 4\pi A(C) \geq 0 \qquad\qquad (1)$$

and the equality holds only for circles. Again we ask whether (1) represents the "whole truth" or whether it can be extended in some significant way.

(C) We start with the simple and well-known identity

$$(1 + 2 + 3 + \cdots + n)^2 = 1^3 + 2^3 + 3^3 + \cdots + n^3, \qquad (2)$$

and with the related integral inequality

$$\left(\int_0^x f(y)\, dy \right)^2 \geq \int_0^x f^3(y)\, dy \qquad\qquad (3)$$

which holds if $f(0) = 0$ and $0 \leq f'(x) \leq 1$. It is observed first that the inequality (3) has a more precise form: under the same conditions on f we have

$$\left(\int_0^x f(y)\,dy\right)^2 - \int_0^x f^3(y)\,dy$$

$$= 2\int_0^x \int_0^y f(u)f(y)[1 - f'(u)]\,du\,dy, \tag{4}$$

as proved in Section 1 of Chapter 2. Next, we produce an inequality corresponding to (2) and (3). If $f(y) = y$ we get the equality in (3):

$$\left(\int_0^x y\,dy\right)^2 = \int_0^x y^3\,dy, \tag{5}$$

and so, by analogy, we manufacture the inequality which is to (2) as (3) is to (5):

$$\left(\sum_{i=0}^n a_i\right)^2 \geq \sum_{i=0}^n a_i^3 \quad \text{if} \quad 0 = a_0 \leq a_1 \leq a_2 \leq \cdots, a_{i+1} - a_i \leq 1. \tag{6}$$

Now we ask: Since in the continuous domain the "full" relation (4) generalizes (3), is there an analogous "full" relation for the discrete inequality (6)?

The first problem, (A), which is the deepest of the three was left purposely in very vague terms whereas the third one, (C), which is quite elementary, was formulated at length and in some detail. This was done because problem (C) being most easily accessible shows rather clearly some common features shared by all three problems. A closer study of it may suggest a joint approach toward generalizing all three problems and, hopefully, some others as well. In (A), (B), and (C) we deal with a quantity which inherently or by the hypotheses is nonnegative:

$$\text{in (A)} \qquad |P(x)| \geq 0,$$
$$\text{in (B)} \qquad L^2(C) - 4\pi A(C) \geq 0,$$
$$\text{in (C)} \qquad \left(\sum_{i=0}^n a_i\right)^2 - \sum_{i=0}^n a_i^3 \geq 0.$$

In all three it is important to isolate the condition when that nonnegative quantity is precisely 0. In (A) $P(x)$ can be 0 only if x is algebraic, but π is not and so $|P(\pi)| > 0$. In (B) the only case of equality is when C is a circle. In (C), on account of the conditions stated in (6) equality occurs only for the sequence $a_i = i$, that is, for (2).

The common ingredient of (A), (B), and (C) is now isolated and all three problems are generalized together when we ask not just that the quantity which is of interest in each case should be nonnegative, and actually positive except when such and such happens, but that, further, this quantity should be more precisely and significantly estimated away from 0.

For (A) this leads to a generalization of the concept of transcendency by means of the so-called transcendency measures. First, we introduce the two basic bounds on the size of the polynomial P: its degree $D(P) = n$ and its height

$$H(P) = \max_{0 \leq i \leq n} |a_i|.$$

A transcendency measure for a transcendental number, here π, is a function $f(x, y)$, defined and positive for all positive integer values of x and y, which measures by how much π fails to be algebraic:

$$|P(\pi)| \geq f(D(P), H(P)).$$

Obviously, such measures exist for every transcendental number but getting a good one for a specific number such as π, can be much harder than the proof of transcendency alone. For π, as Mahler has shown in [107], we have the following: Let x be a real algebraic number satisfying the irreducible equation $P(x) = 0$ and let

$$m = [20.2^{(5D(P)-1)/2}], \quad A = \max (H(P), \ (m+1)^{(m+1)/D(P)}),$$

then

$$|\pi - x| > \left(\frac{m+1}{e}\right)^{-(m+1)} A^{-(m+1)D(P)\log(1+m)}.$$

To pass to an estimate of $|P(x)|$ we observe that

$$|P(\pi)| = |P(\pi) - P(x)| = |\pi - x| \left| \frac{P(\pi) - P(x)}{\pi - x} \right|$$

and the last factor can be bounded away from 0,

$$\left| \frac{P(\pi) - P(x)}{\pi - x} \right| \geq K > 0,$$

by applying the mean-value theorem and observing that $P(\pi)$ and $P'(\pi)$ cannot both be very small.

In the above-quoted paper Mahler also proves the following irrationality measure for π: For *every* pair of positive integers p, q

$$\left| \pi - \frac{p}{q} \right| > q^{-42}.$$

With problem (B) one generalization of our type has been given in vol. 1, p. 42: the Bonnesen extension of the isoperimetric inequality

$$L^2(C) - 4\pi A(C) \geq \pi^2 [R(C) - r(C)]^2 \tag{7}$$

where $R(C)$ is the radius of the smallest circle containing C and $r(C)$ is the radius of the largest circle contained in C. We can still ask here how precise is the Bonnesen extension (7) of (1). One formulation of this loose question is to maximize and minimize the isoperimetric deficit

$$L^2(C) - 4\pi A(C)$$

subject to the side-condition that $R(C)$ and $r(C)$ are given numbers. Another form of the question is, how to extend (7) by means of a nonnegative functional $F(C)$, vanishing if C is a circle and only then, such that

$$L^2(C) - 4\pi A(C) - \pi^2 [R(C) - r(C)]^2 \geq F(C).$$

Of course, once $F(C)$ is found one can then try to estimate further with

$$L^2(C) - 4\pi A(C) - \pi^2 [R(C) - r(C)]^2 - F(C) \geq G(C),$$

and so on. It may also be attempted to obtain by some uniform method the series

$$L^2(C) - 4\pi A(C) \geq \sum_{n=1}^{\infty} F_n(C),$$

where each $F_n(C)$ is a positive functional, vanishing if and only if C is the circle, and with suitable other properties. A certain remote analogy may be noted with respect to both Taylor's series and the absolutely monotonic functions of Section 6 of Chapter 2. This even suggests that $L^2(C) - 4\pi A(C)$ is a sort of an absolutely monotonic functional.

For problem (C), as was shown in Section 1 of Chapter 2, we get the identity

$$\left(\sum_{i=0}^{n} a_i \right)^2 - \sum_{i=0}^{n} a_i^3 = 2 \sum_{i=0}^{n} \sum_{j=0}^{i} a_i \frac{a_j + a_{j-1}}{2} [1 - (a_j - a_{j-1})]$$

where $a_{-1} = a_0 = 0$.

The opportunities for generalization and generation of new problems under the above scheme are extensive; practically any inequality is a candidate. For instance, consider the Weitzenböck inequality

$$a^2 + b^2 + c^2 \geq 4\sqrt{3} A$$

for a triangle T of area A and sides a, b, c, mentioned in this Appendix under Extremization Principles. To convert it under our scheme we should like to have for the triangle T

$$a^2 + b^2 + c^2 - 4\sqrt{3} A = F(T),$$

where $F(T)$ is a function of the parameters of T which is nonnegative and vanishes if and only if T is equilateral. The reader may wish to verify that if x and A_1 are the side and the area of the smallest equilateral triangle T_1 inscriptible into T, then

$$x = \frac{2\sqrt{2} A}{\sqrt{a^2 + b^2 + c^2 + 4\sqrt{3} A}}, \qquad A_1 = \frac{\sqrt{3} x^2}{4}.$$

Therefore

$$a^2 + b^2 + c^2 - 4\sqrt{3} A = \frac{2\sqrt{3} A}{A_1} (A - 4A_1).$$

Is it true that $A - 4A_1 \geq 0$ with the equality if and only if both T and T_1 are equilateral? Yes, and the reader may wish to prove the following generalization of *this*. Let T be an arbitrary triangle of area A and let S be another arbitrary triangle. Let A_1 be the area of the smallest triangle inscribed into T and similar to S. Then $A - 4A_1 \geq 0$ with the equality if and only if T and S are themselves similar. We note that here "similar" does not imply "similarly located." Finally, the inequality $A - 4A_1 \geq 0$ might itself be extended under our scheme.

Our next item is the Mergelyan–Wesler theorem of vol. 1, p. 224: If the unit (circular) disk D_0 in the plane is packed by an infinite sequence $P = D_1, D_2, \ldots$ of open disjoint circular subdisks to within measure 0, that is, if

$$\text{Area } D_0 = \sum_{n=1}^{\infty} \text{Area } D_n \quad \text{or equivalently} \quad \sum_{n=1}^{\infty} r_n^2 = 1, \qquad (8)$$

then

$$\sum_{n=1}^{\infty} r_n \text{ diverges.} \qquad (9)$$

Here also we start by asking whether this represents "the whole truth," we answer "no," and we set out to generalize by introducing the convenient quantity

$$M_x(P) = \sum_{n=1}^{\infty} r_n^x.$$

The convenience is shown by observing that the area condition (8) and the Mergelyan–Wesler theorem (9) itself now read very simply: For any packing P such that $M_2(P) = 1$ we have $M_1(P) = \infty$. Since $r_n < 1$ and therefore r_n^x is a decreasing function of x, we have here something like a Dedekind cut: For every packing P there is a unique number $e(P)$ such that $M_x(P)$ diverges for $x < e(P)$ and converges for $e(P) < x$. By analogy with entire functions the number $e(P)$ may be called the exponent of convergence of the packing P. Although the matter of generalizing the Mergelyan–Wesler theorem is not closed, being still under active research, it appears very likely that the Mergelyan–Wesler estimate is "not the whole truth" being too low: There is a universal constant, heuristically estimated to be about

$$S = 1.306951 \cdots$$

such that $e(P) \geq S$ for every packing P. In other words, if 100% of the area of D_0 is to be exhausted we need such a huge number of tiny disks D_n, that is, the number of disks of radius $\geq \varepsilon$ grows so fast as $\varepsilon \to 0$, that not only the sum (9) diverges but so do all the "higher moments" $M_x(P)$ for any $x < S$.

There appears to be a considerable difficulty in dealing with such higher moments $M_x(P)$. One reason for this difficulty is that our whole problem, in its finer detail, appears to be really a problem concerning nonlinear iteration, which is a distinctly refractory topic (cf. vol. 1, Chapter 2; p. 315, Chapter 5). We recall here, for instance, that the tentative bound $S > 1.18096$ of vol. 1, p. 225 was based on reducing nonlinear to linear iteration. Now, the infinite packing problem appears to be geometrical and it is not immediately obvious that there is any connection with iterating, not to mention that the difficulties arise because we iterate a difficult function. However, the reader may wish here to refer to vol. 1, pp. 74–75 and 225–226 where the connections with iteration are discussed at some length.

An analogous difficulty, of appearing to be about one thing and being really about something else, arises in the case of the still unsolved Borsuk problem (vol. 1, pp. 44, 157; p. 12 in Chapter 1, this volume). This is: To prove that any bounded subset of the Euclidean n-dimensional space E^n is a union of $n + 1$ smaller subsets.

It may appear on the face of it, that the Borsuk problem is about distances, diameters, and subdivisions, hence about the Euclidean metric structure. Yet, it seems that the problem is really about the finer curvature structure of the boundaries of sets of constant width. First, why boundaries and why sets of constant width? This results from the reduction of the problem: to subdivide well a set X into $n+1$ small parts we have the sole parameter Diam (X), its diameter, to work with. Hence it can be assumed without loss of generality that X is adjunction-complete: Any adjunction of even one point not already in X, increases the diameter of X. Now, as was shown in vol. 1, p. 157, a set is adjunction-complete if and only if it is of constant width. Further, a good subdivision of such a set is induced by a corresponding subdivision of its boundary (by taking convex hulls of unions with a fixed point in the interior).

Now we submit some evidence to substantiate our opinion on what the problem of Borsuk is "really" about. First, we recall the proof of vol. 1, p. 44 that the conjecture of Borsuk's problem holds for any set with a smooth boundary. That proof was based on the simple observation that the north and south poles of a smooth solid K cannot be simultaneously seen unobstructed from any point at finite distance. Why not? Because the curvature of the boundary of K necessarily hides the sight of at least one of the two poles.

Further, we note a hierarchy of conjectures, of which we shall cite three, which go beyond that of Borsuk, and which might claim to be closer to "the whole truth." Their general sense is that the rounder the solid K of constant width happens to be, the worse its partitionability $G(K)$. This last quantity is the infimum of all real u such that K is a union of $n+1$ sets of diameter $\leq u$. The first conjecture is: For any set K of constant width d we have $G(K) \leq G(B)$, B being the ball of diameter d; this appears to be supported by the two-dimensional result of vol. 1, pp. 158–159. The second conjecture states that as we round off K, we increase its partitionability $G(K)$:

$$G(\lambda K + (1-\lambda)B) \leq G(B), \qquad 0 \leq \lambda \leq 1,$$

or in stronger form,

$$G(\lambda K + (1-\lambda)B)$$

is an increasing function of λ, for every K. The third conjecture states roughly that any reasonable measure of roundness must have the above monotonicity property.

We return briefly to the problem of infinite packings P and the difficulties of dealing with the moments $M_x(P)$. A possible reason for these difficulties is perhaps partly psychological and might be termed the "natural number myth." We meet that sort of difficulty when dealing with something which, our instinct or habit tells us, ought to be a whole number but is not. For instance, there is the Γ-function as a generalization of the factorial $n!$, fractional differentiation (e.g., the Abel integral equation which is really a differential equation of order $-1/2$), iteration of nonintegral order (vol. 1, p. 59) and the pursuit T_s of real order s (Section 4 of Chapter 5). In the case of infinite packings P we are familiar with the moments $M_1(P)$ and $M_2(P)$ since these correspond to the sum of all circumferences of the packing disks, and to the sum of their areas. But the general moment $M_x(P)$ may be more difficult to grasp and handle.

BIBLIOGRAPHY

[1] Aczel, J., *Vorlesungen über Funktionalgleichungen*, Birkhauser Verlag, Basel, 1961.

[2] Akhiezer, I. N., *Vorlesungen über Approximationstheorie*, Akademie Verlag, Berlin, 1953.

[3] Alexandrov, A. D., *Konvexe Polyeder*, Akademie Verlag, Berlin, 1958.

[4] *Apollonius of Perga* (ed. T. L. Heath), Cambridge, 1896.

[5] Artobolevskii, I. I., *Mechanisms for the Generation of Plane Curves*, Macmillan, New York, 1964.

[6] Attneave, F., *Psych. Rev.*, **61**, 183 (1954).

[7] Auger, L., *Un Savant Méconnu*, Blanchard, Paris, 1962.

[8] Baire, R., *Ann. di Mat.*, **3**, 1 (1899).

[9] Banach, S., *Mechanics*, Monografie Matematyczne, Warsaw, 1951.

[10] Barber, M. N., and Ninham, B. W., *Random and Restricted Walks*, Gordon and Breach, New York, 1970.

[11] Bartlett, M. S., An *Introduction to Stochastic Processes*, Cambridge, 1966.

[12] Bell, E. T., *Amer. J. Math.*, **55**, 50 (1933).

[13] Bell, E. T., *Amer. J. Math.*, **67**, 86 (1945).

[14] Bender, E., and Goldman, J., *Combinatorial Analysis*, Holt, Reinhart, and Winston (in press).

[15] Berlekamp, E., *Algebraic Coding Theory*, McGraw-Hill, New York, 1968.

[16] Blaschke, W., *Kreis und Kugel*, Chelsea, 1949.

[17] Bochner, S., *Fouriersche Integrale*, Chelsea, 1948.

[18] Bonnesen, T., and Fenchel, W., *Konvexe Körper*, Chelsea, 1948.

[19] Booth, R. S., *Derivatives from Fast Sequences* (in press).

[20] Borel, E., *Amer. Math. Monthly*, **38**, 96 (1931).

[21] Born, M., and Wols, E., *Principles of Optics*, 2nd rev. ed., Macmillan, New York, 1964.

[22] Briot, C. A. A., and Bouquet, J., *Analytical Geometry of Two Dimensions*, 14th ed., Chicago, 1896.

[23] Bromwich, T. I., *Theory of Infinite Series*, Cambridge, 1908.

[24] Burks, A. W. (ed.), *Theory of Self-reproducing Automata*, Univ. of Illinois Press, 1966.

400

[25] Butler, S., *Erewhon*, London, 1872.

[26] Cantor, G., *Gesammelte Abhandlungen*, Georg Olms, 1962.

[27] Carleman, T. A. T., *Comptes R. du 5 Congr. Math. Scand.*, 1922, p. 181.

[28] Carlitz, L., *Portugaliae Math.*, **21**, 201 (1962).

[29] Catalan, E., *J. de Liouville, l-er Ser.*, **4**, 323 (1839).

[30] Cauer, W., *Theorie der Linearen Wechselstromschaltungen*, 2nd ed., Akademie Verlag, Berlin, 1954.

[31] Cesaro, E., *Vorlesungen über Natürliche Geometrie*, Teubner, Leipzig, 1901.

[32] Chandrasekhar, S., *Rev. Mod. Phys.*, **15**, 1 (1943).

[33] Cockayne, E., and Melzak, Z. A., *Q. Appl. Math.*, **26**, 213 (1968).

[34] Coddington, E. A., and Levinson, N., *Theory of Ordinary Differential Equations*, McGraw-Hill, New York, 1955.

[35] Cohen, P., *Set Theory and Continuum Hypothesis*, Benjamin, New York, 1965.

[36] Cohn-Vossen, S., *Nachr. Ges. Wiss. Götting.*, 1927, 125.

[37] Coxeter, H. S. M., *Introduction to Geometry*, Wiley, New York, 1961.

[38] Crick, H. C., Griffiths, J. S., and Orgel, L., E., *Proc. Nat. Acad. Sci. U.S.A.*, **43**, 416 (1957).

[39] Crofton, M. W., *Proc. Lond. Math. Soc.*, **10** 13 (1878–79).

[40] Davis, H. T., *Introduction to Nonlinear Differential and Integral Equations*, Dover, 1962.

[41] Denjoy, A., *C. R.*, **173**, 1329 (1921).

[42] Dienes, P., *The Taylor Series*, Dover, 1957.

[43] Dilworth, R. P., *Ann. Math.*, **51**, 161 (1950).

[44] Diogenes, Laertius, *Lives of Eminent Philosophers*, Loeb, 1951.

[45] *Diophantus* (ed. T. L. Heath), Cambridge, 1910.

[46] Disraeli, B., Coningsby, London, 1844.

[47] Drake, R. L., *A General Mathematical Survey of the Coagulation Equation* (mimeo.), Nat. Cent. for Atm. Res., Boulder, Colorado.

[48] Eagle, A., *The Elliptic Functions as They Should Be*, Galloway and Porter, 1958.

[49] Eastman, W. L., *I.E.E.E. Trans.*, **IT-11**, 263 (1965).

[50] Edwards, J., *A Treatise on the Integral Calculus*, 2 vols., Macmillan, 1921.

[51] Eisenhart, L. P., *Differential Geometry*, Ginn, Boston, 1909.

[52] Eliot, G., *Impressions of Theophrastus Such*, Blackwood, London, 1879.

[53] Erdös, P., and Szekeres, G., *Comp. Math.*, **2**, 463 (1935).

[54] Euler, L., *Comment. Acad. Petropol.*, **8**, 66 (1736).

[55] Feller, W., *An Introduction to Probability Theory*, 2 vols., 2nd ed., Wiley, New York, 1966.

[56] Fournier, J., *Israel J. Math.*, **18**, 157 (1974).

[57] Fu, K. S., *Sequential Methods.* Academic Press, New York, 1968.

[58] Gantmacher F. R., *The Theory of Matrices*, 2 vols., Chelsea, 1959.

[59] Gödel, K., *Monatshefte fur Math. Phys.*, **38**, 173 (1931).

[60] Gödel, K., *The Consistency of the Continuum Hypothesis*, Princeton, 1940.

[61] Goldstine, H. H., *The Computer from Pascal to von Neumann*, Princeton, 1972.

[62] Golomb, S. W., Gordon, B., and Welch, L. R., *Canad. Math. J.*, 1958, 202.

[63] Goursat, E., *A Course in Mathematical Analysis*, 3 vols., Ginn, Boston, 1904.

[64] Greenhill, A. G., *The Applications of Elliptic Functions*, Dover, New York, 1959.

[65] Hadley, G., and Kemp, M. C., *Variational Methods in Economics*, North-Holland, 1971.

[66] Hall, M., *Combinatorial Theory*, Blaisdell, 1967.

[66a] Hanan, M., *J.S.I.A.M.* (*Appl. Math.*) **14**, 255 (1966).

[67] Hardy, G. H., and Wright, E. M., *An Introduction to the Theory of Numbers*, Oxford, 1945.

[68] Hardy, G. H., *A Mathematician's Apology*, Cambridge, 1941.

[69] Hartline, H. K., Ratliff, F., and Miller, W. H., in *Nervous Inhibition* (ed. Florey E.), Pergamon, New York, 1961.

[70] Hartline, H. K., and Ratliff, F., *J. Gen. Phys.*, **41**, 1094 (1958).

[71] Hathaway, A. S., *Amer. Math. Monthly*, **28**, 93 (1921).

[72] Herglotz, G., *Abh. Math. Sem. der Hans. Univ.*, **15**, 1943.

[73] Hermes, H., *Enumerability, Decidability, Computability*, Academic Press, New York, 1965.

[74] Hertz, P., *Math. Ann.*, **67**, 387 (1909).

[75] Hilbert, D., *Archiv Math. Phys. 3rd Ser.*, **1**, 44, 213 (1901).

[76] Hobson, E. W., *A Treatise on Plane Trigonometry*, 3rd ed., Cambridge, 1911.

[77] Hobson, E. W., *Squaring the Circle and Other Monographs*, Chelsea, 1950.

[78] Hodge, W. V. D., and Pedoe, D., *Methods of Algebraic Geometry*, vol. 1, Cambridge, 1947.

[78a] Hölder, O., *Math. Ann.*, **28**, 1 (1887).

[79] Holmgren, E., *Ark. Math. Astr. och Fys.*, 1904, 324.

[80] Hubel, D., and Wiesel, T. N., *J. Physiol.* (*Lond.*) **160**, 106 (1962).

[81] Hurewicz, W., and Wallman, H., *Dimension Theory*, Princeton, 1948.

[82] Ingham, A. E., *Proc. Cambr. Phil. Soc.*, **29**, 271 (1933).

[83] Ingham, A. E., *Distribution of Prime Numbers*, Cambridge Tracts, New York, 1964.

[84] Jackson, F. H., *Q. J. Math. Oxf.*, **2**, 1 (1951).

[85] Jeffress, L. A. (ed.), Cerebral Mechanisms in Behavior, Wiley, New York, 1951.

[86] Jessen, B., *Acta Math.*, **63**, 249 (1934).

[87] John, P. W. M., *Lond. Math. Soc.*, **36**, 159 (1961).

[88] Johnson, R. A., *Modern Geometry*, Houghton and Mifflin, Boston, 1929.

[89] Kac, M., *Probability and Related Topics*, Interscience, New York, 1959.

[90] Kac, M., *Comm. Math. Helv.*, **9**, 1 (1936-37).

[91] Katznelson, Y., *Introduction to Harmonic Analysis*, Wiley, New York, 1968.

[92] Keisler, J., and Chang, C. C., *Continuous Model Theory*, Princeton, 1966.

[93] Kempe, A. B., *Proc. Lond. Math. Soc.*, **10**, 229 (1878-79).

[94] Kleene, S. C., *Introduction to Metamathematics*, Van Nostrand, New York, 1952.

[95] Kirchoff, G., *Sitz Preuss. Akad. der Wiss.*, 1882, 641.

[96] Kirchoff, G., *Vorlesungen uber Mathematische Physik.*, Leipzig, 1878.

[97] Lehmer, D., *Canad. J. Math.*, **2**, 267 (1950).

[98] Leveque, W. J., *Topics in Number Theory*, 2 vols., Addison-Wesley, 1955.

[99] Leveque, W. J., *Proc. Amer. Math. Soc.*, **2**, 401 (1951).

[100] Lewis, J. E., *Inf. Control* (in press).

[101] Lighthill, M. J., *Introduction to Fourier Analysis and Generalized Functions*, Cambridge, 1960.

[102] Lotka, A., *Elements of Mathematical Biology*, Dover, 1956.

[103] Lovelace, A. A. (anon.), Annotated translation of [119], signed A. A. L., London, 1843.

[104] Luneburg, R. K., *Mathematical Theory of Optics* (mimeo.), Brown Univ., Providence, 1944.

[105] Lyusternik, L. A., *Shortest Paths*, Pergamon, 1964.

[106] Mahler, K., *Lond. J. Math.*, **15**, 115 (1940).

[107] Mahler, K., *Proc. Kon. Akad. van Wet.*, **38**, 333 (1952).

[108] Mascheroni, G., *La Geometria dell Compasso*, 1797.

[109] Matula, D. W., *Number of Subtrees*, Report AM-68-3, Washington Univ., St Louis, Miss., 1968.

[110] McBride, E. B., *Obtaining Generating Functions*, Springer, 1971.

[111] McClelland, W. J., and Preston, T., *A Treatise on Spherical Trigonometry*, Macmillan, 1897.

[112] Melzak, Z. A., *Canad. Math. Bull.*, **11**, 85 (1968).

[113] Melzak, Z. A., *Trans. Amer. Math. Soc.*, **85**, 547 (1957).

[114] Melzak, Z. A., *Mich. Math. J.*, **4**, 193 (1957).

[115] Melzak, Z. A., *Mich. Math. J.*, **6**, 331 (1959).

[116] Melzak, Z. A., *Proc. Amer. Math. Soc.*, **10**, 438 (1959).

[117] Melzak, Z. A., *J. Math. Anal. Appl.*, **2**, 264 (1961).

[118] Melzak, Z. A., *Inf. Control*, **5**, 163 (1962).

[119] Menabrea, L. F., *Bibliotheque Universelle de Geneve*, Oct. 1842.

[120] Menger, K., *Proc. Nat. Acad. Sci. U.S.A.*, **28**, 535 (1942).

[121] Menger, K., *Proc. Nat. Acad. Sci. U.S.A.*, **37**, 178, 226 (1951).

[122] Menger, K., Schweitzer, B., and Sklar, A., *Czech. Math. J.*, 1959, 459.

[123] Mikusinski, J., *Operational Calculus*, Pergamon, 1959.

[124] Minsky, M., and Papert, S., *Perceptrons*, M.I.T. Press, Cambridge, 1969.

[125] Montroll, E., in *Statistical Mechanics* (Price, S., Freed, K., and Light, J., ed.), Univ. of Chicago Press, 1971.

[126] Montroll, E. et al., *Rev. Mod. Phys.*, **43**, 231 (1971).

[127] Moseley, M., *Irascible Genius*, Hutchison, London, 1964.

[128] Mosevich, J. W., *Geodesic Focussing in Parallel-Plate Systems*, Ph.D. Thesis, Univ. of British Columbia, 1972.

[129] Mosevich, J. W. *Utilitas Math.*, **4**, 129 (1973).

[130] Mullin, R., and Rota, G. C., in *Graph Theory and Applications*, Academic Press, New York, 1970.

[131] Oberhettinger, F., and Magnus, W., *Anwendungen der Elliptischen Funktionen*, Springer, 1949.

[132] Perron, O., *Kettenbrüche*, 3rd., Stuttgart, 1954.

[133] Picard, E., *Rev. Scienza*, **1**, (1907).

[133a] Pinney, E., *Ordinary Difference–Differential Equations*, Univ. of Calif. Press, Berkeley, 1959.

[134] Pogorelov, A. V., *Differential Geometry*, Noordhoof, 1966.

[135] Polya, G., *Math. Ann.*, **74**, 204 (1913).

[136] Polya, G., and Szegö, G., *Aufgaben und Lehrsätze*, 2 vols., Dover, 1945.

[137] Polya, G., and Szegö, G., *Problems and Theorems in Analysis*, vol. 1, Springer, 1972.

[138] Post, E., *J. Symb. Logic*, **1**, 103 (1936).

[139] Prim, R. C., *Bell Syst. Tech. J.*, **36**, 1389 (1957).

[139a] Ramon-Moliner, E., private communication.

[140] Ramsey, F. P., *Econ. J.*, **38**, 543 (1928).

[141] Ratliff, F., *Mach Bands*, Holden-Day, San Francisco, 1965.

[142] Reuleaux, F., The Kinematics of Machinery, Dover, 1963.

[143] Ritt, J. F., *Integration in Finite Terms*, Columbia Univ. Press, New York, 1948.

[144] Roberts, G. E., *J. Opt. Soc. Amer.*, **54**, 1111 (1964).

[145] Rogers, C. A., *Packing and Covering*, Cambridge, 1964.

[146] Rosenblatt, F., *Principles of Neurodynamics*, Spartan Books, New York, 1962.

[147] Rosenthall, E., *Amer. J. Math.*, **65**, 663 (1943).

[148] Rosenthall, E., *Bull. Amer. Math. Soc.*, **54**, 366 (1948).

[149] Rota, G. C., *Zeits. Warch.*, **2**, 340 (1964).

[150] Rota, G. C., et al., *J. Math. Anal. Appl.*, **42**, 685 (1973).

[151] Rudin, W., *Real and Complex Analysis*, McGraw-Hill, New York, 1966.

[152] Schoenberg, I. J., *Acta Math.*, **91**, 143 (1954).

[153] Schubert, H., *Kalkül der Abzählenden Geometrie*, Teubner, Leipzig, 1879.

[154] Schubert, H., *Mitt. Math. Ges. Hamb.*, **4**, (1884).

[155] Schwarzschild, K., *Abh. Kön. Ges. Wiss. Gott.*, **4**, 3 (1905).

[156] Schweitzer, B., and Sklar, A., *J. Lond Math. Soc.*, **38**, 401 (1963).

[157] Siegel, C. L., *Ann. Math.*, **36**, 527 (1935).

[158] Siegel, C. L., *Transcendental Numbers*, Princeton, 1949.

[159] Sierpinski, W., *General Topology*, 2nd ed., University of Toronto Press, 1956.

[160] Sinzow, D. M., *Bull. Phys.-Math. Kazan*, **11**, 13 (1901).

[161] Skolem, T., *Diophantische Gleichungen*, Chelsea, 1950.

[162] Smoluchowski, R., *Phys. Zeitsch.*, **17**, 557 (1916).

[163] Sommerville, D. M. Y., *An Introduction to Geometry of n Dimensions*, Dover, 1958.

[164] Stoker, J. J., *Comm. Pure Appl. Math.*, **21**, 119 (1968).

[165] Struik, D. J., *Lectures on Classical Differential Geometry*, Addison-Wesley, 1950.

[166] Svoboda, A., *Computing Mechanisms and Linkages*, Dover, 1965.

[167] Temple, G., *J. Lond. Math. Soc.*, **28**, 134 (1953).

[168] Temple, G., *Proc. Roy. Soc. Lond.*, Ser. A, 1955, 175.

[169] Thorp, E., *Amer. Math. Soc. Proc.*, **11**, 734 (1960).

[170] Titchmarsh, E. C., *The Theory of Functions*, 2nd ed., Oxford, 1931.

[171] Titchmarsh, E. C., *The Theory of the Riemann Zeta Function*, Oxford, 1951.

[172] Truesdell, C. Y., *Unified Theory of Special Functions*, Princeton, 1948.

[173] Turing, A. M., *Proc. Lond. Math. Soc.*, **42**, 230 (1936–37).

[174] Uspensky, J. V., and Heaslet, M., *Introduction to Number Theory*, McGraw-Hill, New York, 1946.

[175] Verhulst, P. F., *Nouv. Mem. Acad. Roy. Bruxelles*, **18**, (1845).

[176] Volterra, V., *Lecons sur la Theorie Mathematique de la Lutte pour Vie*, Gauthier-Villars, Paris, 1931.

[176a] Volterra, V., *Acta Biotheoretica*, **3**, 1 (1937).

[177] Voronoi, G., *J. Reine Angew. Math.*, **134**, 198 (1908).

[178] Waerden, B. van der, *Modern Algebra*, vol. 1, Ungar, 1949.

[179] Wang, H., *Symp. Math. Th. Automata*, Polytech. Inst. of Brooklyn, 1962, p. 23.

[180] Ward, M., *Amer. J. Math.*, **55**, 67 (1933).

[181] Watson, G. N., *Quart. J. Math. Oxf.*, 1939, 266.

[182] Whittaker, E. T., *Lond. J. Math.*, **16**, 131 (1941).

[183] Whittaker, E. T., and Watson, G. N., *A Course of Modern Analysis*, Amer. ed., 1948.

[184] Whittaker, J. V., *Sections of Convex Polytopes* (in press).

[185] Widder, D. V., *The Laplace Transform*, Princeton, 1946.

[186] Yates, R. C., *Amer. Math. Monthly*, **38**, 573 (1931).

[187] Zadeh, L., *Inf. Control*, **8**, 338 (1965).

[188] Zeeman, E. C., in *Topology of 3-Manifolds* (Fort, M. K., ed.), Prentice-Hall, New Jersey, 1962.

INDEX